Geochemical Reaction Modeling

Geochemical Reaction Modeling

Concepts and Applications

Craig M. Bethke

University of Illinois
Urbana-Champaign

New York   Oxford
OXFORD UNIVERSITY PRESS
1996

Oxford University Press

Oxford   New York
Athens   Auckland   Bangkok   Bombay
Calcutta   Cape Town   Dar es Salaam   Delhi
Florence   Hong Kong   Istanbul   Karachi
Kuala Lumpur   Madras   Madrid   Melbourne
Mexico City   Nairobi   Paris   Singapore
Taipei   Tokyo   Toronto

and associated companies in
Berlin   Ibadan

Library of Congress Cataloging-in-Publication Data
Bethke, Craig.
Geochemical reaction modeling / Craig M. Bethke.
p.   cm.
Includes bibliographical references and indexes.
ISBN 0-19-509475-1
1. Geochemical modeling.   2. Chemical reactions—Simulation methods.   I. Title.
QE515.5.G43B48   1996   95-35674
551.9—dc20

3 5 7 9 8 6 4

Printed in the United States of America
on acid free paper

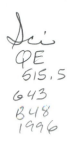

060500 — 5720G2

For A., H., and G.

Preface

Geochemists have long recognized the need for computational models to trace the progress of reaction processes, both natural and artificial. Given a process involving many individual reactions (possibly thousands), some of which yield products that provide reactants for others, how can we know which reactions are important, how far each will progress, what overall reaction path will be followed, and what the path's endpoint will be?

These questions can be answered reliably by hand calculation only in simple cases. Geochemists are increasingly likely to turn to quantitative modeling techniques to make their evaluations, confirm their intuitions, and spark their imaginations.

Computers were first used to solve geochemical models in the 1960s, but the new modeling techniques disseminated rather slowly through the practice of geochemistry. Even today, many geochemists consider modeling to be a "black art," perhaps practiced by digital priests muttering mantras like "Newton-Raphson" and "Runge-Kutta" as they sit before their cathode ray altars. Others show little fear in constructing models but present results in a way that adds little understanding of the problem considered. Someone once told me, "Well, that's what came out of the computer!"

A large body of existing literature describes either the formalism of numerical methods in geochemical modeling or individual modeling applications. Few references, however, provide a perspective of the modeling specialty, and some that do are so terse and technical as to discourage the average geochemist. Hence, there are few resources to which someone wishing to construct a model without investing a career can turn.

I have written this book in an attempt to present in one place both the concepts that underpin modeling studies and the ways in which geochemical models can be applied. Clearly, this is a technical book. I have tried to present enough detail to help the reader understand what the computer does in calculating a model, so that the computer becomes a useful tool rather than an impenetrable black box. At the same time, I have tried to avoid submerging the reader in computational intricacies. Such details I leave to the many excellent articles and monographs on the subject.

I have devoted most of this book to applications of geochemical modeling. I develop specific examples and case studies taken from the literature, my experience, and the experiences over the years of my students and colleagues. In the examples, I have carried through from the initial steps of conceptualizing and constructing a model to interpreting the calculation results. In each case, I present complete input to computer programs so that the reader can follow the calculations and experiment with the models.

The reader will probably recognize that, despite some long forays into hydrologic and basin modeling (a topic for another book, perhaps), I fell in love with geochemical modeling early in my career. I hope that I have communicated the elegance of the underlying theory and numerical methods as well as the value of calculating models of reaction processes, even when considering relatively simple problems.

I first encountered reaction modeling in 1980 when working in Houston at Exxon Production Research Company and Exxon Minerals Company. There, I read papers by Harold Helgeson and Mark Reed and experimented with the programs "EQ3/EQ6," written by Thomas Wolery, and "Path," written by Ernest Perkins and Thomas Brown.

Computing time was expensive then (about a dollar per second!). Computers filled entire rooms but were slow and incapacious by today's standards, and graphical tools for examining results almost nonexistent. A modeler sent a batch job to a central CPU and waited for the job to execute and produce a printout. If the model ran correctly, the modeler paged through the printout to plot the results by hand. But even at this pace, geochemical modeling was fun!

I returned to modeling in the mid-1980s when my graduate students sought to identify chemical reactions that drove sediment diagenesis in sedimentary basins. Computing time was cheaper, graphics hardware more accessible, and patience generally in shorter supply, so I set about writing my own modeling program, GT, which I designed to be fast enough to use interactively. A student programmer, Thomas Dirks, wrote the first version of a graphics program GTPLOT. With the help of another programmer, Jeffrey Biesiadecki, we tied the programs together, creating an interactive, graphical method for tracing reaction paths.

The program was clearly as useful as it was fun to use. In 1987, at the request of a number of graduate students, I taught a course on geochemical reaction modeling. The value of reaction modeling in learning geochemistry by experience rather than rote was clear. This first seminar evolved into a popular course, "Groundwater Geochemistry," which our department teaches each year.

The software also evolved as my group caught the interactive modeling bug. I converted the batch program GT to REACT, which was fully interactive. The user entered the chemical constraints for his problem and then typed "go" to trigger the calculation. Ming-Kuo Lee and I added Pitzer's activity model and a method for tracing isotope fractionation. Twice I replaced GTPLOT with new, more powerful programs. I wrote ACT2 and TACT to produce activity-activity and temperature-activity diagrams, and RXN to balance reactions and compute equilibrium constants and equations.

In 1992, we bundled these programs together into a package called "The Geochemist's Workbench®," which is owned by The Board of Trustees of the University of Illinois and can be licensed inexpensively for educational or commercial purposes. Within a few months of its completion the software was in use at dozens of universities and companies around the world.

We find that the programs allow us to try fresh approaches to teaching aqueous geochemistry. Once a student can reliably balance reactions by hand, the task quickly becomes a chore. After calculating a few Eh-pH diagrams, what does one learn by manually producing more plots? For many students, trees quickly come to obscure a beautiful forest. The computer can take over the mechanics of basic tasks, once they have been mastered, freeing the student to absorb the big picture and find the broad perspective. This approach has proved popular with students and professors. Many examples given in this book were developed originally as class assignments and projects.

I should not, however, give the impression that geochemical modeling is of any greater value in education than in scientific and practical application. The development of our modeling software, as evident in the case studies in this book, reflects the practical needs of petroleum geology and environmental geochemistry expressed to us over nearly a decade by a consortium of industrial and governmental affiliates to the Hydrogeology Program. These affiliates, without whom neither the software nor this book would exist, are: Amoco Production Research; Arco Oil and Gas Company, British Petroleum Research; Chevron Petroleum Technology Company; Conoco, Incorporated; Du Pont Company; Exxon Production Research; Hewlett Packard, Incorporated; Illinois State Geological Survey; Japan National Oil Company; Lawrence Livermore National Laboratory; Marathon Oil Company; Mobil Research and Development; Oak Ridge National Laboratory; Sandia National Laboratories; SiliconGraphics Computer Systems; Texaco, Incorporated; Union Oil Company of California; and the United States Geological Survey.

I can thank just a few of my colleagues and students who helped develop the case studies in this book. John Yecko and William Roy of the Illinois State Geological Survey first modeled degradation of the injection wells at Marshall, Illinois. Rachida Bouhlila provided analyses of the brines at Sebkhat El Melah, Tunisia. Amy Berger helped me write Chapter 8 (Surface Complexation), and Chapter 23 (Acid Drainage) is derived in part from her work. Edward Warren and Richard Worden of British Petroleum's Sunbury lab contributed data for calculating scaling in North Sea oil fields, Richard Wendlandt first modeled the effects of alkali floods on clastic reservoirs, and Kenneth Sorbie helped write Chapter 22 (Petroleum Reservoirs). I borrowed from Elisabeth Rowan's study of the genesis of fluorite ores at the Albigeois district, Wendy Harrison's study of the Gippsland basin, and a number of other published studies, as referenced in the text.

The book benefited enormously from the efforts of a small army of colleagues who served as technical reviewers: Stephen Altaner, Tom Anderson, and Amy Berger (University of Illinois); Greg Anderson (University of Toronto); Paul Barton, Jim Bischoff, Neil Plummer, Geoff Plumlee, and Elisabeth Rowan (U.S. Geological Survey); Bill Bourcier (Livermore); Patrick Brady and Kathy Nagy (Sandia); Ross Brower and Ed Mehnert (Illinois State Geological Survey); David Dzombak (Carnegie Mellon University); Ming-Kuo Lee (Auburn University); Peter Lichtner (Desert Research Institute); Benoit Madé and Jan van der Lee (Ecole des Mines); Mark Reed (University of Oregon); Kenneth Sorbie (Heriot-Watt University); Carl Steefel (Battelle); Jim Thompson (Harvard University); and John Weare (University of California, San Diego). I learned much from them. I also thank Mary Glockner, who read and corrected the entire manuscript; my editor Joyce Berry, and Lisa Stallings at Oxford for their unwavering support; and Bill Bourcier and Randy Cygan, who have always been willing to lend a hand, and often have.

I thank the two institutions that supported me while I wrote this book: the Department of Geology at the University of Illinois and the Centre d'Informatique Géologique at Ecole Nationale Supérieure des Mines de Paris in Fontainebleau, France. I began writing this book in Fontainebleau while on sabbatical leave in 1990 and completed it there under the sponsorship of the Académie des Sciences and Elf Aquitaine in 1995.

Fontainebleau, France   C.M.B.
June 1995

A Note about Software

The geochemical modeler's milieu is software and the computer on which it runs. A number of computer programs have been developed over the past twenty years to facilitate geochemical modeling (sources of the current versions of popular programs are listed in Appendix 1). Each program has its own capabilities, limitations, and indeed, personality. Some programs work on personal computers, others on scientific workstations, and a few run best on supercomputers. Some provide output graphically, others produce printed numbers. There is no best software, only the software that best meets a modeler's needs.

No discussion of geochemical modeling would be fully useful without specific examples showing how models are configured and run. In setting up the examples in this book, I employ a group of interactive programs that my colleagues and I have written over the past ten or so years. The programs, RXN, ACT2, TACT, REACT, and GTPLOT are known collectively as "The Geochemist's Workbench®." The package, which runs on Unix workstations and PCs, can be licensed through the University of Illinois. This book is not intended to serve as documentation for these programs; a separate, comprehensive "User's Guide" is available for that purpose (see Appendix 1).

I chose to use this software for reasons that extend beyond familiarity and prejudice: the programs are interactive and take simple commands as input. As such, I can include within the text of this book scripts that in a few lines show the precise steps taken to calculate each result. Readers can, of course, reproduce the calculations by using any of a number of other modeling programs, such as those listed in Appendix 1. Following the steps shown in the text, they should be able to construct input in the format recognized by the chosen program.

Contents

Part 2 — Reaction Processes

Part 3 — Applied Reaction Modeling

Geochemical Reaction Modeling

1

Introduction

As geochemists, we frequently need to describe the chemical states of natural waters, including how dissolved mass is distributed among aqueous species, and to understand how such waters will react with minerals, gases, and fluids of the Earth's crust and hydrosphere. We can readily undertake such tasks when they involve simple chemical systems, in which the relatively few reactions likely to occur can be anticipated through experience and evaluated by hand calculation. As we encounter more complex problems, we must rely increasingly on quantitative models of solution chemistry and irreversible reaction to find solutions.

The field of geochemical modeling has grown rapidly since the early 1960s, when the first attempt was made to predict by hand calculation the concentrations of dissolved species in seawater. Today's challenges might be addressed by using computer programs to trace many thousands of reactions in order, for example, to predict the solubility and mobility of forty or more elements in buried radioactive waste.

Geochemists now use quantitative models to understand sediment diagenesis and hydrothermal alteration, explore for ore deposits, determine which contaminants will migrate from mine tailings and toxic waste sites, predict scaling in geothermal wells and the outcome of steam-flooding oil reservoirs, solve kinetic rate equations, manage injection wells, evaluate laboratory experiments, and study acid rain, among many examples. Teachers let their students use these models to learn about geochemistry by experiment and experience.

Many hundreds of scholarly articles have been written on the modeling of geochemical systems, giving mathematical, geochemical, mineralogical, and

3

practical perspectives on modeling techniques. Dozens of computer programs, each with its own special abilities and prejudices, have been developed (and laboriously debugged) to analyze various classes of geochemical problems. In this book, I attempt to treat geochemical modeling as an integrated subject, progressing from the theoretical foundations and computational concerns to the ways in which models can be applied in practice. In doing so, I hope to convey, by principle and by example, the nature of modeling and the results and uncertainties that can be expected.

1.1 Development of Chemical Modeling

Hollywood may never make a movie about geochemical modeling, but the field has its roots in top-secret efforts to formulate rocket fuels in the 1940s and 1950s. Anyone who reads cheap novels knows that these efforts involved brilliant scientists endangered by spies, counter-spies, hidden microfilm, and beautiful but treacherous women.

The rocket scientists wanted to be able to predict the thrust that could be expected from a fuel of a certain composition (see historical sketches by Zeleznik and Gordon, 1968; van Zeggeren and Storey, 1970; Smith and Missen, 1982). The volume of gases exiting the nozzle of the rocket motor could be used to calculate the expected thrust. The scientists recognized that by knowing the fuel's composition, the temperature at which it burned, and the pressure at the nozzle exit, they had uniquely defined the fuel's equilibrium volume, which they set about calculating.

Aspects of these early calculations carry through to geochemical modeling. Like rocket scientists, we define a system of known composition, temperature, and pressure in order to calculate its equilibrium state. Much of the impetus for carrying out the calculations remains the same, too. Theoretical models allowed rocket scientists to test fuels without the expense of launching rockets, and even to consider fuels that had been formulated only on paper. Similarly, they allow geoscientists to estimate the results of a hydrothermal experiment without spending time and money conducting it, test a chemical stimulant for an oil reservoir without risking damage to the oil field, or help evaluate the effectiveness of a scheme to immobilize contaminants leaking from buried waste before spending and perhaps wasting millions of dollars and years of effort.

Chemical modeling also played a role in the early development of electronic computers. Early computers were based on analog methods in which voltages represented numbers. Because the voltage could be controlled to within only the accuracy of the machine's components, numbers varied in magnitude over just a small range. Chemical modeling presented special problems because the concentrations of species vary over many orders of magnitude. Even species in small concentrations, such as H^+ in aqueous systems, must be known accurately, since concentrations appear not only added together in mass balance equations, but multiplied by each other in the mass action (equilibrium constant) equations. The mathematical nature of the chemical equilibrium problem helped to demonstrate the limitations of analog methods, providing impetus for the development of digital computers.

Controversy over Free-Energy Minimization

Brinkley (1947) published the first algorithm to solve numerically for the equilibrium state of a multicomponent system. His method, intended for a desk calculator, was soon applied on digital computers. The method was based on evaluating equations for equilibrium constants, which, of course, are the mathematical expression of the minimum point in Gibb's free energy for a reaction.

In 1958, White et al. published an algorithm that used optimization theory to solve the equilibrium problem by "minimizing the free energy directly." Free-energy minimization became a field of study in its own, and the technique was implemented in a number of computer programs. The method had the apparent advantage of not requiring balanced chemical reactions. Soon, the chemical community was divided into two camps, each of which made extravagant claims about guarantees of convergence and the simplicity or elegance of differing algorithms (Zeleznik and Gordon, 1968).

According to Zeleznik and Gordon, tempers became so heated that a panel convened in 1959 to discuss equilibrium computation had to be split in two. Both sides seemed to have lost sight of the fact that the equilibrium constant is a mathematical expression of minimized free energy. As noted by Smith and Missen (1982), the working equations of Brinkley (1947) and White et al. (1958) are suspiciously similar. As well, the complexity of either type of formulation depends largely on the choice of components and independent variables, as described in Chapter 3.

Not surprisingly, Zeleznik and Gordon (1960, 1968) and Brinkley (1960) proved that the two methods were computationally and conceptually equivalent. The balanced reactions of the equilibrium constant method are counterparts to the species compositions required by the minimization technique; in fact, given the same choice of components, the reactions and expressions of species compositions take the same form.

Nonetheless, controversy continues even today among geochemical modelers. Colleagues sometimes take sides on the issue, and claims of simplified formulations and guaranteed convergence by minimization are still heard. In this book, I formalize the discussion in terms of equilibrium constants, which are familiar to geochemists and widely reported in the literature. Quite properly, I treat minimization methods as being computationally equivalent to the equilibrium constant approach, and do not discuss them as a separate group.

Application in Geochemistry

When they calculated the species distribution in seawater, Garrels and Thompson (1962) were probably the first to apply chemical modeling in the field of geochemistry. Modern chemical analyses give the composition of seawater in terms of dissociated ions (Na^+, Ca^{++}, Mg^{++}, HCO_3^-, and so on), even though the solutes are distributed among complexes such as $MgSO_4$(aq) and $CaCl^+$ as well as the free ions. Before advent of the theory of electrolyte dissociation, seawater analyses were reported, with equal validity, in terms of the constituent salts NaCl, $MgCl_2$, and so on. Analyses can, in fact, be reported in many ways, depending on the analyst's choice of chemical components.

Garrels and Thompson's calculation, computed by hand, is the basis for a class of geochemical models that predict species distributions, mineral saturation states, and gas fugacities from chemical analyses. This class of models stems from the distinction between a chemical analysis, which reflects a solution's bulk composition, and the actual distribution of species in a solution. Such *equilibrium models* have become widely applied, thanks in part to the dissemination of reliable computer programs such as SOLMNEQ (Kharaka and Barnes, 1973) and WATEQ (Truesdell and Jones, 1974).

Garrels and Mackenzie (1967) pioneered a second class of models when they simulated the reactions that occur as a spring water evaporates. They began by calculating the distribution of species in the spring water, and then repeatedly removed an aliquot of water and recomputed the species distribution. From concepts of equilibrium and mass transfer, the *reaction path model* was born. This class of calculation is significant in that it extends geochemical modeling from considering state to simulating process.

Helgeson (1968) introduced computerized modeling to geochemistry. Inspired by Garrels and Mackenzie's work, he realized that species distributions and the effects of mass transfer could be represented by general equations that can be coded into computer programs. Helgeson and colleagues (Helgeson et al., 1969, 1970) demonstrated a generalized method for tracing reaction paths, which they automated with their program PATHI ("path-one") and used to study weathering, sediment diagenesis, evaporation, hydrothermal alteration, and ore deposition.

Two conceptual improvements have been made since this early work. First, Helgeson et al. (1970) posed the reaction path problem as the solution to a system of ordinary differential equations. Karpov and Kaz'min (1972) and Karpov et al. (1973) recast the problem algebraically so that a reaction path could be traced by repeatedly solving for a system's equilibrium state as the system varied in composition or temperature. Wolery's (1979) EQ3/EQ6, the first software package for geochemical modeling to be documented and distributed, and Reed's (1977, 1982) SOLVEQ and CHILLER programs used algebraic formulations. This refinement simplified the formulations and codes, separated consideration of mass and heat transfer from the chemical equilibrium calculations, and eliminated the error implicit in integrating differential equations numerically.

Second, modelers took a broader view of the choice of chemical components. Aqueous chemists traditionally think in terms of elements (and electrons) as components, and this choice carried through to the formulations of PATHI and EQ3/EQ6. Morel and Morgan (1972), in calculating species distributions, described composition by using aqueous species for components (much like the seawater analysis described at the beginning of this section; see also Morel, 1983). Reed (1982) formulated the reaction path problem similarly, and Perkins and Brown's (1982) PATH program also used species and minerals as components. Chemical reactions now served double duty by giving the compositions of species and minerals in the system in terms of the chosen component set. This choice, which allowed models to be set up without even acknowledging the existence of elements, simplified the governing equations and provided for easier numerical solutions.

1.2 Scope of This Book

In setting out to write this book, I undertook to describe reaction modeling both in its conceptual underpinnings and its applications. Anything less would not be acceptable. Lacking a thorough introduction to underlying theory, the result would resemble a cookbook, showing the how but not the why of modeling. A book without detailed examples spanning a range of applications, on the other hand, would be sterile and little used.

Of necessity, I limited the scope of the text to discussing reaction modeling itself. I introduce the thermodynamic basis for the equations I derive, but do not attempt a complete development of the field of thermodynamics. A number of texts already present this beautiful body of theory better than I could aspire to in these pages. Among my favorites: Prigogine and Defay (1954), Pitzer and Brewer (1961), Denbigh (1971), Anderson and Crerar (1993), and Nordstrom and Munoz (1994). I present (in Chapter 7) but do not derive models for estimating activity coefficients in electrolyte solutions. The reader interested in more detail may refer to Robinson and Stokes (1968), Helgeson et al. (1981), and Pitzer (1987); Anderson and Crerar (1993, Chapter 17) present a concise but thorough overview of the topic.

I do not discuss questions of the measurement, estimation, evaluation, and compilation of the thermodynamic data upon which reaction modeling depends. Nordstrom and Munoz (1994, Chapters 11 and 12) provide a summary and overview of this topic, truly a specialty in its own right. Haas and Fisher (1976), Helgeson et al. (1978), and Johnson et al. (1991) treat aspects of the subject in detail. Finally, I mention only in passing methods of linking geochemical reaction modeling with hydrologic transport codes. The references in Section 2.2 address this topic in detail.

2

Modeling Overview

A *model* is a simplified version of reality that is useful as a tool. A successful model strikes a balance between realism and practicality. Properly constructed, a model is neither so simplified that it is unrealistic nor so detailed that it cannot be readily evaluated and applied to the problem of interest.

Geologic maps constitute a familiar class of models. To map a sedimentary section, a geologist collects data at certain outcrops. He casts his observations in terms of the local stratigraphy, which is itself a model that simplifies reality by allowing groups of sediments to be lumped together into formations. He then interpolates among his data points (and projects beneath them) to infer positions for formation contacts, faults, and so on across his field area.

The final map is detailed enough to show the general arrangement of formations and major structures, but simplified enough, when drawn to scale, that small details do not obscure the overall picture. The map, despite its simplicity, is without argument a useful tool for understanding the area's geology. To be successful, a geochemical model should also portray the important features of the problem of interest without necessarily attempting to reproduce each chemical or mineralogical detail.

2.1 Conceptual Models

The first and most critical step in developing a geochemical model is conceptualizing the system or process of interest in a useful manner. By *system*, we simply mean the portion of the universe that we decide is relevant. The composition of a *closed system* is fixed, but mass can enter and leave an *open system*. A system has an *extent*, which the modeler defines when he sets the

9

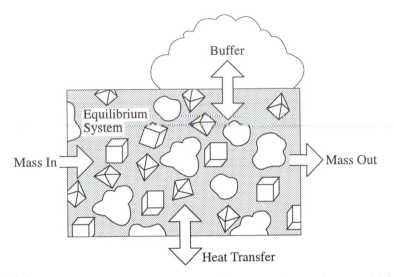

FIG. 2.1 Schematic diagram of a reaction model. The heart of the model is the equilibrium system, which contains an aqueous fluid and, optionally, one or more minerals. The system's constituents remain in chemical equilibrium throughout the calculation. Transfer of mass into or out of the system and variation in temperature drive the system to a series of new equilibria over the course of the reaction path. The system's composition may be buffered by equilibrium with an external gas reservoir, such as the atmosphere.

amounts of fluid and mineral considered in the calculation. A system's extent might be a droplet of rainfall, the groundwater and sediments contained in a unit volume of an aquifer, or the world's oceans.

The "art" of geochemical modeling is conceptualizing the model in a useful way. Figure 2.1 shows schematically the basis for constructing a geochemical model. The heart of the model is the *equilibrium system*, which remains in some form of chemical equilibrium, as described below, throughout the calculation. The equilibrium system contains an aqueous fluid and optionally one or more minerals. The temperature and composition of the equilibrium system are known at the beginning of the model, which allows the system's equilibrium state to be calculated. Pressure also affects the equilibrium state, but usually in a minor way under the near-surface conditions considered in this book (e.g., Helgeson, 1969; but also see Hemley et al., 1986), unless a gas phase is present.

In the simplest class of geochemical models, the equilibrium system exists as a closed system at a known temperature. Such equilibrium models predict the distribution of mass among species and minerals, as well as the species' activities, the fluid's saturation state with respect to various minerals, and the fugacities of different gases that can exist in the chemical system. In this case, the initial equilibrium system constitutes the entire geochemical model.

More complicated models account for the transport of mass or heat into or out of the system, so that its composition or temperature, or both, vary over the course of the calculation. The system's initial equilibrium state provides the starting point for this class of reaction path models. From this point, the model traces how mass entering and leaving the system, or changes in temperature, affects the system's equilibrium state.

Conceptualizing a geochemical model is a matter of defining (1) the nature of equilibrium to be maintained, (2) the initial composition and temperature of the equilibrium system, and (3) the mass transfer or temperature variation to occur over the course of the reaction process envisioned.

Types of Equilibrium

It is useful at this point to differentiate among the ways in which we can define equilibrium. In a classic sense (e.g., Pitzer and Brewer, 1961; Denbigh, 1971), a system is in equilibrium when it occupies a specific region of space within which there is no spontaneous tendency for change to occur. In this case, which we will call *complete equilibrium*, all possible chemical reactions are in equilibrium. Assuming complete equilibrium, for example, we can predict the distribution of dissolved species in a sample of river water, if the water is not supersaturated with respect to any mineral.

Geochemical models can be conceptualized in terms of certain false equilibrium states (Barton et al., 1963; Helgeson, 1968). A system is in *metastable equilibrium* when one or more reactions proceed toward equilibrium at rates that are vanishingly small on the time scale of interest. Metastable equilibria commonly figure in geochemical models. In calculating the equilibrium state of a natural water from a reliable chemical analysis, for example, we may find that the water is supersaturated with respect to one or more minerals. The calculation predicts that the water exists in a metastable state because the reactions to precipitate these minerals have not progressed to equilibrium.

In tracing a reaction path, likewise, we may find a mineral in the calculation results that is unlikely to form in a real system. Quartz, for example, would be likely to precipitate too slowly to be observed in a laboratory experiment conducted at room temperature. A model can be instructed to seek metastable solutions by not considering (*suppressing*, in modeling parlance) certain minerals in the calculation, as would be necessary to model such an experiment.

A system in complete equilibrium is spatially continuous, but this requirement can be relaxed as well. A system can be in internal equilibrium but, like Swiss cheese, have holes. In this case, the system is in *partial equilibrium*. The fluid in a sandstone, for example, might be in equilibrium itself, but may not be in equilibrium with the mineral grains in the sandstone or with just some of the grains. This concept has provided the basis for many published reaction paths, beginning with the work of Helgeson et al. (1969), in which a rock gradually reacts with its pore fluid.

The species dissolved in a fluid may be in partial equilibrium, as well. Many redox reactions equilibrate slowly in natural waters (e.g., Lindberg and Runnells, 1984). The oxidation of methane

$$CH_4(aq) + 2\,O_2(aq) \rightarrow HCO_3^- + H^+ + H_2O \qquad (2.1)$$

is notorious in this regard. Shock (1988), for example, found that although carbonate species and organic acids in oil-field brines appear to be in equilibrium with each other, these species are clearly out of equilibrium with methane. To model such a system, the modeler can *decouple* redox pairs such as HCO_3^-–CH_4 (e.g., Wolery, 1983), denying the possibility that oxidized species react with reduced species.

A third variant is the concept of *local equilibrium*, sometimes called mosaic equilibrium (Thompson, 1959, 1970; Valocchi, 1985; Knapp, 1989). This idea is useful when temperature, mineralogy, or fluid chemistry vary across a system of interest. By choosing a small enough portion of a system, according to this assumption, we can consider that portion to be in equilibrium. The concept of local equilibrium can also be applied to model reactions occurring in systems open to groundwater flow, using the ''flow-through'' and ''flush'' models described in the next section. The various types of equilibrium can sometimes be combined in a single model. A modeler, for example, might conceptualize a system in terms of partial and local equilibrium.

The Initial System

Calculating a model begins by computing the initial equilibrium state of the system at the temperature of interest. By convention but not requirement, the initial system contains a kilogram of water and so, accounting for dissolved species, a somewhat greater mass of fluid. The modeler can alter the system's extent by prescribing a greater or lesser water mass. Minerals may be included as desired, up to the limit imposed by the phase rule, as described in the next chapter. Each mineral included will be in equilibrium with the fluid, thus providing a constraint on the fluid's chemistry.

The modeler can constrain the initial equilibrium state in many ways, depending on the nature of the problem, but the number of pieces of information required is firmly set by the laws of thermodynamics. In general, the modeler sets the temperature and provides one compositional constraint for each chemical component in the system. Useful constraints include

- The mass of solvent water (1 kg by default),

- The amounts of any minerals in the equilibrium system,

- The fugacities of any gases at known partial pressure,

- The amount of any component dissolved in the fluid, such as Na^+ or HCO_3^-, as determined by chemical analysis, and

- The activities of a species such as H^+, as would be determined by *p*H measurement, or the oxidation state given by an Eh determination.

Unfortunately, the required number of constraints is not negotiable. Regardless of the difficulty of determining these values in sufficient number or the apparent desirability of including more than the allowable number, the system is mathematically underdetermined if the modeler uses fewer constraints than components, or overdetermined if he sets more.

Sometimes the calculation predicts that the fluid as initially constrained is supersaturated with respect to one or more minerals, and hence, is in a metastable equilibrium. If the supersaturated minerals are not suppressed, the model proceeds to calculate the equilibrium state, which it needs to find if it is to follow a reaction path. By allowing supersaturated minerals to precipitate, accounting for any minerals that dissolve as others precipitate, the model determines the stable mineral assemblage and corresponding fluid composition. The model output contains the calculated results for the supersaturated system as well as those for the system at equilibrium.

Mass and Heat Transfer: The Reaction Path

Once the initial equilibrium state of the system is known, the model can trace a reaction path. The reaction path is the course followed by the equilibrium system as it responds to changes in composition and temperature (Fig. 2.1). The measure of reaction progress is the variable ξ, which varies from zero to one from the beginning to end of the path. The simplest way to specify mass transfer in a reaction model (Chapter 11) is to set the mass of a reactant to be added or removed over the course of the path. In other words, the reaction rate is expressed in reactant mass per unit ξ. To model the dissolution of feldspar into a stream water, for example, the modeler would specify a mass of feldspar sufficient to saturate the water. At the point of saturation, the water is in equilibrium with the feldspar and no further reaction will occur. The results of the calculation are the fluid chemistry and masses of precipitated minerals at each point from zero to one, as indexed by ξ.

Any number of reactants may be considered, each of which can be transferred at a positive or negative rate. Positive rates cause mass to be added to the system; at negative rates it is removed. Reactants may be minerals, aqueous species (in charge-balanced combinations), oxide components, or gases. Since the role of a reactant is to change the system composition, it is the reactant's composition, not its identity, that matters. In other words, quartz, cristobalite, and SiO_2(aq) behave alike as reactants.

Mass transfer can be described in more sophisticated ways. By taking ξ in the previous example to represent time, the rate at which feldspar dissolves and product minerals precipitate can be set using kinetic rate laws, as discussed in Chapter 14. The model calculates the actual rates of mass transfer at each step of the reaction progress from the rate constants, as measured in laboratory experiments, and the fluid's degree of undersaturation or supersaturation.

The fugacities of gases such as CO_2 and O_2 can be buffered (Fig. 2.1; see Chapter 12) so that they are held constant over the reaction path. In this case, mass transfer between the equilibrium system and the gas buffer occurs as needed to maintain the buffer. Adding acid to a CO_2-buffered system, for example, would be likely to dissolve calcite

$$CaCO_3 + 2\,H^+ \rightarrow Ca^{++} + H_2O + CO_2(g) \qquad (2.2)$$
$$\textit{calcite}$$

Carbon dioxide will pass out of the system into the buffer to maintain the buffered fugacity.

Reaction paths can be traced at steady or varying temperature; the latter case is known as a *polythermal* path. Strictly speaking, heat transfer occurs even at constant temperature, albeit commonly in small amounts, to offset reaction enthalpies. For convenience, modelers generally define polythermal paths in terms of changes in temperature rather than heat fluxes.

2.2 Configurations of Reaction Models

Reaction models, despite their simple conceptual basis (Fig. 2.1), can be configured in a number of ways to represent a variety of geochemical processes. Each type of model imposes on the system some variant of equilibrium, as described in the previous section, but differs from others in the manner in which mass and heat transfer are specified. This section summarizes the configurations that are commonly applied in geochemical modeling.

Closed-System Models

Closed-system models are those in which no mass transfer occurs. Equilibrium models, the simplest of this class, describe the equilibrium state of a system composed of a fluid, any coexisting minerals, and, optionally, a gas buffer. Such models are not true reaction models, however, because they describe state instead of process.

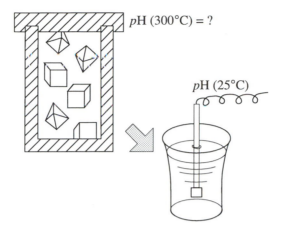

FIG. 2.2 Example of a polythermal path. Fluid from a hydrothermal experiment is sampled at 300°C and analyzed at room temperature. To reconstruct the fluid's *p*H at high temperature, the calculation equilibrates the fluid at 25°C and then carries it as a closed system to the temperature of the experiment.

Polythermal reaction models (Section 12.1), however, are commonly applied to closed systems, as in studies of groundwater geothermometry (Chapter 17), and interpretations of laboratory experiments. In hydrothermal experiments, for example, researchers sample and analyze fluids from runs conducted at high temperature, but can determine *p*H only at room temperature (Fig. 2.2). To reconstruct the original *p*H (e.g., Reed and Spycher, 1984), assuming that gas did not escape from the fluid before it was analyzed, an experimentalist can calculate the equilibrium state at room temperature and follow a polythermal path to estimate the fluid chemistry at high temperature.

There is no restriction against applying polythermal models in open systems. In this case, the modeler defines mass transfer as well as the heating or cooling rate in terms of ξ. Realistic models of this type can be hard to construct (e.g., Bowers and Taylor, 1985), however, because the heating or cooling rates need to be balanced somehow with the rates of mass transfer.

Titration Models

The simplest open-system model involves a reactant which, if it is a mineral, is undersaturated in an initial fluid. The reactant is gradually added into the equilibrium system over the course of the reaction path (Fig. 2.3). The reactant dissolves irreversibly. The process may cause minerals to become saturated and precipitate or drive minerals that already exist in the system to dissolve. The equilibrium system continues to evolve until the fluid reaches saturation with the

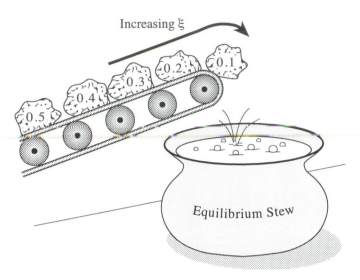

FIG. 2.3 Configuration of a reaction path as a titration model. One or more reactants are gradually added to the equilibrium system, as might occur as the grains in a rock gradually react with a pore fluid.

reactant or the reactant is exhausted. A model of this nature can be constructed with several reactants, in which case the process proceeds until each reactant is saturated or exhausted.

This type of calculation is known as a *titration model* because the calculation steps forward through reaction progress ξ, adding an aliqout of the reactant at each step $\Delta\xi$. To predict, for example, how the a rock will react with its pore fluid, we can titrate the minerals that make up the rock into the fluid. The solubility of most minerals in water is rather small, so the fluid in such a calculation is likely to become saturated after only a small amount of the minerals has reacted. Reacting on the order of 10^{-3} moles of a silicate mineral, for example, is commonly sufficient to saturate a fluid with respect to the mineral.

In light of the small solubilities of many minerals, the extent of reaction predicted by this type of calculation may be smaller than expected. Considerable amounts of diagenetic cements are commonly observed, for example, in sedimentary rocks, and crystalline rocks can be highly altered by weathering or hydrothermal fluids. A titration model may predict that the proper cements or alteration products form, but explaining the quantities of these minerals observed in nature will probably require that the rock react repeatedly as its pore fluid is replaced. Local equilibrium models of this nature are described later in this section.

Fixed-Fugacity and Sliding-Fugacity Models

Many geochemical processes occur in which a fluid remains in contact with a gaseous phase. The gas, which could be the Earth's atmosphere or a subsurface gas reservoir, acts to buffer the system's chemistry. By dissolving gas species from the buffer or exsolving gas into it, the fluid will, if reaction proceeds slowly enough, maintain equilibrium with the buffer.

Reaction paths in which the fugacities of one or more gases are buffered by an external reservoir (Fig. 2.1) are known as *fixed-fugacity paths* (Section 12.2). Because Garrels and Mackenzie (1967) assumed a fixed CO_2 fugacity when they calculated their pioneering reaction path by hand (see Chapter 1), they also calculated a fixed-fugacity path. Numerical modelers (e.g., Delany and Wolery, 1984) have more recently programmed buffered gas fugacities as options in their software.

The results of fixed-fugacity paths can differ considerably from those of simple titration models. Consider, for example, the oxidation of pyrite to goethite

$$\underset{pyrite}{FeS_2} + \frac{5}{2}H_2O + \frac{15}{4}O_2(aq) \rightarrow \underset{goethite}{FeOOH} + 4\,H^+ + 2\,SO_4^{--} \qquad (2.3)$$

in a surface water. In a simple titration model, pyrite dissolves until the water's dissolved oxygen is consumed. Water equilibrated with the atmosphere contains about 10 mg/kg O_2(aq), so the amount of pyrite consumed is small. In a fixed-fugacity model, however, the concentration of O_2(aq) remains in equilibrium with the atmosphere, allowing the reaction to proceed almost indefinitely.

Reaction paths in which the fugacities of one or more gases vary along ξ instead of remaining fixed are called *sliding fugacity paths* (Section 12.3). This type of path is useful when changes in total pressure allow gases to exsolve from the fluid (e.g., Leach et al., 1991). For example, layers of dominantly carbonate cement are observed along the tops of geopressured zones in the U.S. Gulf of Mexico basin (e.g., Hunt, 1990). The cement is apparently precipitated (Fig. 2.4) as fluids slowly migrate from overpressured sediments in overlying strata at hydrostatic pressure. The pressure drop allows CO_2 to exsolve,

$$Ca^{++} + 2\,HCO_3^- \rightarrow \underset{calcite}{CaCO_3} + H_2O + CO_2(g) \qquad (2.4)$$

causing calcite to precipitate.

Fixed activity and *sliding activity paths* (Sections 12.2-12.3) are analogous to their counterparts in fugacity, except that they apply to aqueous species instead of gases. Fixed activity paths are useful for simulating, for example, a laboratory

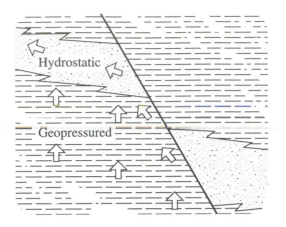

FIG. 2.4 Example of a sliding fugacity path. Deep groundwaters of a geopressured zone in a sedimentary basin migrate upward to lower pressures. During migration, CO_2 exsolves from the water so that its fugacity follows the variation in total pressure. The loss of CO_2 causes carbonate cements to form.

experiment controlled by a pH-stat, a device that holds pH constant. Sliding activity paths make easy work of calculating speciation diagrams, as described in Chapter 12.

Kinetic Reaction Models

In *kinetic reaction paths* (discussed in Chapter 14), the rates at which minerals dissolve into or precipitate from the equilibrium system are set by kinetic rate laws. In this class of models, reaction progress is measured in time instead of by the nondimensional variable ξ. According to the rate law, as would be expected, a mineral dissolves into fluids in which it is undersaturated and precipitates when supersaturated. The rate of dissolution or precipitation in the calculation depends on the variables in the rate law: the reaction's rate constant, the mineral's surface area, the degree to which the mineral is undersaturated or supersaturated in the fluid, and the activities of any catalyzing and inhibiting species.

Kinetic and equilibrium-controlled reactions can be readily combined into a single model. The two descriptions might seem incompatible, but kinetic theory (Chapter 14) provides a conceptual link: the equilibrium point of a reaction is the point at which dissolution and precipitation rates balance. For practical purposes, mineral reactions fall into three groups: those in which reaction rates may be so slow relative to the time period of interest that the reaction can be ignored altogether; those in which the rates are fast enough to maintain equilibrium; and the remaining reactions. Only those in the latter group require a kinetic description.

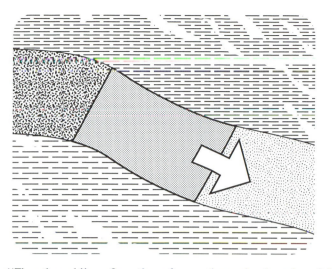

FIG. 2.5 "Flow-through" configuration of a reaction path. A packet of fluid reacts with an aquifer as it migrates. Any minerals that form as reaction products are left behind and, hence, isolated from further reaction.

Local Equilibrium Models

Reaction between rocks and the groundwaters migrating through them is most appropriately conceptualized by using a model configuration based on the assumption of local equilibrium (Section 11.3).

In a *flow-through* reaction path, the model isolates from the system minerals that form over the course of the calculation, preventing them from reacting further. Garrels and Mackenzie (1967) suggested this configuration, and Wolery (1979) implemented it in the EQ3/EQ6 code. In terms of the conceptual model (Fig. 2.1), the process of isolating product minerals is a special case of transferring mass out of the system. Rather than completely discarding the removed mass, however, the software tracks the cumulative amount of each mineral isolated from the system over the reaction path.

Using a flow-through model, for example, we can follow the evolution of a packet of fluid as it traverses an aquifer (Fig. 2.5). Fresh minerals in the aquifer react to equilibrium with the fluid at each step in reaction progress. The minerals formed by this reaction are kept isolated from the fluid packet, as if the packet has moved farther along the aquifer and is no longer able to react with the minerals produced previously.

In a second example of a flow-through path, we model the evaporation of seawater (Fig. 2.6). The equilibrium system in this case is a unit mass of seawater. Water is titrated out of the system over the course of the path,

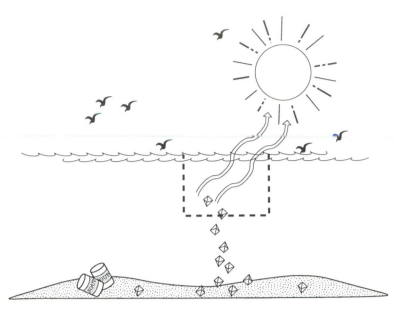

FIG. 2.6 Example of a "flow-through" path. Titrating water from a unit volume of seawater increases the seawater's salinity until evaporite minerals form. The product minerals sink to the sea floor, where they are isolated from further reaction.

concentrating the seawater and causing minerals to precipitate. The minerals sink to the sea floor as they form, and so are isolated from further reaction. We carry out such a calculation in Chapter 18.

In a *flush* model, on the other hand, the model tracks the evolution of a system through which fluid migrates (Fig. 2.7). The equilibrium system in this case might be a specified volume of an aquifer, including rock grains and pore fluid. At each step in reaction progress, an increment of unreacted fluid is added to the system, displacing the existing pore fluid. The model is analogous to a "mixed-flow reactor" as applied in chemical engineering (Levenspeil, 1972; Hill, 1977).

Flush models are useful for applications such as studying the diagenetic reactions resulting from groundwater flow in sedimentary basins (see Chapter 19) and predicting formation damage in petroleum reservoirs and injection wells (Fig. 2.8; see Chapters 21 and 22). Stimulants intended to increase production from oil wells (including acids, alkalis, and hot water) as well as the industrial wastes pumped into injection wells commonly react strongly with geologic formations (e.g., Hutcheon, 1984). Reaction models are likely to find increased application as well operators seek to minimize damage from caving and the loss of permeability due to the formation of oxides, clay minerals, and zeolites in the formation's pore space.

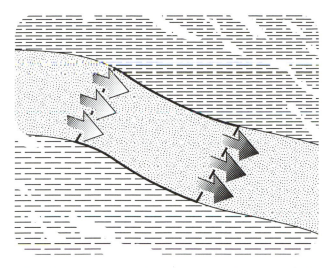

FIG. 2.7 "Flush" configuration of a reaction model. Unreacted fluid enters the equilibrium system, which contains a unit volume of an aquifer and its pore fluid, displacing the reacted fluid.

Flush models can also be configured to simulate the effects of dispersive mixing. Dispersion is the physical process by which groundwaters mix in the subsurface (Freeze and Cherry, 1979). With mixing, the groundwaters react with each other and the aquifer through which they flow (e.g., Runnells, 1969). In a flush model, two fluids can flow into the equilibrium system, displacing the mixed and reacted fluid (Fig. 2.9).

A final variant of local equilibrium models is the *dump* option (Wolery, 1979). Here, once the equilibrium state of the initial system is determined, the minerals in the system are jettisoned. The minerals present in the initial system, then, are not available over the course of the reaction path. The dump option differs from the flow-through model in that while the minerals present in the initial system are prevented from back-reacting, those that precipitate over the reaction path are not.

As an example of how the dump option might be used, consider the problem of predicting whether scale will form in the wellbore as groundwater is produced from a well (Fig. 2.10). The fluid is in equilibrium with the minerals in the formation, so the initial system contains both fluid and minerals. The dump option simulates movement of a packet of fluid from the formation into the well bore, since the minerals in the formation are no longer available to the packet. As the packet ascends the well bore, it cools, perhaps exsolves gas as it moves toward lower pressure, and leaves behind any scale produced. The reaction model, then, is a polythermal, sliding-fugacity, and flow-through path combined with the dump option.

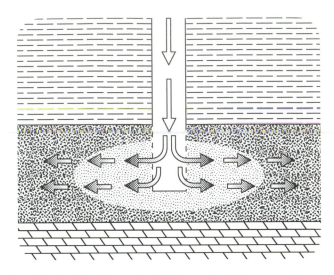

FIG. 2.8   Example of a "flush" model. Fluid is pumped into a petroleum reservoir as a stimulant, or industrial waste is pumped into a disposal well. Unreacted fluid enters the formation, displacing the fluid already there.

Continuum Models

In the late 1980s, several groups developed *continuum* models of geochemical reaction in systems open to groundwater flow. Continuum models are a natural marriage (Rubin, 1983; Bahr and Rubin, 1987) of the local equilibrium and kinetic models already discussed with the mass transport models traditionally applied in hydrology and various fields of engineering (e.g., Bird et al., 1960; Bear, 1972). By design, this class of models predicts the distribution in space and time of the chemical reactions that occur along a groundwater flow path. Published formulations, either for systems involving dissolution and precipitation of minerals or those accounting for surface adsorption reactions, include those of Lichtner (1985, 1988) and Lichtner et al. (1986), Cederberg et al. (1985), Ortoleva et al. (1987), Liu and Narasimhan (1989a, 1989b), and Steefel and Lasaga (1992, 1994).

The configuration of a continuum model can be thought of as a series of flush models placed end to end (Fig. 2.11). Fluid enters one end of the series, reacts with the medium, and discharges at the far end. Much of the reaction occurs along fronts that migrate through the medium until they either traverse to the outlet or reach a steady-state position (Lichtner, 1988). As noted by Lichtner (1988), continuum models predict that reactions occur in the same sequence as reaction path models. The advantage of continuum models is that they predict how the positions of reaction fronts migrate through time, provided that reliable

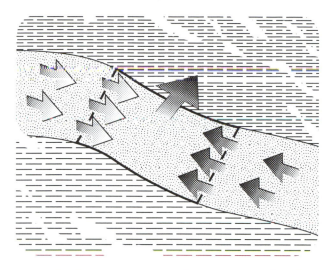

FIG. 2.9 Use of a "flush" model to simulate dispersive mixing. Two fluids enter a unit volume of an aquifer where they react with each other and minerals in the aquifer, displacing the mixed and reacted fluid.

input is available about flow rates, the permeability and dispersivity of the medium, and reaction rate constants.

Models of this configuration are more difficult to set up and are more challenging to compute than are reaction path models. In addition, the reaction distributions predicted by continuum models primarily reflect the assumed hydrologic properties and kinetic rate constants, which normally comprise the most poorly known parameters in a natural system. Since a valid reaction model is a prerequisite for a continuum model, the first step in any case is to construct a successful reaction path model for the problem of interest. The reaction path model provides the modeler with an understanding of the nature of the chemical process before he undertakes more complex modeling. The scope of this book is restricted to reaction models like those already described; we will not consider the added complexities of continuum calculations.

2.3 Uncertainty in Geochemical Modeling

Calculating a geochemical model provides not only results, but uncertainty about the accuracy of the results. Uncertainty, in fact, is an integral part of modeling that deserves as much attention as any other aspect of a study. To evaluate the sources of error in a study, a modeler should consider a number of questions:

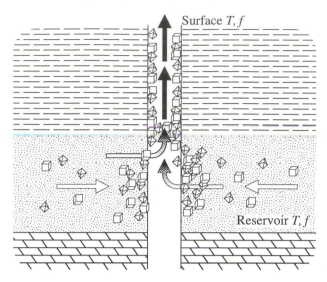

FIG. 2.10 Use of the "dump" option to simulate scaling. The pore fluid is initially in equilibrium with minerals in the formation. As the fluid enters the well bore, the minerals are isolated (dumped) from the system. The fluid then follows a polythermal, sliding fugacity path as it ascends the wellbore toward lower temperatures and pressures, depositing scale.

• *Is the chemical analysis used sufficiently accurate to support the modeling study?* The chemistry of the initial system in most models is constrained by a chemical analysis, including perhaps a pH determination and some description of the system's oxidation state. The accuracy and completeness of available chemical analyses, however, vary widely. Routine tests made of drinking water supplies and formation fluids from oil wells are commonly too rough and incomplete to be of much use to a modeler. Sets of analyses retrieved unselectively from water quality databases such as WATSTORE at the U.S. Geological Survey are generally not suitable for modeling applications (Hem, 1985). Careful analyses such as those of Iceland's geothermal waters made for scientific purposes (Arnorsson et al., 1983; see Chapter 17), on the other hand, are invaluable.

As Hem (1985) notes, a chemical analysis with concentrations reported to two or three, and sometimes four or five, significant figures can be misleadingly authoritative. Analytical accuracy and precision are generally in the range of ± 2 to $\pm 10\%$, but depend on the technique used, the skill of the analyst, and on whether or not the constituent was present near the detection limit of the analytical method. The third digit in a reported concentration is seldom meaningful, and confidence should not necessarily be placed on the second.

Care should be taken in interpreting reported pH values, which may have been determined in the field or in the laboratory after the sample had been stored

FIG. 2.11 Configuration of a continuum model of water-rock interaction in a system open to groundwater flow, showing positions of reaction fronts as they migrate through the system.

for an unknown period of time. Only the field measurement of pH is meaningful and, in the case of a groundwater, even the field measurement is reliable only if it is made immediately after sampling, before the water can exchange CO_2 with the atmosphere.

Significant error is introduced when a sample is acidified to "preserve" it, if the sample is not first carefully filtered to remove sediment and colloids (as illustrated in Section 6.2). Until the 1950s, it was normal procedure to sample unfiltered waters, and this practice continues in some organizations today. Even today's common practice of passing samples through a .45-μm filter in the field fails to remove colloidal aluminum and iron (e.g., Kennedy et al., 1974); a .10 μm filter is usually required to separate these colloids. By adding acid, the sampler dissolves any colloids and some of any suspended sediments, the constituents of which will appear in the chemical analysis as if they had originally been in solution.

Samples of formation water taken from drill-stem tests during oil exploration are generally contaminated by drilling fluids. The expense of keeping a drilling rig idle generally precludes pumping the formation fluid long enough to produce uncontaminated fluid, a procedure that might require weeks. As only rough knowledge of groundwater composition is needed in exploration, there is little impetus to improve procedures. Samples obtained at the well head after a field has been in production (e.g., Lico et al., 1982) may be preferable to analyses made during drill-stem tests, but care must be taken: samples obtained in this way may have already exsolved CO_2 or other gases before sampling.

• *Does the thermodynamic dataset contain the species and minerals likely to be important in the study?* A set of thermodynamic data, especially one intended to span a range of temperatures, is by necessity a balance between completeness and accuracy. The modeler is responsible for assuring that the database includes the predominant species and important minerals in the problem of interest.

The following example shows why this is important. The calculations in this book make use of the dataset compiled by Thomas Wolery, Ken Jackson, and numerous co-workers at Lawrence Livermore National Laboratory (the LLNL dataset; Delany and Lundeen, 1989), which is based in part on a dataset developed by Helgeson et al. (1978). The dataset includes a number of Cu-bearing species and minerals, including the cupric species Cu^{++} and $Cu(OH)^+$ that are dominant at room temperature under oxidized conditions in acidic and neutral solutions.

At pH values greater than about 9.5, the species $Cu(OH)_2$, $Cu(OH)_3^-$, and $Cu(OH)_4^{--}$ dominate the solubility of cupric copper by some orders of magnitude (Baes and Mesmer, 1976); these species, however, are not included in the database version used in this book. To construct a valid model of copper chemistry in an oxidizing, alkaline solution, the modeler would need to extend the database to include these species.

The same requirement extends to the minerals considered in the calculation. Minerals in nature occur as solid solutions in which elements substitute for one another in the mineral's crystal structure, but thermodynamic datasets generally contain data for pure minerals of fixed composition. A special danger arises in considering the chemistry of trace metals. In nature, these would be likely to occur as ions substituted into the framework of common minerals or sorbed onto mineral or organic surfaces, but the chemical model would consider only the possibility that the species occur as dissolved species or as the minerals of these elements that are seldom observed in nature.

• *Are the equilibrium constants for the important reactions in the thermodynamic dataset sufficiently accurate?* The collection of thermodynamic data is subject to error in the experiment, chemical analysis, and interpretation of the experimental results. Error margins, however, are seldom reported and never seem to appear in data compilations. Compiled data, furthermore, have generally been extrapolated from the temperature of measurement to that of interest (e.g., Helgeson, 1969). The stabilities of many aqueous species have been determined only at room temperature, for example, and mineral solubilities many times are measured at high temperatures where reactions approach equilibrium most rapidly. Evaluating the stabilities and sometimes even the stoichiometries of complex species is especially difficult and prone to inaccuracy.

For these reasons, the thermodynamic data on which a model is based vary considerably in quality. At the minimum, data error limits the resolution of a geochemical model. The energetic differences among groups of silicates, such as the clay minerals, is commonly smaller than the error implicit in estimating mineral stability. A clay mineralogist, therefore, might find less useful information in the results of a model than expected.

• *Can the species' activity coefficients be calculated accurately?* An activity coefficient relates each dissolved species' concentration to its activity. Most commonly, a modeler uses an extended form of the Debye-Hückel equation to estimate values for the coefficients. Helgeson (1969) correlated the activity coefficients to this equation for dominantly NaCl solutions having concentrations up to 3 molal. The resulting equations are probably reliable for electrolyte solutions of general composition (i.e., those dominated by salts other than NaCl) where ionic strength is less than about 1 molal (Wolery, 1983; see Chapter 7). Calculated activity coefficients are less reliable in more concentrated solutions. As an alternative to the Debye-Hückel method, the modeler can use virial equations (the "Pitzer equations") designed to predict activity coefficients for electrolyte brines. These equations have their own limitations, however, as discussed in Chapter 7.

• *Do the kinetic rate constants and rate laws apply well to the system being studied?* Using kinetic rates laws to describe the dissolution and precipitation rates of minerals adds an element of realism to a geochemical model but can be a source of substantial error. Much of the difficulty arises because a measured rate constant reflects the dominant reaction mechanism in the experiment from which the constant was derived, even though an entirely different mechanism may dominate the reaction in nature (see Chapter 14).

Rate constants for the dissolution and precipitation of quartz, for example, have been measured in deionized water (Rimstidt and Barnes, 1980). Dove and Crerar (1990), however, found that reactions rates increased by as much as one and a half orders of magnitude when the reaction proceeded in dilute electrolyte solutions. As well, reaction rates determined in the laboratory from hydrothermal experiments on "clean" systems differ substantially from those that occur in nature, where clay minerals, oxides, and other materials may coat mineral surfaces and hinder reaction.

• *Is the assumed nature of equilibrium appropriate?* The modeler defines an equilibrium system that forms the core of a geochemical model, using one of the equilibrium concepts already described. The modeler needs to ask whether the reactions considered in an equilibrium system actually approach equilibrium. If not, it may be necessary to decouple redox reactions, suppress minerals from the system, or describe mineral dissolution and precipitation using a kinetic rate law in order to calculate a realistic chemical model.

• *Most importantly, has the modeler conceptualized the reaction process correctly?* The modeler defines a reaction process on the basis of a concept of how the process occurs in nature. Many times the apparent failure of a calculation indicates a flawed concept of how the reaction occurs rather than error in a chemical analysis or the thermodynamic data. The "failed" calculation, in this case, is more useful than a successful one because it points out a basic error in the modeler's understanding.

Errors in conceptualizing a problem are easy to make but can be hard to discover. A modeler, distracted by the intricacies of his tools and the complexities of his results, can too easily lose sight of the nature of the conceptual model used or the assumptions implicit in deriving it. A mistake in the study of sediment diagenesis, for example, is to try to explain the origin of cements in a marine orthoquartzite by heating the original quartz grains and seawater along a polythermal path, to simulate burial.

The rock in question might contain a large amount of calcite cement, but the reaction path predicts that only a trace of calcite forms during burial. Considering this contradiction, the modeler realizes that this model could not have been successful in the first place: there is not enough calcium or carbonate in seawater to have formed that amount of cement. The model in this case was improperly conceptualized as a closed rather than open system.

Given this array of error sources, how can a geochemical modeler cope with the uncertainties implicit in his calculations? The best answer is probably that the modeler should begin work by integrating experimental results and field observations into the study. Having successfully explained the experimental or field data, the modeler can extrapolate to make predictions with greater confidence.

The modeler should also take heart that his work provides an impetus to determine more accurate thermodynamic data, derive better activity models for electrolyte solutions, and measure reaction rates under more realistic conditions.

3

The Equilibrium State

Aqueous geochemists work daily with equations that describe the equilibrium points of chemical reactions among dissolved species, minerals, and gases. To study an individual reaction, a geochemist writes the familiar expression, known as the *mass action equation*, relating species' activities to the reaction's equilibrium constant. In this chapter we carry this type of analysis a step farther by developing expressions that describe the conditions under which not just one but all of the possible reactions in a geochemical system are at equilibrium.

We consider a geochemical system comprising at least an aqueous solution in which the species of many elements are dissolved. We generally have some information about the fluid's bulk composition, perhaps directly because we have analyzed it in the laboratory. The system may include one or more minerals, up to the limit imposed by the phase rule (see Section 3.4), that coexist with and are in equilibrium with the aqueous fluid. The fluid's composition might also be buffered by equilibrium with a gas reservoir (perhaps the atmosphere) that contains one or more gases. The gas buffer is large enough that its composition remains essentially unchanged if gas exsolves from or dissolves into the fluid.

How can we express the equilibrium state of such a system? A direct approach would be to write each reaction that could occur among the system's species, minerals, and gases. To solve for the equilibrium state, we would determine a set of concentrations that simultaneously satisfy the mass action equation corresponding to each possible reaction. The concentrations would also have to add up, together with the mole numbers of any minerals in the system, to give the system's bulk composition. In other words, the concentrations would also need to satisfy a set of *mass balance equations*.

Such an approach, however, is unnecessarily difficult to carry out. Dissolving even a few elements in water produces many tens of species that need

29

be considered, and complex solutions contain many hundreds of species. Each species represents an independent variable, namely its concentration, in our scheme. For any but the simplest of chemical systems, the problem would contain too many unknown values to be solved conveniently.

Fortunately, few of these variables are truly independent. Geochemists have developed a variety of numerical schemes to solve for equilibrium in multicomponent systems, each of which features a reduction in the number of independent variables carried through the calculation. The schemes are alike in that each solves sets of mass action and mass balance equations. They vary, however, in their choices of thermodynamic components and independent variables, and how effectively the number of independent variables has been reduced.

In this chapter we develop a description of the equilibrium state of a geochemical system in terms of the fewest possible variables and show how the resulting equations can be applied to calculate the equilibrium states of natural waters. We reserve for the next two chapters discussion of how these equations can be solved by using numerical techniques.

3.1 Thermodynamic Description of Equilibrium

To this point we have used a number of terms familiar to geochemists without giving the terms rigorous definitions. We have, for example, discussed thermodynamic components without considering their meaning in a strict sense. Now, as we begin to develop an equilibrium model, we will be more careful in our use of terminology. We will not, however, develop the basic equations of chemical thermodynamics, which are broadly known and clearly derived in a number of texts (as mentioned in Chapter 1).

Phases and Species

A geochemical system can be thought of as an assemblage of one or more phases of given bulk composition. A *phase* is a region of space that is physically distinct, mechanically separable, and homogeneous in its composition and properties. Phases are separated from one another by very thin regions known as *surfaces* over which properties and commonly composition change abruptly (e.g., Pitzer and Brewer, 1961; Nordstrom and Munoz, 1994).

The ancient categories of water, earth, and air persist in classifying the phases that make up geochemical systems. For purposes of constructing a geochemical model, we assume that our system will always contain a fluid phase composed of water and its dissolved constituents, and that it may include the phases of one or more minerals and be in contact with a gas phase. If the fluid phase occurs alone, the system is *homogeneous*; the system when composed of more than one phase is *heterogeneous*.

Species are the molecular entities, such as the gases CO_2 and O_2 in a gas, or the electrolytes Na^+ and SO_4^{--} in an aqueous solution, that exist within a phase. Species, unlike phases, do not have clearly identifiable boundaries. In addition, species may exist only for the most fleeting of moments. Arriving at a precise definition of what a species is, therefore, can be less than straightforward.

In aqueous solutions, geochemists generally recognize dissociated electrolytes and their complexes as species. For example, we can take Ca^{++} as a species in itself, rather than combined with its sphere of hydration as $Ca^{++} \cdot nH_2O$. Similarly, we can represent the neutral species of dissolved silica $SiO_2(aq)$ as H_4SiO_4, or silicic acid, just by adjusting our concept of the species' boundaries. In electrolyte brines, cations and anions occur so close together that the degree of complexing among ions, and hence the extent of a species in solution, is difficult to determine. Keeping in mind the unlikeliness of arriving at a completely unambiguous definition, we will define (following, e.g., Smith and Missen, 1982) a *species* as a chemical entity distinguishable from other entities by molecular formula and structure, and by the phase within which it occurs.

Components and the Basis

The overall composition of any system can be described in terms of a set of one or more chemical *components*. We can think of components as the ingredients in a recipe. A certain number of moles of each component goes into making up a system, just as the amount of each ingredient is specified in a recipe. By combining the components, each in its specified mass, we reproduce the system's bulk composition.

Whereas species and phases exist as real entities that can be observed in nature, components are simply mathematical tools for describing composition. Expressed another way, a component's stoichiometry but not identity matters: water, ice, and steam serve equally well as component H_2O. Since a component needs no identity, it may be either fictive or a species or phase that actually exists in the system. When we express the composition of a fluid in terms of elements or the composition of a rock in terms of oxides, we do not imply that elemental sodium occurs in the fluid, or that calcium oxide is found in the rock. These are fictive components. If we want, we can invent components that exist nowhere in nature.

On the other hand, when we express the chemical analysis of a fluid in terms of the ions Na^+, Ca^{++}, HCO_3^-, and so on, we use a set of components with the stoichiometries of species that really appear in the fluid. In this case, the distinction between species and component is critical. The bicarbonate component, for example, is distributed among a number of species: HCO_3^-, $CO_2(aq)$, CO_3^{--}, $CaHCO_3^+$, $NaCO_3^-$, etc. Hence, the number of moles of component HCO_3^- in the system would differ, perhaps by orders of magnitude, from that of the HCO_3^- species. Similarly, the other components would be distributed among the various species Na^+, NaCl, $CaSO_4$, and so on.

As a second example, we choose quartz (or any silica polymorph) as a component for a system containing an aqueous fluid and quartz. Now the mole number for the quartz component includes not only the silica in the quartz mineral, the real quartz, but the silica in solution in species such as $SiO_2(aq)$ and $H_3SiO_4^-$. Again, the mole numbers of component quartz and real quartz are not the same. A common mistake in geochemical modeling is confusing the components used to describe the composition of a system with the species and phases that are actually present.

The set of components used in a geochemical model is the calculation's *basis*. The basis is the coordinate system chosen to describe composition of the overall system of interest, as well as the individual species and phases that make up the system (e.g., Greenwood, 1975). There is no single basis that describes a given system. Rather, the basis is chosen for convenience from among an infinite number of possibilities (e.g., Morel, 1983). Any useful basis can be selected, and the basis may be changed at any point in a calculation to a more convenient one. We discuss the choice of basis species in the next section.

Chemical Potentials, Activities, and Fugacities

The tools for calculating the equilibrium point of a chemical reaction arise from the definition of the chemical potential. If temperature and pressure are fixed, the equilibrium point of a reaction is the point at which the Gibb's free energy function G is at its minimum (Fig. 3.1). As with any convex-upward function, finding the minimum G is a matter of determining the point at which its derivative vanishes.

To facilitate this analysis, we define the *chemical potential* μ of each species that makes up a phase. The chemical potential of a species B

$$\mu_B \equiv \frac{\partial G_B}{\partial n_B} \tag{3.1}$$

is the derivative of the species' free energy G_B with respect to its mole number n_B. The value of μ depends on temperature, pressure, and the mole numbers of each species in the phase. Since μ is defined as a partial derivative, we take its value holding constant each of these variables except n_B.

Knowing the chemical potential function for each species in a reaction defines the reaction's equilibrium point. Consider a hypothetical reaction

$$bB + cC \rightleftarrows dD + eE \tag{3.2}$$

among species B, C, etc., where b, c, and so on are the reaction coefficients. The free energy is at a minimum at the point where driving the reaction by a small

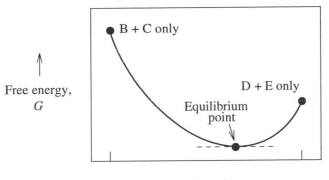

FIG. 3.1 Variation in free energy G with reaction progress for the reaction $b\text{B} + c\text{C} \rightleftarrows d\text{D} + e\text{E}$. The reaction's equilibrium point is the minimum along the free energy curve.

amount forward or backward has no effect on G. From the definition of chemical potential (Eqn. 3.1), the point of minimum G satisfies

$$d\mu_D + e\mu_E - b\mu_B - c\mu_C = 0 \tag{3.3}$$

since d moles of D and e moles of E are produced in the reaction for each b moles of B and c moles of C consumed.

We can find the reaction's equilibrium point from Eqn. 3.3 as soon as we know the form of the function representing chemical potential. The theory of ideal solutions (e.g., Pitzer and Brewer, 1961; Denbigh, 1971) holds that the chemical potential of a species can be calculated from the potential μ_B^o of the species in its pure form at the temperature and pressure of interest. According to this result, a species' chemical potential is related to its standard potential by

$$\mu_B = \mu_B^o + RT_K \ln X_B \tag{3.4}$$

Here, R is the gas constant, T_K is absolute temperature, and X_B is the mole fraction of B in the solution phase. Using this equation, we can calculate the equilibrium point of reactions in ideal systems directly from tabulated values of standard potentials μ^o.

Unfortunately, phases of geochemical interest are not ideal. As well, aqueous species do not occur in a pure form, since their solubilities in water are limited, so a new choice for the standard state is required. For this reason, the chemical potentials of species in solution are expressed less directly (Stumm and Morgan, 1981, and Nordstrom and Munoz, 1994, e.g., give complete discussions), although the form of the ideal solution equation (Eqn. 3.4) is retained.

Aqueous species

The chemical potential of an aqueous species A_i is given

$$\mu_i = \mu_i^o + RT_K \ln\ a_i \qquad (3.5)$$

The mole fraction X in the previous equation is replaced with a new variable a_i, the species' *activity*. The standard potentials μ_i^o are tabulated at a new standard state: a hypothetical one-molal solution of the species in which activity and molality are equal, and in which the species properties have been extrapolated to infinite dilution.

This choice of a standard state seems like impossible mental gymnastics, but it allows activity to follow a molal scale, so that activity and molality are numerically equivalent in dilute solutions. A species' molality m_i, the number of moles of the species per kilogram of solvent, is related to its activity by

$$a_i = \gamma_i m_i \qquad (3.6)$$

The constant of proportionality γ_i is the species' *activity coefficient*, which accounts for the nonideality of the aqueous solution. The species' activity coefficients approach unity in very dilute solutions

$$\gamma_i \rightarrow 1 \ \text{ and } \ a_i \rightarrow m_i \qquad (3.7)$$

so that the species' activities and molalities become nearly equal.

Minerals

Chemical potentials for the constituents of minerals are defined in a similar manner. All minerals contain substitutional impurities that affect their chemical properties. Impurities range from trace substitutions, as might be found in quartz, to widely varying fractions of the end-members of solid solutions series. Solid solutions of geologic significance include clay minerals, zeolites, and plagioclase feldspars, which are important components in most geochemical models.

The chemical potential of each end-member component of a mineral,

$$\mu_k = \mu_k^o + RT_K \ln\ a_k \qquad (3.8)$$

is given in terms of a standard potential μ_k^o, representing the end-member in pure form at the temperature and pressure of interest, and an activity a_k. A geochemical model constructed in the most general manner would account for the activities of all of the constituents in each stable solid solution.

Models can be constructed in this manner (e.g., Bourcier, 1985), but most modelers choose for practical reasons to consider only minerals of fixed composition. The data needed to calculate activities in even binary solid solutions are, for the most part, lacking at temperatures of interest. The solid solutions can sometimes be assumed to be ideal, so that activities equate to mole fractions, but in many cases this assumption leads to errors more severe than those produced by ignoring the solutions altogether. As well, there are several conceptual and theoretical problems (e.g., Glynn et al., 1990) that increase the difficulty of incorporating solid solution theory into reaction modeling in a meaningful way.

In our models, we will consider only minerals of fixed composition. Each mineral, then, exists in its standard state, so that its chemical potential and standard potential are the same

$$\mu_k = \mu_k^o \tag{3.9}$$

and its activity is unity

$$a_k = 1 \tag{3.10}$$

This equality will allow us to eliminate the a_k terms from the governing equations. We will carry these variables through the mathematical development, however, so that the results can be readily extended to account for solid solutions, even though we will not apply them in this manner.

Gases

The chemical potential of a gas species

$$\mu_m = \mu_m^o + RT_K \ln\, f_m \tag{3.11}$$

is given in terms of a standard potential of the pure gas at 1 atm and the temperature of interest, and the gas' fugacity f_m. Fugacity is related to partial pressure

$$f_m = \chi_m P_m \tag{3.12}$$

by a fugacity coefficient χ_m. At low pressures,

$$\chi_m \rightarrow 1 \;\; \text{and} \;\; f_m \rightarrow P_m \tag{3.13}$$

so that fugacity and partial pressure become numerically equivalent.

The Equilibrium Constant

The equilibrium constant expresses the point of minimum free energy for a chemical reaction, as set forth in Eqn. 3.3, in terms of the chemical potential functions above. The criterion for equilibrium becomes

$$d\mu_D^o + e\mu_E^o - b\mu_B^o - c\mu_C^o = -RT_K\Big[d \ln\, a_D + e \ln\, a_E - b \ln\, a_B - c \ln\, a_C\Big] \tag{3.14}$$

when we substitute the chemical potential functions into Eqn. 3.3. (If the reaction involves a gas species, we would replace the appropriate activity with the gas' fugacity.)

The left side of this equation is the reaction's standard free energy

$$\Delta G^\circ = d\mu_D^\circ + e\mu_E^\circ - b\mu_B^\circ - c\mu_C^\circ \qquad (3.15)$$

The equilibrium constant is defined in terms of the standard free energy as

$$\ln\ K = -\frac{\Delta G^\circ}{RT_K} \qquad (3.16)$$

Eqn. 3.14 can be written

$$\ln\ K = d \ln\ a_D + e \ln\ a_E - b \ln\ a_B - c \ln\ a_C \qquad (3.17)$$

or, equivalently,

$$K = \frac{a_D^d a_E^e}{a_B^b a_C^c} \qquad (3.18)$$

which is the familiar mass action equation.

3.2 Choice of Basis

The first decision to be made in constructing a geochemical model is how to choose the basis, the set of thermodynamic components used to describe composition. Thermodynamics provides little guidance in our choice. Given this freedom, we choose a basis for convenience, subject to three rules:

- We must be able to form each species and phase considered in our model from some combination of the components in the basis.

- The number of components in the basis is the minimum necessary to satisfy the first rule.

- The components must be linearly independent of one another. In other words, we should not be able to write a balanced reaction to form one component in terms of the others.

The third rule is, in fact, a logical consequence of the first and second, but we write it out separately because it provides a useful test of a basis choice.

The way we select components to make up the basis is similar to the way a restaurant chef might decide what foodstuffs to buy. The chef needs to be able to prepare each item on the menu from a pantry of ingredients. For various reasons (to simplify ordering, account for limited storage, minimize costs, allow the menu to be changed from day to day, and keep the ingredients fresh), the chef keeps only the minimum number of ingredients on hand. Therefore, the pantry contains no ingredient that can be prepared from the other ingredients on hand. There is no need to store cake mix, since a mixture of flour, sugar, eggs, and so on, serves the same purpose. The chef chooses foodstuffs the same way we choose chemical components.

A straightforward way to choose a basis is to select elements as components. In the presence of redox reactions, the basis also includes the electron or some measure of oxidation state. Clearly, this choice satisfies the three rules mentioned, since any species or phase is composed of elements, and reactions converting one element to another is the stuff of alchemy or nuclear physics, both of which are beyond the scope of this book.

Such a straightforward choice, although commonly used, is seldom the most convenient way to formulate a geochemical model. The chef in our restaurant, if talented enough, could prepare any dish from such ingredients as elemental carbon, hydrogen, and oxygen, instead of flour, eggs, sugar, and so on. (But our analogy is not perfect, since there are more basic ingredients in a kitchen than chemical elements in the foodstuffs; the reader should not take it too literally.) Like the chef's work, our job gets easier if we pick as components certain species or phases that actually go into making up the system of interest.

Convention for Choosing the Basis

Throughout this book, we will choose the following species and phases as components:

- Water, the solvent species,
- Each mineral in equilibrium with the system of interest,
- Each gas species set at known fugacity in the external buffer, and
- Enough aqueous species, preferably those abundantly present in solution, to complete the basis set.

The aqueous species included in the basis are known as *basis species*, while the remaining species in solution comprise the set of *secondary species*.

This choice of basis follows naturally from the steps normally taken to study a geochemical reaction by hand. An aqueous geochemist balances a reaction between two species or minerals in terms of water, the minerals that would be formed or consumed during the reaction, any gases such as O_2 or CO_2 that remain at known fugacity as the reaction proceeds, and, as necessary, the predominant aqueous species in solution. We will show later that formalizing our basis choice in this way provides for a simple mathematical description of equilibrium in multicomponent systems and yields equations that can be evaluated rapidly.

Choosing the basis in this manner sometimes leads to some initial confusion, because we select species present in the system to serve as components. There is a risk of confusing the amount of a component, which describes bulk composition but not the actual state of the system, with the amount of a species or mineral that exists in reality.

Components with Negative Masses

Perhaps the most clear-cut distinction between components and species occurs when a component is present at negative mass. To see how this can occur, we return to our restaurant analogy. The dessert menu includes cakes, which contain whole eggs, and meringue pies made from egg whites. The chef could stock both egg yolks and whites in his pantry, but this would hardly be convenient. He would prefer to stock whole eggs and discard the yolks when necessary. His cake and meringue, then, contain a positive number of eggs, but the meringue contains negative egg yolks.

The same principle applies to a chemical system. Let us consider an alkaline water and assume a component set that includes H_2O and H^+. Each hydroxyl ion

$$H_2O - H^+ \rightarrow OH^- \tag{3.19}$$

is made up of a water molecule less a hydrogen ion. Since the solution contains more hydroxyl than hydrogen ions, the overall solution composition is described in terms of a positive amount of water component and a negative amount of the H^+ component. The molality of the H^+ species itself, of course, is positive.

As a second example, consider a solution rich in dissolved H_2S. If our basis includes SO_4^{--}, H^+, and $O_2(aq)$, then H_2S is formed

$$SO_4^{--} + 2\,H^+ - 2\,O_2(aq) \rightarrow H_2S(aq) \tag{3.20}$$

from a negative amount of $O_2(aq)$. The overall solution composition might well include a negative amount of $O_2(aq)$ component, although the $O_2(aq)$ species would be present at a small but positive concentration.

3.3 Governing Equations

At this point we can derive a set of governing equations that fully describes the equilibrium state of the geochemical system. To do this we will write the set of independent reactions that can occur among species, minerals, and gases in the system and set forth the mass action equation corresponding to each reaction. Then we will derive a mass balance equation for each chemical component in the system. Substituting the mass action equations into the mass balance equations gives a set of governing equations, one for each component, that can be solved directly for the system's equilibrium state.

Independent Reactions

To derive the governing equations we need to identify each independent chemical reaction that can occur in the system. It is possible to write many more reactions than are independent in a geochemical system. The remaining or

TABLE 3.1 Constituents of a geochemical model: the cast of characters

A_w	Water, the solvent
A_i	Aqueous species in the basis, the basis species
A_j	Other aqueous species, the secondary species
A_k	Minerals in the system
A_l	All minerals, even those that do not exist in the system
A_m	Gases of known fugacity
A_n	All gases

dependent reactions, however, are linear combinations of the independent reactions and need not be considered.

Since a geochemical model needs to be cast in general form, the species occurring in reactions are represented symbolically (Table 3.1). Depending on the nature of the problem, we have chosen a basis

$$\mathbf{B} = \left[A_w, A_i, A_k, A_m \right] \tag{3.21}$$

according to the convention in the previous section. Here, A_w is water, A_i are the aqueous species, A_k the minerals, and A_m the gases in the basis. These variables are labels rather than numerical values. For example, A_w is "H_2O" and A_i might be "Na^+."

The independent reactions are those between the secondary species and the basis. In general form, the reactions are

$$A_j \rightleftarrows \nu_{wj} A_w + \sum_i \nu_{ij} A_i + \sum_k \nu_{kj} A_k + \sum_m \nu_{mj} A_m \tag{3.22}$$

Here, ν represents the reaction coefficients: ν_{wj} is the number of moles of water in the reaction to form A_j, ν_{ij} is the number of moles of the basis species A_i, and so on for the minerals and gases.

Need we consider any other reactions? From the previous section, according to the criteria for a valid basis set, we know that no reactions can be written among the basis species themselves. We could write reactions among the secondary species, but such reactions would not be independent. In other words, since we have already written Eqn. 3.22 specifying equilibrium between each secondary species and the basis, any reaction written among the secondary species is redundant (Fig. 3.2).

Taking a basis that contains H_2O, H^+, and HCO_3^-, for example, the reactions for secondary species $CO_2(aq)$ and CO_3^{--} are

$$CO_2(aq) \rightleftarrows HCO_3^- + H^+ - H_2O \tag{3.23}$$

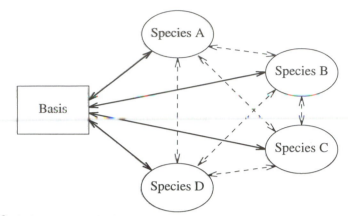

FIG. 3.2 Independent (solid lines) and dependent (dashed lines) reactions in a chemical system composed of a basis and four secondary species A through D. Only the independent reactions need be considered.

and

$$CO_3^{--} \rightleftarrows HCO_3^- - H^+ \tag{3.24}$$

The reaction between these two secondary species

$$CO_2(aq) + H_2O \rightleftarrows CO_3^{--} + 2H^+ \tag{3.25}$$

is simply the first reaction above less the second, and so need not be considered independently.

It is fortunate that we do not have to consider the dependent reactions. Given N_j secondary species, there are just N_j reactions with the basis, but $(N_j^2 - N_j)/2$ reactions could be written among the secondary species. The formula for the latter number, for example, is the number of handshakes if everyone in a group shook everyone else's hand. This is practical at a small party, but impossible at a convention. In chemical systems with many hundreds of species, taking the dependent reactions into account might tax even the most powerful computers.

It is worth noting that the Reaction 3.22 serves two purposes. First, it defines the compositions of all the species A_j in terms of the current component set, the basis **B**. Second, it represents the chemical reactions, each of which has its own equilibrium constant, between the secondary species A_j and the basis species A_w, A_i, A_k, and A_m.

If we had chosen to describe composition in terms of elements, we would need to carry the elemental compositions of all species, minerals, and gases, as well as the coefficients of the independent chemical reactions. Our choice of components, however, allows us to store only one array of reaction coefficients, thereby reducing memory use on the computer and simplifying the forms of the governing equations and their solution. In fact, it is possible to build a complete chemical model (excluding isotope fractionation) without acknowledging the existence of elements in the first place!

Mass Action Equations

Each independent Reaction 3.22 in the system has an associated equilibrium constant K_j at the temperature of interest and, hence, a mass action equation of the form

$$K_j = \frac{a_w^{v_{wj}} \cdot \prod_{i} (\gamma_i m_i)^{v_{ij}} \cdot \prod_{k} a_k^{v_{kj}} \cdot \prod_{m} f_m^{v_{mj}}}{\gamma_j m_j} \qquad (3.26)$$

Here, we have represented the activities of aqueous species with the product $\gamma \cdot m$ of the species' activity coefficients and molalities, according to Eqn. 3.6. The symbol $\prod$ in this equation is the product function, the analog in multiplication to the summation Σ. Table 3.2 lists the meaning of each variable in this and following equations.

A goal in deriving the governing equations is to reduce the number of independent variables by eliminating the molalities m_j of the secondary species. To this end, we can rearrange the equation above to give the value of m_j

$$m_j = \frac{1}{K_j \gamma_j} \left[a_w^{v_{wj}} \cdot \prod_{i} (\gamma_i m_i)^{v_{ij}} \cdot \prod_{k} a_k^{v_{kj}} \cdot \prod_{m} f_m^{v_{mj}} \right] \qquad (3.27)$$

in terms of the molality and activity coefficient of each aqueous species in the basis and the activity or fugacity of each of the other basis entries. This expression is the mass action equation in its final form.

Mass Balance Equations

The mass balance equations express conservation of mass in terms of the components in the basis. The mass of each chemical component is distributed among the species and minerals that make up the system. The water component, for example, is present in free water molecules of the solvent and as the water required to make up the secondary species. According to Eqn. 3.22, each mole of species A_j is composed of v_{wj} moles of the water component. The mole number M_w of water component is given

$$M_w = n_w \left(55.5 + \sum_j v_{wj} m_j \right) \qquad (3.28a)$$

where 55.5 (more precisely, 55.5087) is the number of moles of H_2O in a kilogram of water. Multiplying molality units by the mass n_w of solvent water gives result in moles, as desired.

Similar logic gives the mass balance equation for the species components. The mass of the ith component is distributed among the single basis species A_i

TABLE 3.2 Variables in the governing equations

Bulk composition, moles

M_w	Water component
M_i	Species components
M_k	Mineral components
M_m	Gas components

Solvent mass, molalities, mole numbers

n_w	Solvent mass, kg
m_i	Molalities of basis species
m_j	Molalities of secondary species
n_k	Mole numbers of minerals

Activities and fugacities

a_w	Water activity
a_i	Activities of basis species
a_j	Activities of secondary species
a_k	Mineral activities
f_m	Gas fugacities

Activity coefficients

γ_i	Basis species
γ_j	Secondary species

Reaction coefficients

$\nu_{wj}, \nu_{ij}, \nu_{kj}$	Secondary species
$\nu_{wl}, \nu_{il}, \nu_{kl}$	Minerals
$\nu_{wn}, \nu_{in}, \nu_{kn}$	Gases

Equilibrium constants

K_j	Secondary species
K_l	Minerals
K_n	Gases

---

and the secondary species in the system. By Eqn. 3.22, there are ν_{ij} moles of component i in each mole of secondary species A_j. There is one mole of Na^+ component, for example, per mole of the basis species Na^+, one per mole of the ion pair $NaCl$, and two per mole of the aqueous complex Na_2SO_4, and so on. Mass balance for species component i, then, is expressed

$$M_i = n_w \left(m_i + \sum_j \nu_{ij} m_j \right) \qquad (3.28b)$$

in terms of the solvent mass n_w and the molalities m_i and m_j.

Mineral components are distributed among the mass of actual mineral in the system and the amount required to make up the dissolved species. In a system containing a mole of quartz, for example, there is (in the absence of other silica-bearing components) somewhat more than a mole of component quartz. The additional component mass is required to make up species such as $SiO_2(aq)$ and $H_3SiO_4^-$. Since v_{kj} moles of mineral component k go into making up each mole of secondary species j, mass balance is expressed

$$M_k = n_k + n_w \sum_j v_{kj} m_j \qquad (3.28c)$$

where n_k is the mole number of the mineral corresponding to the component.

Mass balance on gas components is somewhat less complicated because the gas buffer is external to the system. In this case, we need only consider the gas components that make up secondary species:

$$M_m = n_w \sum_j v_{mj} m_j \qquad (3.28d)$$

where v_{mj} is the reaction coefficient from Eqn. 3.22.

Substituted Equations

The final form of the governing equations is given by substituting the mass action equation (Eqn. 3.27) for each occurrence of m_j in the mass balance equations (Eqns. 3.28a–d). The substituted equations are

$$M_w - n_w \left\{ 55.5 + \sum_j \frac{v_{wj}}{K_j \gamma_j} \left[a_w^{v_{wj}} \cdot \prod^i (\gamma_i m_i)^{v_{ij}} \cdot \prod^k a_k^{v_{kj}} \cdot \prod^m f_m^{v_{mj}} \right] \right\} \qquad (3.29a)$$

$$M_i = n_w \left\{ m_i + \sum_j \frac{v_{ij}}{K_j \gamma_j} \left[a_w^{v_{wj}} \cdot \prod^i (\gamma_i m_i)^{v_{ij}} \cdot \prod^k a_k^{v_{kj}} \cdot \prod^m f_m^{v_{mj}} \right] \right\} \qquad (3.29b)$$

$$M_k = n_k + n_w \sum_j \frac{v_{kj}}{K_j \gamma_j} \left[a_w^{v_{wj}} \cdot \prod^i (\gamma_i m_i)^{v_{ij}} \cdot \prod^k a_k^{v_{kj}} \cdot \prod^m f_m^{v_{mj}} \right] \qquad (3.29c)$$

$$M_m = n_w \sum_j \frac{v_{mj}}{K_j \gamma_j} \left[a_w^{v_{wj}} \cdot \prod^i (\gamma_i m_i)^{v_{ij}} \cdot \prod^k a_k^{v_{kj}} \cdot \prod^m f_m^{v_{mj}} \right] \qquad (3.29d)$$

Writing the appropriate governing equation for each chemical component produces a set of equations that completely describes the equilibrium state of the chemical system. As such, the set will include Eqn. 3.29a written once, Eqn. 3.29b written individually for each species component, and Eqns. 3.29c and 3.29d for each mineral and gas component.

Like all formulations of the multicomponent equilibrium problem, these equations are nonlinear by nature because the unknown variables appear in product functions raised to the values of the reaction coefficients. (Nonlinearity also enters the problem because of variation in the activity coefficients.) Such nonlinearity, which is an unfortunate fact of life in equilibrium analysis, arises from the differing forms of the mass action equations, which are product functions, and the mass balance equations, which appear as summations. The equations, however, occur in a straightforward form that can be evaluated numerically, as discussed in the Chapter 5.

We have considered a large number of values (including the molality of each aqueous species, the mole number of each mineral, and the mass of solvent water) to describe the equilibrium state of a geochemical system. In Eqns. 3.29a–d, however, this long list has given way to a much smaller number of values that constitute the set of independent variables. Since there is only one independent variable per chemical component, and hence per equation, we have succeeded in reducing the number of unknowns in the equation set to the minimum possible. In addition, Eqn. 3.29c is linear with respect to n_k and, as discussed below, Eqn. 3.29d need only be evaluated for M_m and hence is linear in its unknown. As we will discuss in the next chapter, the partial linearity of the governing equations leaves them especially easy to evaluate.

To see how the governing equations might be solved, we consider a system that contains an aqueous fluid and several minerals but has no gas buffer. If we know the system's bulk composition in terms of M_w, M_i, and M_k, we can evaluate Eqns. 3.29a–c to give values for the unknown variables: the solvent mass n_w, the basis species' molalities m_i, and the mineral mole numbers n_k.

The other variables in Eqns. 3.29a–c are either known values, such as the equilibrium constants K and reaction coefficients ν, or, in the case of the activity coefficients γ_i, γ_j and activities a_w, a_k, values that can be considered to be known. In practice, the model updates the activity coefficients and activities during the numerical solution so that their values have been accurately determined by the time the iterative procedure is complete.

In solving the equations, we can consider the set of bulk compositions (M_w, M_i, M_k) to be the "boundary conditions" from which we determine the system's equilibrium state. The result is given in terms of the values of (n_w, m_i, n_k). Once these values are known, the dependent variables m_j can be set immediately using Eqn. 3.27. Note that we have demonstrated the conjecture of the first chapter: that the equilibrium state of any system at known temperature and pressure can be calculated once the system's bulk composition in known.

It is commonly convenient, however, to apply some of the governing equations in the reverse manner. A modeler may specify the value for one or more of the variables n_w, m_i, or n_k that we considered independent in the previous paragraph. Such situations are quite common. The modeler may know the mass of solvent water or of the minerals in the system, or the molalities or activities of certain species. He may wish to constrain a_{H^+}, for example, on the basis of a pH measurement.

In these cases, the equation in question is evaluated to give the mole number M_w, and so on, of the corresponding component. In the presence of a gas buffer, the values of one or more fugacities f_m are fixed. Now, the mole number M_m of the gas component remains to be determined. In general, the value of either M_w or n_w needs to be set to evaluate Eqn. 3.29a, and either M_i or m_i is required for each Eqn. 3.29b. Each Eqn. 3.29c can be solved knowing either M_k or n_k, whereas Eqn. 3.29d is generally evaluated directly for M_m.

Charge Balance

The principle of electroneutrality requires that the ionic species in an electrolyte solution remain charge balanced on a macroscopic scale. The requirement of electroneutrality arises from the large amount of energy required to separate oppositely charged particles by any significant distance against Coulombic forces (e.g., Denbigh, 1971). Because of this requirement, we cannot obtain a flask of sodium ions at the chemistry supply room, nor can we measure the activity coefficients of individual ions directly.

The electroneutrality condition can be expressed by the condition of charge balance among the species in solution, according to

$$\sum_i z_i m_i + \sum_j z_j m_j = 0 \tag{3.30}$$

Here, z_i and z_j are the ionic charges on basis and secondary species. It is useful to note, however, that electroneutrality is assured when the components in the basis are charge balanced.

To see this, we use can Eqn. 3.22 to write the ionic charge on a secondary species

$$z_j = \sum_i \nu_{ij} z_i \tag{3.31}$$

in terms of the charges on the basis species. Substituting, the electroneutrality condition becomes

$$\sum_i z_i \left(m_i + \sum_j \nu_{ij} m_j \right) = 0 \tag{3.32}$$

According to the mass balance Eqn. 3.28a, the expression in parentheses is M_i. Further, the charge Z_i on a species component is the same as the charge z_i on the corresponding basis species, since components and species share the same stoichiometry. Substituting, the electroneutrality condition becomes

$$\sum_i Z_i M_i = 0 \qquad (3.33)$$

which requires that components be charge balanced.

This relation is useful because it effectively removes the requirement that M_i be known for one of the basis species. Instead of setting this value directly, it can be determined by balance from the mole numbers of the other basis species. When charged species appear in the basis, in fact, it is customary for equilibrium models to force charge balance by adjusting M_i for a component chosen either by the modeler or the computer program.

The electroneutrality condition is almost always used to set the bulk concentration of the species in abundant concentration for which the greatest analytic uncertainty exists. In practice, this component is generally Cl^- because most commercial labs, unless instructed otherwise, report a chloride concentration calculated by a rough charge balance (i.e., one that excludes the H^+ component and perhaps others) rather than a value resulting from direct analysis. If we were to use the reported chloride content to constrain by rigorous charge balance the concentration of another component, we would at best be propagating the error in the laboratory's rough calculation.

A special danger of the automatic implementation of the electroneutrality condition within computer codes is that the feature can be used to give calculated values for species in small concentration, pH, and even oxidation state. Such values are, however, almost always meaningless because they merely reflect analytical error or even the rough charge balance calculation made by the analytical lab. To see why such calculations fail, consider an attempt to back-calculate the pH of a neutral groundwater. Hydrogen ions are present at a concentration of about 10^{-7} molal. In order to resolve such a small concentration by charge balance, the analyses of all of the other components would have to be accurate to at least 10^{-8} molal, which is of course impossible. Even worse, if the lab reported a chloride concentration calculated to give the appearance of charge balance, then the computed pH would merely reflect the rounding error in the lab's calculation!

Mineral Saturation States

Once we have calculated the distribution of species in the fluid, we can determine the degree to which it is undersaturated or supersaturated with respect to various minerals. For any mineral A_l in the thermodynamic database, we can write a reaction

$$A_l \rightleftarrows \nu_{wl} A_w + \sum_i \nu_{il} A_i + \sum_k \nu_{kl} A_k + \sum_m \nu_{ml} A_m \qquad (3.34)$$

in terms of the basis. Here, A_l is a mineral that can be formed by combining components in the basis. We could not, for example, write a reaction to form muscovite in a system devoid of potassium.

Reaction 3.34 has an activity product Q_l in the form

$$Q_l = \frac{a_w^{\nu_{wl}} \cdot \prod_i^i (\gamma_i m_i)^{\nu_{il}} \cdot \prod_k^k a_k^{\nu_{kl}} \cdot \prod_m^m f_m^{\nu_{ml}}}{a_l} \tag{3.35}$$

Since this equation has the same form as the mass action equation, the reaction is in equilibrium if Q_l equals the reaction's equilibrium constant K_l. In this case, the fluid is saturated with respect to A_l. As a test of our calculations, for example, we would expect the fluid to be saturated with respect to any mineral in the equilibrium system.

The fluid is undersaturated if Q_l is less than K_l. This condition indicates that Reaction 3.34 has not proceeded to the right far enough to reach the saturation point, either because the water has not been in contact with sufficient amounts of the mineral or has not reacted with the mineral long enough. Values of Q_l greater than K_l, on the other hand, indicate that the reaction needs to proceed to the left to reach equilibrium. In this case, the fluid is supersaturated with respect to the mineral.

A fluid's saturation with respect to a mineral A_l is commonly expressed in terms of the *saturation index*

$$SI_l = \log Q_l - \log K_l = \log (Q_l/K_l) \tag{3.36}$$

which is the ratio of activity product to equilibrium constant, expressed as a logarithm. From this equation, an undersaturated mineral has a negative saturation index, a supersaturated mineral has a positive index, and a mineral at the point of saturation has an index of zero. A positive saturation index indicates that the calculated state of the system is a metastable equilibrium because of the thermodynamic drive for Reaction 3.34 to precipitate the supersaturated mineral.

Gas Fugacities

Having determined the distribution of species in solution, we can also calculate the fugacity of the various gases with respect to the fluid. For any gas A_n in the database that can be composed from the component set, we can write a reaction

$$A_n \rightleftarrows \nu_{wn} A_w + \sum_i \nu_{in} A_i + \sum_k \nu_{kn} A_k + \sum_m \nu_{mn} A_m \tag{3.37}$$

in terms of the basis. By mass action, the fugacity f_n of this gas is

$$f_n = \frac{a_w^{\nu_{wn}} \cdot \prod_i^i (\gamma_i m_i)^{\nu_{in}} \cdot \prod_k^k a_k^{\nu_{kn}} \cdot \prod_m^m f_m^{\nu_{mn}}}{K_n} \tag{3.38}$$

where K_n is the equilibrium constant for Reaction 3.37.

The fugacities calculated in this way are those that would be found in a gas phase that is in equilibrium with the system, if such a gas phase were to exist.

Whether a gas phase exists or is strictly hypothetical depends on how the modeler has defined the system, but not on the gas fugacities given by Eqn. 3.38.

The pe and Eh

The pe and Eh are equivalent electrochemical descriptions of oxidation state for a system in equilibrium. For an aqueous solution, any half-cell reaction

$$n\, e^- \rightleftharpoons \nu_{wn} A_w + \sum_i \nu_{in} A_i + \sum_k \nu_{kn} A_k + \sum_m \nu_{mn} A_m \qquad (3.39)$$

where e^- is the electron and n its reaction coefficient, sets the pe and Eh. The electron, of course, does not exist as a free species in solution (e.g., Thorstenson, 1984) and so has no concentration. Species in the fluid can donate and accept electrons from a metallic electrode, however, so we can define and measure the electron's free energy and equilibrium activity (Hostettler, 1984). For example, the reaction

$$e^- + \tfrac{1}{4}\, O_2(aq) + H^+ \rightleftharpoons \tfrac{1}{2}\, H_2O \qquad (3.40)$$

which has a log equilibrium constant K_{e^-} of about 25.5 at 25°C, fixes the equilibrium electron activity when pH and the oxygen and water activities are known.

The pe, by analogy to pH, is defined as

$$pe = -\log a_{e^-}$$

$$= -\frac{1}{n} \log \frac{Q_{e^-}}{K_{e^-}} \qquad (3.41)$$

where n is the number of electrons consumed in the half-cell reaction and Q_{e^-} is the activity product for the half-cell reaction, calculated accounting for each species except the electron. The analogy to pH is imperfect because whereas H^+ is a species that exists in solution, e^- does not. The Nernst equation

$$Eh = -\frac{2.303\, RT_K}{nF} \log \frac{Q_{e^-}}{K_{e^-}}$$

$$= \frac{2.303\, RT_K}{F}\, pe \qquad (3.42)$$

gives the Eh value corresponding to any half-cell reaction. Here, R is the gas constant, T_K absolute temperature, and F the Faraday constant.

Many natural waters, including most waters at low temperature, do not achieve redox equilibrium (e.g., Lindberg and Runnells, 1984; see Section 6.4). In this case, no single value of pe or Eh can be used to represent the redox state. Instead, there is a distinct value for each redox couple in the system. Applying the Nernst equation to Reaction 3.40 gives a pe or Eh representing the hydrolysis of water. Under disequilibrium conditions, this value differs from those calculated from reactions such as

$$8\,e^- + SO_4^{--} + 9\,H^+ \rightleftarrows HS^- + 4\,H_2O \tag{3.43}$$

and

$$e^- + FeOOH + 3\,H^+ \rightleftarrows Fe^{++} + 2\,H_2O \tag{3.44}$$
$$\text{goethite}$$

which represent redox couples for sulfur and iron. The variation among the resulting values of *pe* or Eh provides a measure of the extent of disequilibrium in a system. Techniques for modeling waters in redox disequilibrium are discussed in Chapter 6.

3.4 Number of Variables and the Phase Rule

The most broadly recognized theorem of chemical thermodynamics is probably the phase rule derived by Gibbs in 1875 (see Guggenheim, 1967; Denbigh, 1971). Gibbs' phase rule defines the number of pieces of information needed to determine the state, but not the extent, of a chemical system at equilibrium. The result is the number of degrees of freedom N_F possessed by the system.

The phase rule says that for each phase beyond the first that occurs at equilibrium in a system, N_F decreases by one. Expressed in general form, the phase rule is

$$N_F = N_C - N_\phi + 2 \tag{3.45}$$

where N_C is the number of chemical components in the system, and N_ϕ is the number of phases. If temperature and pressure in the system are fixed (i.e., they have equilibrated with some external medium), as we have assumed here, the rule takes the simplified form

$$N_F = N_C - N_\phi \tag{3.46}$$

The proof of the phase rule is actually implicit in the derivation of the governing equations (Eqns. 3.29a–d), and is not repeated here. It is interesting, nonetheless, to compare this well-known result to the governing equations, if only to demonstrate that we have reduced the problem to the minimum number of independent variables.

The number of components in our geochemical system is given

$$N_C = 1 + N_i + N_k + N_m \tag{3.47}$$

where 1 accounts for water, N_i is the number of aqueous species serving as components (the basis species), and N_k and N_m are the numbers of mineral and gas components. Phases in the system include the fluid, each mineral, and each gas at known fugacity, so

$$N_\phi = 1 + N_k + N_m \tag{3.48}$$

Since the gases are buffered independently, each counts as a separate phase.

The phase rule (Eqn. 3.46), then, predicts that our system has $N_F = N_i$ degrees of freedom. In other words, given a constraint on the concentration or activity of each basis species, we could determine the system's equilibrium state. To constrain the governing equations, however, we need N_C pieces of information, somewhat more than the degrees of freedom predicted by the phase rule.

The extra pieces of information describe the extent of the system — the amounts of fluid and minerals that are present. It is not necessary to know the system's extent to determine its equilibrium state, but in reaction modeling (see Chapter 11) we generally want to track the masses of solution and minerals in the system; we also must know these masses to search for the system's stable phase assemblage (as described in Section 5.4).

Providing an additional piece of information about the size of each phase predicts that a total of $N_i + N_\phi$, or N_C, values is needed to constrain the system's state and extent. This total matches the number of variables we must supply in order to solve the governing equations. Hence, although we can make no claim that we have cast the governing equations in simplest form, we can say that we have reduced the number of independent variables to the minimum allowed by thermodynamics.

4

Changing the Basis

To this point we have assumed the existence of a basis of chemical components that corresponds to the system to be modeled. The basis, as discussed in the previous chapter, includes water, each mineral in the equilibrium system, each gas at known fugacity, and certain aqueous species. The basis serves two purposes: each chemical reaction considered in the model is written in terms of the members of the basis set, and the system's bulk composition is expressed in terms of the components in the basis.

Since we could not possibly store each possible variation on the basis, it is important for us to be able at any point in the calculation to adapt the basis to match the current system. It may be necessary to change the basis (make a *basis swap*, in modeling vernacular) for several reasons. This chapter describes how basis swaps can be accomplished in a computer model, and Chapter 9 shows how this technique can be applied to automatically balance chemical reactions and calculate equilibrium constants.

The modeler first encounters basis swapping in setting up a model, when it may be necessary to swap the basis to constrain the calculation. The thermodynamic dataset contains reactions written in terms of a preset basis that includes water and certain aqueous species (Na^+, Ca^{++}, K^+, Cl^-, HCO_3^-, SO_4^{--}, H^+, and so on) normally encountered in a chemical analysis. Some of the members of the original basis are likely to be appropriate for a calculation. When a mineral appears at equilibrium or a gas at known fugacity appears as a constraint, however, the modeler needs to swap the mineral or gas in question into the basis in place of one of these species.

Over the course of a reaction model, a mineral may dissolve away completely or become supersaturated and precipitate. In either case, the modeling software must alter the basis to match the new mineral assemblage before continuing the calculation. Finally, the basis sometimes must be changed in response to numerical considerations (e.g., Coudrain-Ribstein and Jamet, 1989). Depending on the numerical technique employed, the model may have trouble converging to a solution for the governing equations when one of the basis species occurs at small concentration. Including such a species in the basis can lead to numerical instability because making small corrections to its molality leads to large deviations in the molalities of the secondary species, when they are calculated using the mass action equations. In such a case, the modeling software may swap a more abundant species into the basis.

Fortunately, the process of changing the basis is straightforward and quickly performed on a computer using linear algebra, as we will see in this chapter. Modeling software, furthermore, performs basis changes automatically, so that the details need be of little concern in practice. Nonetheless, the nature of the process should be clear to a modeler. There are four steps in changing the basis: finding the transformation matrix, rewriting reactions to reflect the new basis, altering the equilibrium constants for the reactions, and reexpressing bulk composition in terms of the new basis. The process is familiar to mathematicians, who will recognize it as a linear, nonorthogonal transformation of coordinates.

4.1 Determining the Transformation Matrix

Initially, a basis vector

$$\mathbf{B} = (A_w, A_i, A_k, A_m) \tag{4 1}$$

describes a system. We wish to transform $\mathbf{B}$ to a new basis $\mathbf{B}'$. The new basis, which also contains water, is

$$\mathbf{B}' = (A_w, A'_i, A'_k, A'_m) \tag{4.2}$$

Each entry A'_i, A'_k, or A'_m is a species or mineral that can be formed according to a *swap reaction* as a combination of the entries in the original basis. If A_j, a secondary species under the original basis, is to be swapped into basis position A'_i, the corresponding swap reaction is

$$A'_i = A_j = v_{wj}A_w + \sum_i v_{ij}A_i + \sum_k v_{kj}A_k + \sum_m v_{mj}A_m \tag{4.3}$$

Alternatively, the reaction for swapping a mineral A_l or gas A_n into position A'_k or A'_m is

$$A'_k = A_l = v_{wl}A_w + \sum_i v_{il}A_i + \sum_k v_{kl}A_k + \sum_m v_{ml}A_m \tag{4.4}$$

$$A'_m = A_n = v_{wn}A_w + \sum_i v_{in}A_i + \sum_k v_{kn}A_k + \sum_m v_{mn}A_m \tag{4.5}$$

An equilibrium constant K^{sw} is associated with each swap reaction. The swap reactions, written ensemble, form a matrix equation

$$
\begin{bmatrix} A_w \\ A_i' \\ A_k' \\ A_m' \end{bmatrix} = \begin{bmatrix} \beta \end{bmatrix} \begin{bmatrix} A_w \\ A_i \\ A_k \\ A_m \end{bmatrix}
\tag{4.6}
$$

that gives the new basis in terms of the old. Here, (β) is a matrix of the stoichiometric coefficients

$$
\begin{bmatrix} \beta \end{bmatrix} = \begin{bmatrix} 1 & 0 & 0 & 0 \\ \nu_{wj} & \nu_{ij} & \nu_{kj} & \nu_{mj} \\ \nu_{wl} & \nu_{il} & \nu_{kl} & \nu_{ml} \\ \nu_{wn} & \nu_{in} & \nu_{kn} & \nu_{mn} \end{bmatrix}
\tag{4.7}
$$

from the swap reactions. Writing the old basis in terms of the new, then, is simply a matter of reversing the equation to give

$$
\begin{bmatrix} A_w \\ A_i \\ A_k \\ A_m \end{bmatrix} = \begin{bmatrix} \beta \end{bmatrix}^{-1} \begin{bmatrix} A_w \\ A_i' \\ A_k' \\ A_m' \end{bmatrix}
\tag{4.8}
$$

Here $(\beta)^{-1}$, the inverse of (β), is the *transformation matrix*, which is applied frequently in petrology (e.g., Brady, 1975; Greenwood, 1975; Thompson, 1982), but somewhat less commonly in aqueous geochemistry.

Example: Calculating the Transformation Matrix

These equations are abstract, but an example makes their meaning clear. In calculating a model, we might wish to convert a basis

$$
\mathbf{B} = \begin{bmatrix} H_2O, \ Ca^{++}, \ HCO_3^-, \ H^+ \end{bmatrix}
\tag{4.9}
$$

to a new basis

$$
\mathbf{B}' = \begin{bmatrix} H_2O, \ Ca^{++}, \ Calcite, \ CO_2(g) \end{bmatrix}
\tag{4.10}
$$

in order to take advantage of a known CO_2 fugacity and the assumption of equilibrium with calcite. The swap reactions (including "null" swaps for H_2O and Ca^{++}) are

$$
\begin{aligned}
H_2O &= H_2O \\
Ca^{++} &= Ca^{++} \\
Calcite &= Ca^{++} - H^+ + HCO_3^- \\
CO_2(g) &= -H_2O + H^+ + HCO_3^-
\end{aligned}
\tag{4.11}
$$

or, in matrix form,

$$
\begin{pmatrix} H_2O \\ Ca^{++} \\ Calcite \\ CO_2(g) \end{pmatrix} =
\begin{pmatrix}
1 & 0 & 0 & 0 \\
0 & 1 & 0 & 0 \\
0 & 1 & 1 & -1 \\
-1 & 0 & 1 & 1
\end{pmatrix}
\begin{pmatrix} H_2O \\ Ca^{++} \\ HCO_3^- \\ H^+ \end{pmatrix}
\tag{4.12}
$$

Inverting the matrix of coefficients gives reactions to form the old basis in terms of the new

$$
\begin{pmatrix} H_2O \\ Ca^{++} \\ HCO_3^- \\ H^+ \end{pmatrix} =
\begin{pmatrix}
1 & 0 & 0 & 0 \\
0 & 1 & 0 & 0 \\
\tfrac{1}{2} & -\tfrac{1}{2} & \tfrac{1}{2} & \tfrac{1}{2} \\
\tfrac{1}{2} & \tfrac{1}{2} & -\tfrac{1}{2} & \tfrac{1}{2}
\end{pmatrix}
\begin{pmatrix} H_2O \\ Ca^{++} \\ Calcite \\ CO_2(g) \end{pmatrix}
\tag{4.13}
$$

Written in standard form, the four reactions represented in the matrix equation may appear unusual because of the choice of components, but can be verified to balance. The transformation matrix for this change of basis is

$$
\left[\beta \right]^{-1} =
\begin{pmatrix}
1 & 0 & 0 & 0 \\
0 & 1 & 0 & 0 \\
\tfrac{1}{2} & -\tfrac{1}{2} & \tfrac{1}{2} & \tfrac{1}{2} \\
\tfrac{1}{2} & \tfrac{1}{2} & -\tfrac{1}{2} & \tfrac{1}{2}
\end{pmatrix}
\tag{4.14}
$$

Test for a Valid Basis

The process of determining the transformation matrix provides a chance to check that the current basis is thermodynamically valid. In the previous chapter we noted that if a basis is valid, it is impossible to write a balanced reaction to form one entry in terms of the other entries in the basis.

An equivalent statement is that no row of the coefficient matrix (β) can be formed as a linear combination of the other rows. Since the matrix's determinant is nonzero when and only when this statement is true, we need only evaluate the determinant of (β) to demonstrate that a new basis $\mathbf{B'}$ is valid. In practice, this test can be accomplished using a linear algebra package, or implicitly by testing for error conditions produced while inverting the matrix, since a square matrix has an inverse if and only if its determinant is not zero.

4.2 Rewriting Reactions

Each time we change the basis, we must rewrite each chemical reaction in the system in terms of the new basis. This task, which might seem daunting, is quickly accomplished on a computer, using the transformation matrix. Consider the reaction to form an aqueous species

$$A_j = A_w \nu_{wj} + \sum_i \nu_{ij} A_i + \sum_k \nu_{kj} A_k + \sum_m \nu_{mj} A_m \qquad (4.15)$$

The reaction can be written in vector form

$$A_j = \begin{bmatrix} \nu_{wj}, \nu_{ij}, \nu_{kj}, \nu_{mj} \end{bmatrix} \begin{bmatrix} A_w \\ A_i \\ A_k \\ A_m \end{bmatrix} \qquad (4.16)$$

in terms of the old basis. Substituting the transformation to the new basis (Eqn. 4.6) gives

$$A_j = \begin{bmatrix} \nu_{wj}, \nu_{ij}, \nu_{kj}, \nu_{mj} \end{bmatrix} \begin{bmatrix} \beta \end{bmatrix}^{-1} \begin{bmatrix} A_w \\ A_i' \\ A_k' \\ A_m' \end{bmatrix} \qquad (4.17)$$

The new reaction coefficients for the species A_j, then, are simply the matrix products of the old coefficients and the transformation matrix:

$$\begin{bmatrix} \nu_{wj}', \nu_{ij}', \nu_{kj}', \nu_{mj}' \end{bmatrix} = \begin{bmatrix} \nu_{wj}, \nu_{ij}, \nu_{kj}, \nu_{mj} \end{bmatrix} \begin{bmatrix} \beta \end{bmatrix}^{-1} \qquad (4.18)$$

Similarly, the products

$$\begin{bmatrix} \nu_{wl}', \nu_{il}', \nu_{kl}', \nu_{mk}' \end{bmatrix} = \begin{bmatrix} \nu_{wl}, \nu_{il}, \nu_{kl}, \nu_{ml} \end{bmatrix} \begin{bmatrix} \beta \end{bmatrix}^{-1} \qquad (4.19)$$

$$\begin{bmatrix} \nu_{wn}', \nu_{in}', \nu_{kn}', \nu_{mn}' \end{bmatrix} = \begin{bmatrix} \nu_{wn}, \nu_{in}, \nu_{kn}, \nu_{mn} \end{bmatrix} \begin{bmatrix} \beta \end{bmatrix}^{-1} \qquad (4.20)$$

give the revised reaction coefficients for minerals A_l and gases A_m.

Following our example from above, we write the reaction to form, as an example, carbonate ion

$$CO_3^{--} = \begin{bmatrix} 0, 0, 1, -1 \end{bmatrix} \begin{bmatrix} H_2O \\ Ca^{++} \\ HCO_3^- \\ H^+ \end{bmatrix}$$

$$= \begin{bmatrix} 0, 0, 1, -1 \end{bmatrix} \begin{bmatrix} 1 & 0 & 0 & 0 \\ 0 & 1 & 0 & 0 \\ \frac{1}{2} & -\frac{1}{2} & \frac{1}{2} & \frac{1}{2} \\ \frac{1}{2} & \frac{1}{2} & -\frac{1}{2} & \frac{1}{2} \end{bmatrix} \begin{bmatrix} H_2O \\ Ca^{++} \\ Calcite \\ CO_2(g) \end{bmatrix} \tag{4.21}$$

using the transformation matrix that we already calculated (Eqn. 4.14). Multiplying the numeric terms gives the reaction to form this species in terms of the new basis:

$$CO_3^{--} = \begin{bmatrix} 0, -1, 1, 0 \end{bmatrix} \begin{bmatrix} H_2O \\ Ca^{++} \\ Calcite \\ CO_2(g) \end{bmatrix} \tag{4.22}$$

or simply

$$CO_3^{--} = Calcite - Ca^{++} \tag{4.23}$$

4.3 Altering Equilibrium Constants

The third step in changing the basis is to set the equilibrium constants for the revised reactions. The new equilibrium constant K'_j for a species reaction can be found from its value K_j before the basis swap according to

$$\log K'_j = \log K_j - \sum_i v'_{ij} \log K_i^{sw} - \sum_k v'_{kj} \log K_k^{sw} - \sum_m v'_{mj} \log K_m^{sw} \tag{4.24}$$

Here, K_i^{sw}, K_k^{sw}, and K_m^{sw} are the equilibrium constants for the reactions by which we swap species, minerals, and gases into the basis.

In matrix form, the equation is

$$\log K'_j = \log K_j - \begin{bmatrix} v'_{wj}, v'_{ij}, v'_{kj}, v'_{mj} \end{bmatrix} \begin{bmatrix} 0 \\ \log K_i^{sw} \\ \log K_k^{sw} \\ \log K_m^{sw} \end{bmatrix} \tag{4.25}$$

The equilibrium constants for mineral and gas reactions are calculated from their revised reaction coefficients in similar fashion as

$$\log K'_l = \log K_l - \begin{bmatrix} v'_{wl}, v'_{il}, v'_{kl}, v'_{ml} \end{bmatrix} \begin{bmatrix} 0 \\ \log K_i^{sw} \\ \log K_k^{sw} \\ \log K_m^{sw} \end{bmatrix} \tag{4.26}$$

$$\log K'_n = \log K_n - \left[v'_{wn}, v'_{in}, v'_{kn}, v'_{mn} \right] \begin{bmatrix} 0 \\ \log K_i^{sw} \\ \log K_k^{sw} \\ \log K_m^{sw} \end{bmatrix} \qquad (4.27)$$

Basis entries that do not change over the swap have no effect in these equations, since they are represented by null swap reactions (e.g., $H_2O = H_2O$) with equilibrium constants of unity.

What is the equilibrium constant for CO_3^{--} in the example from the previous section? The value at 25°C for the reaction written in terms of the original basis is $10^{10.34}$, and the equilibrium constants of the swap reactions for calcite and $CO_2(g)$ are $10^{1.71}$ and $10^{-7.82}$. The new value according to the equation above is

$$\log K'_{CO_3^-} = (10.34) - \left[0, -1, 1, 0 \right] \begin{bmatrix} 0 \\ 0 \\ 1.71 \\ -7.82 \end{bmatrix} \qquad (4.28)$$

or $10^{8.63}$. To verify this result, we can calculate the equilibrium constant directly by elimination

$$CO_3^{--} = HCO_3^- - H^+ \qquad\qquad \log K = 10.34$$
$$HCO_3^- + Ca^{++} - H^+ = \text{Calcite} \qquad \log K = -1.71$$

$$CO_3^{--} = \text{Calcite} - Ca^{++} \qquad \log K = 8.63 \qquad (4.29)$$

and arrive at the same value.

4.4 Reexpressing Bulk Composition

The final step in changing the basis is to recalculate the system's bulk composition in terms of the new component set. Composition in terms of the old and new bases are related by the stoichiometric coefficients

$$\begin{bmatrix} M_w \\ M_i \\ M_k \\ M_m \end{bmatrix} = (\beta)^T \begin{bmatrix} M'_w \\ M'_i \\ M'_k \\ M'_m \end{bmatrix} \qquad (4.30)$$

where $(\beta)^T$ is the transpose of (β)

$$
\left[\beta\right]^{T} = \begin{pmatrix} 1 & v_{wj} & v_{wl} & v_{wn} \\ 0 & v_{ij} & v_{il} & v_{in} \\ 0 & v_{kj} & v_{kl} & v_{kn} \\ 0 & v_{mj} & v_{ml} & v_{mn} \end{pmatrix} \tag{4.31}
$$

(i.e., the coefficient matrix flipped on its diagonal). This relationship holds because reaction coefficients give the amounts of the old basis entries that go into making up the new entries, as noted in the previous chapter.

Reversing this equation shows that the composition in terms of the new basis is given immediately by the transpose of the transformation matrix

$$
\begin{pmatrix} M'_w \\ M'_i \\ M'_k \\ M'_m \end{pmatrix} = \left[\beta\right]^{-1 \, T} \begin{pmatrix} M_w \\ M_i \\ M_k \\ M_m \end{pmatrix} \tag{4.32}
$$

(because the inverse of a transposed matrix is the same as the transposed inverse). In other words, the entries in the transformation matrix can be simply flipped (or, in practice, their subscripts reversed) to revise the bulk composition.

Continuing our example from the previous section, we want to use this formula to revise bulk composition in a system in which Calcite and $CO_2(g)$ have been swapped for HCO_3^- and H^+. The old and new compositions are related by

$$
\begin{pmatrix} M_{H_2O} \\ M_{Ca^{++}} \\ M_{HCO_3^-} \\ M_{H^+} \end{pmatrix} = \begin{pmatrix} 1 & 0 & 0 & -1 \\ 0 & 1 & 1 & 0 \\ 0 & 0 & 1 & 1 \\ 0 & 0 & -1 & 1 \end{pmatrix} \begin{pmatrix} M_{H_2O} \\ M_{Ca^{++}} \\ M_{Calcite} \\ M_{CO_2(g)} \end{pmatrix} \tag{4.33}
$$

which we write by transposing the coefficient matrix for the swap reactions (from Eqn. 4.12).

To find the new composition, we need only flip the elements in the transformation matrix about its diagonal, giving

$$
\begin{pmatrix} M_{H_2O} \\ M_{Ca^{++}} \\ M_{Calcite} \\ M_{CO_2(g)} \end{pmatrix} = \begin{pmatrix} 1 & 0 & \tfrac{1}{2} & \tfrac{1}{2} \\ 0 & 1 & -\tfrac{1}{2} & \tfrac{1}{2} \\ 0 & 0 & \tfrac{1}{2} & -\tfrac{1}{2} \\ 0 & 0 & \tfrac{1}{2} & \tfrac{1}{2} \end{pmatrix} \begin{pmatrix} M_{H_2O} \\ M_{Ca^{++}} \\ M_{HCO_3^-} \\ M_{H^+} \end{pmatrix} \tag{4.34}
$$

In interpreting this equation, it is important to recall that the total mole numbers M of the components provide a mathematical tool for describing the system's composition, but do not give the amount of species, minerals, or gases actually present in the system.

According to the equation, we expect more water component when the system is defined in terms of the new basis than the old because hydrogen occurs in the new basis only in H_2O, whereas it was also found in the H^+ and HCO_3^- components of the old basis. (Of course, the amount of actual water remains the same, since we are expressing the same bulk composition in different terms.) Similarly, the calcium in the original system will be distributed between the Ca^{++} and Calcite components of the new system, and so on.

To complete the example numerically, we take as our system a solution of 10^{-3} moles calcium bicarbonate dissolved in one kg (55.5 moles) of water. For simplicity, we set the H^+ component to zero moles, which corresponds to a *p*H of about 8. The system's composition expressed in terms of the two bases is

$$\begin{pmatrix} M_{H_2O} \\ M_{Ca^{++}} \\ M_{HCO_3^-} \\ M_{H^+} \end{pmatrix} = \begin{pmatrix} 55.5 \\ .001 \\ .002 \\ 0 \end{pmatrix} \qquad \begin{pmatrix} M_{H_2O} \\ M_{Ca^{++}} \\ M_{Calcite} \\ M_{CO_2(g)} \end{pmatrix} = \begin{pmatrix} 55.501 \\ 0 \\ .001 \\ .001 \end{pmatrix} \qquad (4.35)$$

where we have calculated the second set of values from the first according to Eqn. 4.34. We can quickly verify that each composition is charge balanced, and that the two vectors contain the same mole numbers of the elements hydrogen, oxygen, calcium, and carbon. The compositions expressed in terms of the two component sets are, in fact, equivalent.

5

Solving for the Equilibrium State

In Chapter 3, we developed equations that govern the equilibrium state of an aqueous fluid and coexisting minerals. The principal unknowns in these equations are the mass of water n_w, the concentrations m_i of the basis species, and the mole numbers n_k of the minerals.

If the governing equations were linear in these unknowns, we could solve them directly using linear algebra. However, some of the unknowns in these equations appear raised to exponents and multiplied by each other, so the equations are nonlinear. Chemists have devised a number of numerical methods to solve such equations (e.g., van Zeggeren and Storey, 1970; Smith and Missen, 1982). All the techniques are iterative and, except for the simplest chemical systems, require a computer. The methods include optimization by steepest descent (White et al., 1958; Boynton, 1960) and gradient descent (White, 1967), back substitution (Kharaka and Barnes, 1973; Truesdell and Jones, 1974), and progressive narrowing of the range of the values allowed for each variable (the monotone sequence method; Wolery and Walters, 1975).

Geochemists, however, seem to have reached a consensus (e.g., Karpov and Kaz'min, 1972; Morel and Morgan, 1972; Crerar, 1975; Reed, 1982; Wolery, 1983) that Newton-Raphson iteration is the most powerful and reliable approach, especially in systems where mass is distributed over minerals as well as dissolved species. In this chapter, we consider the special difficulties posed by the nonlinear forms of the governing equations and discuss how the Newton-Raphson method can be used in geochemical modeling to solve the equations rapidly and reliably.

5.1 Governing Equations

The governing equations are composed of two parts: mass balance equations that require mass to be conserved, and mass action equations that prescribe chemical equilibrium among species and minerals. Water A_w, a set of species A_i, the minerals in the system A_k, and any gases A_m of known fugacity make up the basis **B**:

$$\mathbf{B} = \left[A_w, A_i, A_k, A_m \right] \tag{5.1}$$

The remaining aqueous species are related to the basis entries by the reaction

$$A_j = v_{wj} A_w + \sum_i v_{ij} A_i + \sum_k v_{kj} A_k + \sum_m v_{mj} A_m \tag{5.2}$$

which has an equilibrium constant K_j.

The mass balance equations corresponding to the basis entries are

$$M_w = n_w \left[55.5 + \sum_j v_{wj} m_j \right] \tag{5.3a}$$

$$M_i = n_w \left[m_i + \sum_j v_{ij} m_j \right] \tag{5.3b}$$

$$M_k = n_k + n_w \sum_j v_{kj} m_j \tag{5.3c}$$

$$M_m = n_w \sum_j v_{mj} m_j \tag{5.3d}$$

Here, M_w, M_i, M_k, and M_m give the system's composition in terms of the basis **B**, and m_j is the concentration of each secondary species A_j.

At equilibrium, m_j is given by the mass action equation

$$m_j = \frac{1}{K_j \gamma_j} \left[a_w^{v_{wj}} \cdot \prod^i (\gamma_i m_i)^{v_{ij}} \cdot \prod^k a_k^{v_{kj}} \cdot \prod^m f_m^{v_{mj}} \right] \tag{5.4}$$

Here the a_w and a_k are the activities of water and minerals, and the f_m are gas fugacities. We assume that each a_k equals one, and that a_w and the species' activity coefficients γ can be evaluated over the course of the iteration and thus can be treated as known values in posing the problem.

By substituting this equation for each occurrence of m_j in the mass balance equations, we find a set of equations, one for each basis entry, that describes the equilibrium state in terms of the principal variables. The form of the substituted equations appears in Chapter 3 (Eqns. 3.29a–d), but in this chapter we will carry the variable m_j with the understanding that it represents the result of evaluating Eqn. 5.4.

In some cases, we set one or more of the values of the principal variables as constraints on the system. For example, specifying pH sets m_{H^+}, and setting the mass of solvent water or quartz in the system fixes n_w or n_{Quartz}. To set the bulk compositions M_w and so on, in these cases, we need only evaluate the corresponding equations after the values of the other variables have been determined. Gases appear in the basis as a constraint on fugacity f_m, so Eqn. 5.3d is always evaluated in this manner.

We pose the problem for the remaining equations by specifying the total mole numbers M_w, M_i, and M_k of the basis entries. Our task in this case is to solve the equations for the values of n_w, m_i, and n_k. The solution is more difficult now because the unknown values appear raised to their reaction coefficients and multiplied by each other in the mass action Eqn. 5.4. In the next two sections we discuss how such nonlinear equations can be solved numerically.

5.2 Solving Nonlinear Equations

There is no general method for finding the solution of nonlinear equations directly. Instead, such problems need to be solved indirectly by iteration. The set of values that satisfies a group of equations is called the group's *root*. An iterative solution begins with a guess of the root's value, which the solution procedure tries to improve incrementally until the guess satisfies the governing equations to the desired accuracy.

Of such schemes, two of the most robust and powerful are Newton's method for solving an equation with one unknown variable, and Newton-Raphson iteration, which treats systems of equations in more than one unknown. I will briefly describe these methods here before I approach the solution of chemical problems. Further details can be found in a number of texts on numerical analysis, such as Carnahan et al. (1969).

Newton's Method

In Newton's method, we seek a value of x that satisfies

$$f(x) = a \qquad (5.5)$$

where f is an arbitrary function that we can differentiate, and a is a constant. To start the iteration, we provide a guess $x^{(o)}$ at the value of the root. Unless we are incredibly lucky (or have picked an easy problem that we have already solved), our guess will not satisfy the equation. The inequality between the two sides of the equation is the *residual*

$$R(x) = f(x) - a \qquad (5.6)$$

We can think of the residual as a measure of the "badness" of our guess.

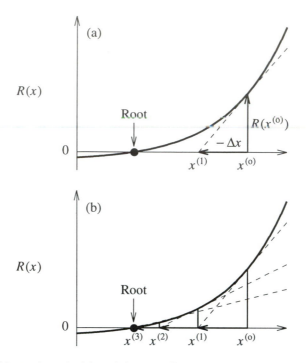

FIG. 5.1 Newton's method for solving a nonlinear equation with one unknown variable. The solution, or root, is the value of x at which the residual function $R(x)$ crosses zero. In (a), given an initial guess $x^{(o)}$, projecting the tangent to the residual curve to zero gives an improved guess $x^{(1)}$. By repeating this operation (b), the iteration approaches the root.

The method's goal is to make the residual vanish by successively improving our guess. To find an improved value $x^{(1)}$, we take the tangent line to the residual function at point $x^{(o)}$ and project it to the zero line (Fig. 5.1). We repeat the projection from $x^{(1)}$ to give $x^{(2)}$, and so on. The process continues until we reach a value $x^{(q)}$ on the (q)-th iteration that satisfies our equation to within a small tolerance.

The method can be expressed mathematically by noting that the slope of the residual function plotted against x is df/dx. Geometrically, the slope is rise over run, so

$$\frac{df}{dx} = -\frac{R(x^{(q)})}{x^{(q+1)} - x^{(q)}} = -\frac{R(x^{(q)})}{\Delta x} \tag{5.7}$$

At any iteration (q), then, the correction Δx is

$$\Delta x = -\frac{R(x^{(q)})}{df/dx} \tag{5.8}$$

As an example, we seek a root to the function

$$2x^3 - x^2 = 1 \tag{5.9}$$

which we can see is satisfied by $x=1$. We iterate from an initial guess of $x=10$ following the Fortran program

```
implicit real*8 (a-h,o-z)
x = 10.
do iter = 1, 99
   resid = 2.*x**3 - x**2 - 1.
   write (6,*) iter-1, x, resid
   if (abs(resid).lt.1.d-10) stop
   dfdx = 6.*x**2 - 2.*x
   x = x - resid/dfdx
end do
end
```

with the results

Iteration, (q)	$x^{(q)}$	$R(x^{(q)})$
0	10.	1900.
1	6.72	560.
2	4.55	170.
3	3.10	49.
4	2.15	14.
5	1.54	3.9
6	1.19	0.94
7	1.033	0.14
8	1.0013	5.2×10^{-3}
9	1.0000021	8.5×10^{-6}
10	1.0000000000057	2.3×10^{-11}

Some words of caution are in order. Many nonlinear functions have more than one root. The choice of an initial guess controls which root will be identified by the iteration. As well, the method is likely to work only when the function is somewhat regular between initial guess and root. A rippled function, for example, would produce tangent lines that project along various positive and negative slopes. Iteration in this case might never locate a root. In fact, the Newton's method can diverge or cycle indefinitely, given a poorly chosen function or initial guess.

Newton-Raphson Iteration

The multidimensional counterpart to Newton's method is Newton-Raphson iteration. A mathematics professor once complained to me, with apparent sincerity, that he could visualize surfaces in no more than twelve dimensions. My perspective on hyperspace is less incisive, as perhaps is the reader's, so we will consider first a system of two nonlinear equations $f = a$ and $g = b$ with unknowns x and y.

To solve the equations, we want to find x and y such that the residual functions

$$R_1(x,y) = f(x,y) - a$$
$$R_2(x,y) = g(x,y) - b \tag{5.10}$$

nearly vanish. Imagine that x and y lie along a table top, and z normal to it with the table surface representing z of zero. Plotting the values of R_1 and R_2 along z produces two surfaces that might extend above and below the table surface (Fig. 5.2). The surfaces intersect along a curved line that, if the problem has a solution, passes through the table top at one or more points. Each such point is a root (x, y) to our problem.

To improve an initial guess $(x^{(o)}, y^{(o)})$, we reach above this point and project tangent planes from the surfaces of R_1 and R_2. The improved guess is the point $(x^{(1)}, y^{(1)})$ where these tangent planes intersect each other and the table top. We repeat this process until each residual function is less than a negligible value.

The corrections Δx and Δy to x and y are those that will project to $R = 0$ along the tangent planes, according to

$$\frac{\partial R_1}{\partial x} \Delta x + \frac{\partial R_1}{\partial y} \Delta y = -R_1$$

$$\frac{\partial R_2}{\partial x} \Delta x + \frac{\partial R_2}{\partial y} \Delta y = -R_2 \tag{5.11}$$

Written in matrix form,

$$\begin{bmatrix} \dfrac{\partial R_1}{\partial x} & \dfrac{\partial R_1}{\partial y} \\[2ex] \dfrac{\partial R_2}{\partial x} & \dfrac{\partial R_2}{\partial y} \end{bmatrix} \begin{bmatrix} \Delta x \\ \Delta y \end{bmatrix} = \begin{bmatrix} -R_1 \\ -R_2 \end{bmatrix} \tag{5.12}$$

the corrections Δx and Δy are given by the solution to two linear equations. This equation can be seen to be the counterpart in two dimensions to the equation (Eqn. 5.8) giving Δx in Newton's method. The equations differ in that there are now vectors of values for the residuals and corrections, and a matrix of partial derivatives replacing the ordinary derivative df/dx.

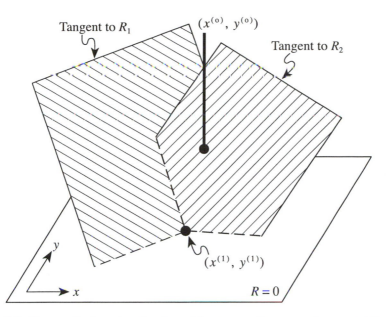

FIG. 5.2 Newton-Raphson iteration for solving two nonlinear equations containing the unknown variables x and y. Planes are drawn tangent to the residual functions R_1 and R_2 at an initial estimate $(x^{(o)}, y^{(o)})$ to the value of the root. The improved guess $(x^{(1)}, y^{(1)})$ is the point at which the tangent planes intersect each other and the plane $R = 0$.

In general, the matrix, known as the *Jacobian*, contains entries for the partial derivative of each residual function R_i with respect to each unknown variable x_j. For a system of n equations in n unknowns, the Jacobian is an $n \times n$ matrix with n^2 entries:

$$\left[\mathbf{J} \right] = \begin{bmatrix} \dfrac{\partial R_1}{\partial x_1} & \dfrac{\partial R_1}{\partial x_2} & \cdots & \dfrac{\partial R_1}{\partial x_n} \\[2ex] \dfrac{\partial R_2}{\partial x_1} & \dfrac{\partial R_2}{\partial x_2} & \cdots & \dfrac{\partial R_2}{\partial x_n} \\[2ex] \cdots & \cdots & & \cdots \\[2ex] \dfrac{\partial R_n}{\partial x_1} & \dfrac{\partial R_n}{\partial x_2} & \cdots & \dfrac{\partial R_n}{\partial x_n} \end{bmatrix} \tag{5.13}$$

Writing the residual functions and corrections as vectors

$$\left[\mathbf{R} \right] = \left[R_1, R_2, \cdots R_n \right] \tag{5.14}$$

$$\left[\Delta \mathbf{x}\right] = \left[\Delta x_1, \ \Delta x_2, \ \cdots \ \Delta x_n\right] \tag{5.15}$$

gives the general equation for determining the correction

$$\left[\mathbf{J}\right]\left[\Delta \mathbf{x}\right] = -\left[\mathbf{R}\right] \tag{5.16}$$

in a Newton-Raphson iteration.

Like Newton's method, the Newton Raphson procedure has just a few steps. Given an estimate of the root to a system of equations, we calculate the residual for each equation. We check to see if each residual is negligibly small. If not, we calculate the Jacobian matrix and solve the linear Eqn. 5.16 for the correction vector. We update the estimated root with the correction vector

$$\left[\mathbf{x}\right]^{(q+1)} = \left[\mathbf{x}\right]^{(q)} + \left[\Delta \mathbf{x}\right] \tag{5.17}$$

and return to calculate new values for residuals. The solution procedure, then, reduces to the repetitive solution of a system of linear equations, a task well suited to modern computers.

5.3 Solving the Governing Equations

In this section we consider how Newton-Raphson iteration can be applied to solve the governing equations listed in Section 5.1. There are three steps to setting up the iteration: (1) reducing the complexity of the problem by reserving the equations that can be solved linearly, (2) computing the residuals, and (3) calculating the Jacobian matrix. Because reserving the equations with linear solutions reduces the number of basis entries carried in the iteration, the solution technique described here is known as the "reduced basis method."

The Reduced Problem

The computing time required to evaluate Eqn. 5.16 in a Newton-Raphson iteration increases with the cube of the number of equations considered (Dongarra et al., 1979). The numerical solution to Eqns. 5.3a–d, therefore, can be found most rapidly by reserving from the iteration any of these equations that can be solved linearly. There are four cases in which equations can be reserved:

- If the mass of water n_w is a constraint on the system, Eqn. 5.3a can be evaluated directly for M_w.

- When the system chemistry is constrained by the concentration m_i (or activity a_i) of a basis species, Eqn. 5.3b gives M_i directly.

- Eqn. 5.3c can *always* be reserved, because the mineral mass n_k is linear in the equation.

- Eqn. 5.3d can also be reserved because the gas fugacities f_m are known.

The nonlinear portion of the problem, then, consists of just two parts:

- Eqn. 5.3a, when n_w is unknown, and
- Eqn. 5.3b, for each basis species at unknown concentration m_i.

The basis entries corresponding these two cases are given by the "reduced basis"

$$\mathbf{B}_r = \left[A_w, A_i \right]_r \tag{5.18}$$

We will carry the subscript r (for "reduced") to indicate that a vector or matrix includes only entries conforming to one of the two nonlinear cases.

Residual Functions

The residual functions measure how well a guess $(n_w, m_i)_r$ satisfies the governing Eqns. 5.3a–b. The form of the residuals can be written

$$R_w = n_w \left[55.5 + \sum_j \nu_{wj} m_j \right] - M_w \tag{5.19a}$$

$$R_i = n_w \left[m_i + \sum_j \nu_{ij} m_j \right] - M_i \tag{5.19b}$$

to represent the inequalities involved in evaluating these equations. The vector of residuals corresponding to the reduced basis $\mathbf{B}_r$ is

$$\mathbf{R}_r = \left[R_w, R_i \right]_r \tag{5.20}$$

Jacobian Matrix

The Jacobian matrix contains the partial derivatives of the residuals with respect to each of the unknown values $(n_w, m_i)_r$. To derive the Jacobian, it is helpful to note that

$$\frac{\partial m_j}{\partial n_w} = 0 \quad \text{and} \quad \frac{\partial m_j}{\partial m_i} = \nu_{ij} \, m_j / m_i \tag{5.21}$$

as can be seen by differentiating Eqn. 5.4. The Jacobian entries, given by differentiating 5.19a–b, are

$$J_{ww} = \frac{\partial R_w}{\partial n_w} = 55.5 + \sum_j \nu_{wj} m_j \tag{5.22a}$$

Concurrent loops

```
do 3   i1 = 1, nbasis_spec
do 2   i2 = 1, nbasis_spec
```

Vector loop

```
      sum = 0.0
      do 1  j = 1, n_species
1     sum = sum + v_i1,j  * v_i2,j  * m_j
```

```
2   J_i1,i2 = n_w * sum/m_i2
3   J_i1,i1 = J_i1,i1 + n_w
```

FIG. 5.3 Calculation of the entries $J_{ii'}$ in the Jacobian matrix on a vector-parallel computer, using a concurrent-outer, vector-inner (COVI) scheme. Each summation in the Jacobian can be calculated as a vector pipeline as separate processors calculate the entries in parallel.

$$J_{wi} = \frac{\partial R_w}{\partial m_i} = \frac{n_w}{m_i} \sum_j v_{wj} v_{ij} m_j \tag{5.22b}$$

$$J_{iw} = \frac{\partial R_i}{\partial n_w} = m_i + \sum_j v_{ij} m_j \tag{5.22c}$$

$$J_{ii'} = \frac{\partial R_i}{\partial m_{i'}} = n_w \left[\delta_{ii'} + \sum_j v_{ij} v_{i'j} m_j / m_{i'} \right] \tag{5.22d}$$

where $\delta_{ii'}$ is the Kronecker delta function

$$\delta_{ii'} = \begin{cases} 1 & \text{if } i=i' \\ 0 & \text{if } i \neq i' \end{cases} \tag{5.23}$$

From a computational point of view, the forms of the Jacobian entries above are welcome because they conform to the architectural requirements of vector, parallel, and vector-parallel computers (Fig. 5.3)

Newton-Raphson Iteration

The Newton-Raphson iteration works by incrementally improving an estimate to values $(n_w, m_i)_r$ of the unknown variables in the reduced basis. The procedure begins with a guess at the variables' values. The first guess might be supplied by the modeler, but more commonly the model sets the guess using an *ad hoc* procedure such as assigning 90% of the mole numbers M_w and M_i to the basis species. If the procedure is invoked while tracing a reaction path, the result for the previous step along the path probably provides the best first guess. The guess can be optimized to better assure convergence in difficult cases, as described later in this section.

From Eqn. 5.16, the correction to an estimated solution is given by solving the system

$$\begin{bmatrix} J_{ww} & J_{wi} \\ J_{iw} & J_{ii'} \end{bmatrix}_r \begin{bmatrix} \Delta n_w \\ \Delta m_i \end{bmatrix}_r = \begin{bmatrix} -R_w \\ -R_i \end{bmatrix}_r \qquad (5.24)$$

for $(\Delta n_w, \Delta m_i)_r$. The equation system can be solved by a variety of methods using widely available software, such as the Linpack library (Dongarra et al., 1979). The correction is added to the current values of the unknown variables, and the iteration continues until the magnitudes of the residual functions fall below a prescribed tolerance.

At this point we can see the advantage of working with the reduced problem. Most published algorithms carry a nonlinear variable for each chemical component plus one for each mineral in the system. The number of nonlinear variables in the method presented here, on the other hand, is the number of components minus the number of minerals. Depending on the size of the problem, the savings in computing effort in evaluating Eqn. 5.24 can be dramatic (Fig. 5.4).

Non-Negativity

There are, in fact, a number of solutions to the governing equations, but usually (see Chapter 10) only one with positive mole numbers and concentrations. Fortunately, the latter answer is of interest to all but the most abstract-thinking geochemist. The requirement that the iteration produce positive masses is known in chemical modeling as the non-negativity constraint.

One method of assuring positive values is to carry the logarithms of the unknown variables through the solution, since this function is not defined for negative numbers (van Zeggeren and Storey, 1970; Wolery, 1979). An alternative and perhaps more straightforward method is to begin the iteration with positive values for the unknown variables and then to scale back corrections that would drive any value negative. This technique in numerical analysis is called *under-relaxation*. The method is the computational equivalent of the childhood riddle: If Marco Polo traveled each day half way to China, how long would it take to get there? Marco's journey would shorten by half each day so that he would never quite reach his destination. Similarly, values in the under-relaxed iteration can approach zero but never become negative.

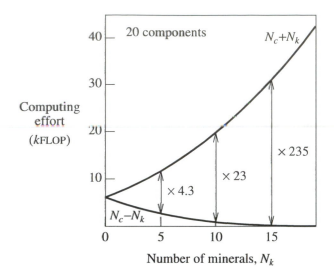

FIG. 5.4 Comparison of the computing effort, expressed in thousands of floating point operations (kFLOP), required to factor the Jacobian matrix for a 20-component system ($N_c = 20$) during a Newton-Raphson iteration. For a technique that carries a nonlinear variable for each chemical component and each mineral in the system (top line), the computing effort increases as the number of minerals increases. For the reduced basis method (bottom line), however, less computing effort is required as the number of minerals increases.

We force non-negativity upon a Newton-Raphson iteration by defining an under-relaxation factor

$$\frac{1}{\delta_{UR}} = \max\left[1, \ -\frac{\Delta n_w}{\frac{1}{2}\, n_w^{(q)}} \ , \ -\frac{\Delta m_i}{\frac{1}{2}\, m_i^{(q)}}\right]_r \tag{5.25}$$

The updated values are calculated according to

$$\begin{bmatrix} n_w \\ m_i \end{bmatrix}_r^{(q+1)} = \begin{bmatrix} n_w \\ m_i \end{bmatrix}_r^{(q)} + \delta_{UR}\begin{bmatrix} \Delta n_w \\ \Delta m_i \end{bmatrix}_r \tag{5.26}$$

By these equations, the correction is allowed to reduce any variable by no more than half its value.

Examples of Convergence

The resulting iteration scheme converges strongly, with each iteration likely to reduce the residuals by an order of magnitude or more. Figure 5.5 shows how

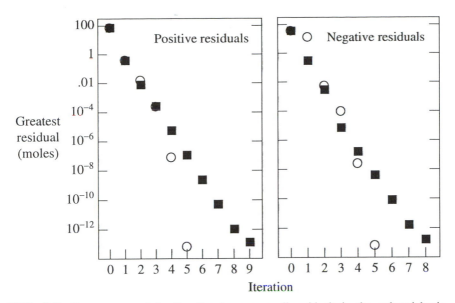

FIG. 5.5 Convergence of the iteration to very small residuals in the reduced basis method, for systems contrived to have large positive and negative residuals at the start of the iteration. The tests assume Debye-Hückel activity coefficients (■) and an ideal solution in which the activity coefficients are unity (O).

the iteration converges when the initial residuals take on large negative and positive values. The figure also shows the convergence when the activity coefficients and water activity are set to one. The extra nonlinearity introduced by the activity coefficients slows convergence by two or three iterations in these tests. From experience, the iteration can be expected to converge within about ten iterations; solutions requiring more than a hundred iterations are rather uncommon.

Optimizing the Starting Guess

The Newton-Raphson iteration usually converges to a solution rapidly enough that the choice of a starting guess is of little practical importance. Sometimes, however, an aqueous species that will end up at extremely low concentration appears in the basis. For example, the basis used to solve for the species distribution in a reduced fluid might contain $O_2(aq)$. The oxidized species begins at high concentration, so by the mass action equations the reduced species such as $H_2(aq)$ and $H_2S(aq)$ start with impossibly large molalities. Hence, the residual for this basis entry becomes extremely large, sometimes in excess of 10^{100}.

Such a situation is dangerous because even though the iteration many times converges nicely, seemingly against impossible odds, the algorithm sometimes diverges to the pseudoroot $n_w = 0$ to Eqn. 5.19b. An effective strategy is to repeatedly halve the starting guess for the basis species corresponding to very large positive residuals until the residuals reach a manageable size, perhaps less than 10^3.

Activity Coefficients

To this point, we have assumed that the activity coefficients γ_i and γ_j as well as the activity of water a_w are known values. In fact, these values vary with m_i. Our strategy is to ignore this variation while calculating the Jacobian matrix and then update the activity coefficients and water activity after each step in the iteration.

Such a scheme is sometimes called a "soft" Newton-Raphson formulation because the partial derivatives in the Jacobian matrix are incomplete. We could, in principle, use a "hard" formulation in which the Jacobian accounts for the deviatives $\partial\gamma/\partial m_i$ and $\partial a_w/\partial m_i$. The hard formulation sometimes converges in fewer iterations, but in tests, the advantage was more than offset by the extra effort in computing the Jacobian. The soft method also allows us to keep the method for calculating activity coefficients (see Chapter 7) separate from the Newton-Raphson formulation, which simplifies programming.

Charge Balance

Our numerical solution must also honor charge balance among the dissolved species. As shown in Chapter 3, species are charge balanced when the components balance. In calculations constrained by known values of M_i, charge balance can be checked before beginning the solution.

A modeler, however, sometimes constrains one or more components in terms of a basis species' free concentration m_i (i.e., by specifying pH). Charge balance cannot be assured a priori because the system's bulk composition is not known until the iteration has converged. To force electrical neutrality, the model adjusts the mole number M_i of a charged component such as Cl^- after each iteration. This adjustment may be of little practical importance, because laboratories commonly report chloride concentrations computed from charge balance rather than from direct analysis of the element.

Mineral Masses

As we have noted, the mole numbers n_k of minerals in the system appear as linear terms in Eqn. 5.3c. For this reason, these equations are omitted from the reduced basis. After the iteration is complete, the values of n_k, when unknown, are calculated according to

$$n_k = M_k - n_w \sum_j \nu_{kj} m_j \tag{5.27}$$

which is obtained by reversing Eqn. 5.3c. Note that there is nothing in the solution procedure that prevents negative values of n_k, a useful feature in determining the stable mineral assemblage (see Section 5.4).

Bulk Composition

The remaining step is to compute the system's bulk composition, if it is not fully known, according to the mass balance equations. The mole numbers M_w, M_i, and M_k are not known when the modeler has constrained the corresponding variable n_w, m_i, or n_k. In these cases, the mole numbers are determined directly from Eqns. 5.3a–c. Where gases appear in the basis, the mole numbers M_m of gas components are similarly calculated from Eqn. 5.3d.

5.4 Finding the Stable Phase Assemblage

The calculation described to this point does not predict the assemblage of minerals that is stable in the current system. Instead, the assemblage is assumed implicitly by setting the basis **B** before the calculation begins. A solution to the governing equations constitutes the equilibrium state of the system if two conditions are met: (1) no mineral in the system has dissolved away and become undersaturated, and (2) the fluid is not supersaturated with respect to any mineral.

A calculation procedure could, in theory, predict at once the distribution of mass within a system and the equilibrium mineral assemblage. Brown and Skinner (1974) undertook such a calculation for petrologic systems. For an n-component system, they calculated the shape of the free energy surface for each possible solid solution in a rock. They then raised an n-dimensional hyperplane upward, allowing it to rotate against the free energy surfaces. The hyperplane's resting position identified the stable minerals and their equilibrium compositions. Inevitably, the technique became known as the "crane plane" method.

Such a method seldom has been used with systems containing an aqueous fluid, probably because the complexity of the solution's free energy surface and the wide range in aqueous solubilities of the elements complicate the numerics of the calculation (e.g., Harvie et al., 1987). Instead, most models employ a procedure of elimination. If the calculation described fails to predict a system at equilibrium, the mineral assemblage is changed to swap undersaturated minerals out of the basis or supersaturated minerals into it, following the steps in the previous chapter; the calculation is then repeated.

Undersaturated Minerals

Minerals that have become undersaturated are revealed in the iteration results by negative mole numbers n_k. A negative mass, of course, is not meaningful physically beyond demonstrating that the mineral was completely consumed, perhaps to form another mineral, in the approach to equilibrium.

Minerals that develop negative masses are removed from the basis one at a time, and the solution is then recalculated. When a mineral is removed, an aqueous species must be selected from among the secondary species A_j to replace it in the basis. The species selected should be in high concentration to assure numerical stability in the iterative scheme described here and must, in combination with the other basis entries, form a valid component set (see Section 3.2). The species best fitting these criteria satisfies

$$\max_{j} \left[m_j \cdot |\nu_{kj}| \right] \tag{5.28}$$

where ν_{kj} is the reaction coefficient for the undersaturated mineral in the reaction to form the secondary species.

Supersaturated Minerals

The saturation state of each mineral that can form in a model must be checked once the iteration is complete to identify supersaturated minerals. A mineral A_l, which is not in the basis, forms by the reaction

$$A_l = \nu_{wl}A_w + \sum_{i} \nu_{il}A_i + \sum_{k} \nu_{kl}A_k + \sum_{m} \nu_{ml}A_m \tag{5.29}$$

which has an associated equilibrium constant K_l. The saturation state is given by the ratio Q_l/K_l, where Q_l is the mineral's activity product

$$Q_l = a_w^{\nu_{wl}} \cdot \prod_{i} (\gamma_i m_i)^{\nu_{il}} \cdot \prod_{k} a_k^{\nu_{kl}} \cdot \prod_{m} f_m^{\nu_{ml}} \tag{5.30}$$

Ratios greater than one identify supersaturated minerals that need to be swapped into the basis and allowed to precipitate.

Choosing the location in the basis for the new mineral is a matter of identifying a basis species A_i that is similar in composition to the mineral to be removed and preferably in small concentration. The best species to be displaced from the basis satisfies

$$\max_{i} \left[\frac{|\nu_{il}|}{m_i} \right] \tag{5.31}$$

where ν_{il} are the coefficients for the basis species in the reaction to form the supersaturated mineral.

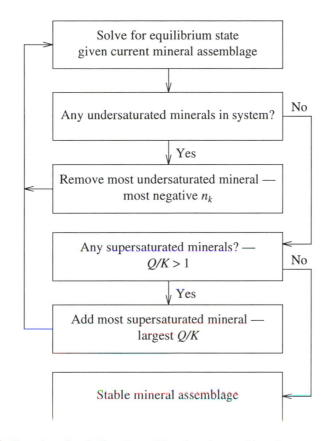

FIG. 5.6 Procedure for finding the stable mineral assemblage in a system of known composition.

Swap Procedure

A calculated solution may have just one supersaturated or undersaturated mineral, in which case calculating the solution for the new mineral assemblage will give the equilibrium state. Not uncommonly, however, more than one mineral appears under- or supersaturated. Such a situation is best handled one swap at a time, according to a set algorithm coded into the model.

No algorithm is guaranteed to arrive at the equilibrium state in this manner (Wolery, 1979), but the procedure outlined in Fig. 5.6 gives good results in solving a wide range of problems. The procedure first checks for undersaturated minerals. If any exist, the one with the most negative n_k is removed from the basis and the governing equations solved once again.

Once there are no undersaturated minerals, the procedure checks for supersaturated minerals. If any exist, the most supersaturated mineral, identified by the largest Q_l/K_l, is swapped into the basis and the governing equations are solved. Precipitating a new mineral, however, may dissolve another away, so now the process begins anew by checking for undersaturated minerals. Once a solution has been found that includes neither undersaturated or supersaturated minerals, the true equilibrium state has been located.

Apparent Violation of the Phase Rule

Sometimes when a mineral becomes supersaturated, there is no logical aqueous species in the basis with which to swap the mineral. Such a situation occurs when no species appear in the reaction to form the mineral. Wolery (1979) and Reed (1982) refer to such a situation as an ''apparent violation of the phase rule,'' because adding the mineral to the basis would produce more phases in the system than there are components.

To include the supersaturated mineral in the basis in this case, another mineral must be removed. Although different schemes have been suggested for identifying the mineral to be removed, the most straightforward is to recognize that the reaction to form the supersaturated mineral

$$v_{wl}A_w + \sum_k v_{kl}A_k + \sum_m v_{ml}A_m \to A_l \tag{5.32}$$

will proceed until a mineral A_k in the basis is exhausted. The mineral satisfying

$$\min_k \left[\frac{n_k}{v_{kl}} \right] \tag{5.33}$$

is the first mineral exhausted and hence the entry to be swapped out of the basis

6

Equilibrium Models of Natural Waters

Having derived a set of equations describing the equilibrium state of a multicomponent system and devised a scheme for solving them, we can begin to model the chemistries of natural waters. In this chapter we construct four models, each posing special challenges, and look in detail at the meaning of the calculation results.

In each case, we use program REACT and employ an extended form of the Debye-Hückel equation for calculating species' activity coefficients, as discussed in Chapter 7. In running REACT, you work interactively following the general procedure:

- **Swap** into the basis any needed species, minerals, or gases. Table 6.1 shows the basis in its original configuration (as it exists when you start the program). You might want to change the basis by replacing $SiO_2(aq)$ with quartz so that equilibrium with this mineral can be used to constrain the model. Or to set a fugacity buffer you might swap $CO_2(g)$ for either H^+ or HCO_3^-.

- **Set** a constraint for each basis member that you want to include in the calculation. For instance, the constraint might be the total concentration of sodium in the fluid, the free mass of a mineral, or the fugacity of a gas. You may also set temperature (25°C, by default) or special program options.

- **Run** the program by typing go.

- **Revise** the basis or constraints and reexecute the program as often as you wish.

In this book, input scripts for running the various programs are set in a "typewriter" typeface. Unless a script is marked as a continuation of the

TABLE 6.1 Basis species in the LLNL database

H_2O	Cr^{+++}	Li^+	Ru^{+++}
Ag^+	Cs^+	Mg^{++}	SO_4^{--}
Al^{+++}	Cu^+	Mn^{++}	SeO_3^{--}
Am^{+++}	Eu^{+++}	NO_3^-	$SiO_2(aq)$
$As(OH)_4^-$	F^-	Na^+	Sn^{++++}
Au^+	Fe^{++}	Ni^{++}	Sr^{++}
$B(OH)_3$	H^+	Np^{++++}	TcO_4^-
Ba^{++}	HCO_3^-	$O_2(aq)$	Th^{++++}
Br^-	HPO_4^{--}	Pb^{++}	U^{++++}
Ca^{++}	Hg^{++}	PuO_2^+	V^{+++}
Cl^-	I^-	Ra^{++}	Zn^{++}
Co^{++}	K^+	Rb^+	

previous script, you should start the program anew or type `reset` to clear your previous configuration.

6.1 Chemical Model of Seawater

For a first chemical model, we calculate the distribution of species in surface seawater, a problem first undertaken by Garrels and Thompson (1962; see also Thompson, 1992). We base our calculation on the major element composition of seawater (Table 6.2), as determined by chemical analysis. To set pH, we assume equilibrium with CO_2 in the atmosphere (Table 6.3). Since the program will determine the HCO_3^- and water activities, setting the CO_2 fugacity (about equal to partial pressure) fixes pH according to the reaction

$$H^+ + HCO_3^- \rightleftarrows CO_2(g) + H_2O \qquad (6.1)$$

Similarly, we define oxidation state according to

$$O_2(aq) \rightleftarrows O_2(g) \qquad (6.2)$$

by specifying the fugacity of $O_2(g)$ in the atmosphere.

The latter two assumptions are simplistic, considering the number of factors that affect pH and oxidation state in the oceans (e.g., Sillén, 1967; Holland, 1978; McDuff and Morel, 1980). Consumption and production of CO_2 and O_2 by plant and animal life, reactions among silicate minerals, dissolution and precipitation of carbonate minerals, solute fluxes from rivers, and reaction between convecting seawater and oceanic crust all affect these variables. Nonetheless, it will be interesting to compare the results of this simple calculation to observation.

TABLE 6.2 Major element composition
of seawater (Drever, 1988)

	Concentration (mg/kg)
Cl^-	19,350
Na^+	10,760
SO_4^{--}	2,710
Mg^{++}	1,290
Ca^{++}	411
K^+	399
HCO_3^-	142
$SiO_2(aq)$	.5-10
$O_2(aq)$	.1-6

To calculate the model, we swap $CO_2(g)$ and $O_2(g)$ into the basis in place of H^+ and $O_2(aq)$, and constrain each basis member. The procedure is

```
swap CO2(g)  for H+
swap O2(g)  for O2(aq)
log f CO2(g)  =  -3.5
f O2(g)       =   0.2

TDS = 35080
Cl-        = 19350 mg/kg
Ca++       =   411 mg/kg
Mg++       =  1290 mg/kg
Na+        = 10760 mg/kg
K+         =   399 mg/kg
SO4--      =  2710 mg/kg
HCO3-      =   142 mg/kg
SiO2(aq)  =     6 mg/kg

print species=long
go
```

Here, we define the total dissolved solids (in mg/kg) so that the program can correctly convert our input constraints from mg/kg to molal units, as carried internally (i.e., variables m_i and m_j). The print command causes the program to list in the output all of the aqueous species, not just those in greatest concentration. Typing go triggers the model to begin calculations and write its results to the output dataset.

TABLE 6.3 Partial pressures of some
gases in the atmosphere (Hem, 1985)

Gas	Pressure (atm)
N_2	.78
O_2	.21
H_2O	.001–.23
CO_2	.0003
CH_4	1.5×10^{-6}
CO	$(.06–1) \times 10^{-6}$
SO_2	1×10^{-6}
N_2O	5×10^{-7}
H_2	$\sim 5 \times 10^{-7}$
NO_2	$(.05–2) \times 10^{-8}$

Species Distribution

The program produces in its output dataset a block of results that shows the concentration, activity coefficient, and activity calculated for each aqueous species (Table 6.4), the saturation state of each mineral that can be formed from the basis, the fugacity of each such gas, and the system's bulk composition. The extent of the system is 1 kg of solvent water and the solutes dissolved in it; the solution mass is 1.0364 kg.

In the calculation results, we can quickly identify the input constraints: the fugacities of $CO_2(g)$ and $O_2(g)$ and the bulk composition expressed in terms of components Cl^-, Ca^{++}, and so on. Note that the free species concentrations do not satisfy the input constraints, which are bulk or total values. The free concentration of the species Ca^{++}, in other words, accounts for just part of the solution's calcium content.

The predicted *p*H is 8.34, a value lying within but toward the alkaline end of the range 7.8 to 8.5 observed in seawater (Fig. 6.1). The dissolved oxygen content predicted by the calculation is 215 μmol/kg, or 6.6 mg/kg. This value compares well with values measured near the ocean surface (Fig. 6.2).

It is clear from the species distribution that the dissolved components in seawater react to varying extents to form complex species (Table 6.5). Components Na^+ and Cl^- are present almost entirely as free ions. Only a few percent of their masses appear in complexes, most notably the ion pairs $MgCl^+$, $NaSO_4^-$, $CaCl^+$, and $NaCl$ (Table 6.4). Components Ca^{++}, SO_4^{--}, and HCO_3^-, on the other hand, complex strongly; complex species account for a third to a half of their total concentrations.

TABLE 6.4 Calculated molalities (m), activity coefficients (γ), and log activities (a) of the most abundant species in seawater

Species	m	γ	log a
Cl^-	.5500	.6276	−.4619
Na^+	.4754	.6717	−.4958
Mg^{++}	$.3975\times10^{-1}$	.3160	−1.9009
SO_4^{--}	$.1607\times10^{-1}$	.1692	−2.5657
K^+	$.1033\times10^{-1}$	.6276	−2.1881
$MgCl^+$	$.9126\times10^{-2}$	.6717	−2.2125
$NaSO_4^-$	$.6386\times10^{-2}$	.6717	−2.3676
Ca^{++}	$.5953\times10^{-2}$	.2465	−2.8334
$MgSO_4$	$.5767\times10^{-2}$	1.0	−2.2391
$CaCl^+$	$.3780\times10^{-2}$	.6717	−2.5953
$NaCl$	$.2773\times10^{-2}$	1.0	−2.5571
HCO_3^-	$.1498\times10^{-2}$	.6906	−2.9851
$CaSO_4$	$.8334\times10^{-3}$	1.0	−3.0792
$NaHCO_3$	$.4447\times10^{-3}$	1.0	−3.3519
$O_2(aq)$	$.2151\times10^{-3}$	1.1735	−3.5980
$MgHCO_3^+$	$.1981\times10^{-3}$	.6717	−3.8760
KSO_4^-	$.1869\times10^{-3}$	.6717	−3.9013
$MgCO_3$	$.1068\times10^{-3}$	1.0	−3.9715
$SiO_2(aq)$	$.8188\times10^{-4}$	1.1735	−4.0174
KCl	$.5785\times10^{-4}$	1.0	−4.2377
CO_3^{--}	$.5437\times10^{-4}$	.1891	−4.9879

Some of the species concentrations predicted by the mathematical model are too small to be physically meaningful. The predicted concentration of $H_2(aq)$, for example, is 4×10^{-45} molal. Multiplying this value by Avogrado's number (6.02×10^{23} mol^{-1}) and the volume of the Earth's oceans (1400×10^{18} l), the concentration is equivalent to just three molecules in all of the world's seas!

We can calculate a more realistic $H_2(aq)$ concentration from the partial pressure of $H_2(g)$ in the atmosphere (Table 6.3) and the equilibrium constant for the reaction

$$H_2(g) \rightleftarrows H_2(aq) \tag{6.3}$$

which is $10^{-3.1}$. In this case, the molality of $H_2(aq)$ in equilibrium with the atmosphere is about 5×10^{-10}. This value, while small, is tens of orders of magnitude greater than the value calculated at equilibrium. Clearly, the equilibrium value is a mathematical abstraction.

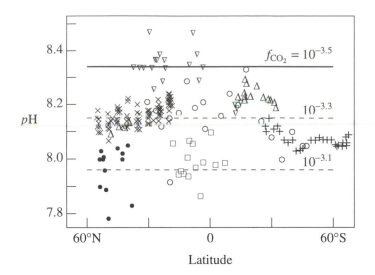

FIG. 6.1 *p*H of surface seawater from the western Pacific Ocean (Skirrow, 1965), as measured *in situ* during oceanographic cruises (various symbols). Line shows *p*H predicted by the model for seawater in equilibrium with atmospheric CO_2 at a fugacity of $10^{-3.5}$. Dashed lines show *p*H values that result from assuming larger fugacities of $10^{-3.3}$ and $10^{-3.1}$.

Mineral Saturation

Thirteen minerals appear supersaturated in the first block of results produced by the chemical model (Table 6.6). These results, therefore, represent an equilibrium achieved internally within the fluid but metastable with respect to mineral precipitation. It is quite common in modeling natural waters, especially when working at low temperature, to find one or more minerals listed as supersaturated. Unfortunately, the error sources in geochemical modeling are large enough that it can be difficult to determine whether or not a water is in fact supersaturated.

Many natural waters are supersaturated at low temperature, primarily because less stable minerals dissolve more quickly than more stable minerals precipitate. Relatively unstable silica phases such as chalcedony or amorphous silica, for example, may control a fluid's SiO_2(aq) concentration because quartz, the most stable silica mineral, precipitates slowly.

Uncertainty in the calculation, however, affects the reliability of values reported for saturation indices. Reaction log K's for many minerals are determined by extrapolating the results of experiments conducted at high temperature to the conditions of interest. The error in this type of extrapolation shows up directly in the denominator of log Q/K. Error in calculating activity

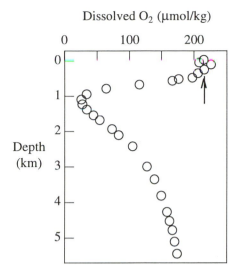

FIG. 6.2 Profile of dissolved oxygen vs. depth at GEOSECS site 226 (Drever, 1988, p. 267). Arrow marks oxygen content predicted by the model, assuming equilibrium with oxygen in the atmosphere.

coefficients (see Chapter 7), on the other hand, directly affects the computed activity product Q. The effect is pronounced for reactions with large coefficients, such as those for clay minerals.

Error in the input data can also be significant. The saturation state calculated for an aluminosilicate mineral, for example, depends on the analytical concentrations determined for aluminum and silicon. These analyses are difficult to perform accurately. As discussed in the next section, the presence of colloids and suspended particles in solution often affects the analytical results profoundly.

Averaging analytical data from different fluids commonly leads to inflated saturation indices. Scientists generally regard averaging as a method for reducing uncertainty in measured values, but the practice can play havoc in geochemical modeling. Waters with a range of Ca^{++} and SO_4^{--} concentrations, for example, can be in equilibrium with anhydrite ($CaSO_4$), so long as the activity product $a_{Ca^{++}} \times a_{SO_4^{--}}$ matches the equilibrium constant for the dissolution reaction. The averaged composition of two fluids (A and B) with differing Ca^{++} to SO_4^{--} ratios is, in the absence of activity coefficient effects, invariably supersaturated. This point can be shown quickly:

$$Q_A = (a_{Ca^{++}})_A \, (a_{SO_4^{-}})_A = K \tag{6.4}$$

$$Q_B = (a_{Ca^{++}})_B \, (a_{SO_4^{-}})_B = K \tag{6.5}$$

TABLE 6.5 Extent of complexing of major cations and anions in seawater

Species	Total concentration mg/kg	Total concentration molal	Free concentration molal	Free concentration % of total	Complexes % of total
Na^+	10760	.4850	.4754	98	2
Mg^{++}	1290	.05501	.03975	72	28
Ca^{++}	411	.01063	.005953	56	44
Cl^-	19350	.5658	.5500	97	3
SO_4^{--}	2710	.02924	.01607	55	45
HCO_3^-	142	.002412	.001498	62	38

$$Q_{ave} = \frac{1}{2}\left[(a_{Ca^{++}})_A + (a_{Ca^{++}})_B\right] \times \frac{1}{2}\left[(a_{SO_4^-})_A + (a_{SO_4^-})_B\right]$$

$$= \frac{1}{2}K + \frac{1}{4}(a_{Ca^{++}})_A(a_{SO_4^-})_B + \frac{1}{4}(a_{Ca^{++}})_B(a_{SO_4^-})_A \geq K \qquad (6.6)$$

The activity product Q_{ave} corresponding to the averaged analysis (ignoring variation in activity coefficients) equals the equilibrium constant K only when fluids A and B are identical; otherwise Q_{ave} exceeds K and anhydrite is reported to be supersaturated. To demonstrate this inequality, we can assume arbitrary values for $a_{Ca^{++}}$ and $a_{SO_4^-}$ that satisfy Eqns. 6.4-5 and substitute them into Eqn. 6.6.

The physical analogy to the averaging problem occurs when a sample consists of a mixture of fluids, as can occur when a well draws water from two or more producing intervals. In this case, the mixture may be supersaturated even when the individual fluids are not.

In the seawater example (Table 6.6), the saturation indices are inflated somewhat by the choice of a rather alkaline pH, reflecting equilibrium with atmospheric CO_2. If we had chosen a more acidic pH within the range observed in seawater, the indices would be smaller. The choice of large formula units for the phyllosilicate minerals, as discussed later in this section, also serves to inflate the saturation indices reported for these minerals.

Calculating the saturation state of dolomite presents an interesting problem. Geochemists have generally believed that ordered dolomite is supersaturated in seawater but prevented from precipitating by kinetic factors (e.g., Kastner, 1984; Hardie, 1987). The stability of dolomite at 25°C, however, is poorly known (Carpenter, 1980). The mineral might be more soluble, and hence less supersaturated in seawater, than shown in the LLNL database, which uses an estimate by Helgeson et al. (1978) derived by extrapolating data from a metamorphic reaction to room temperature. On the basis of careful experiments at lower temperatures, Lafon et al. (1992) suggest that seawater is close to

TABLE 6.6 Calculated saturation indices of various minerals in seawater and the mass of each precipitate in the stable phase assemblage

Mineral	Composition	Initial SI (log Q/K)	Amount formed (mg)
Antigorite	$Mg_{24}Si_{17}O_{42.5}(OH)_{31}$	44.16	—
Tremolite	$Ca_2Mg_5Si_8O_{22}(OH)_2$	7.73	—
Talc	$Mg_3Si_4O_{10}(OH)_2$	6.68	—
Chrysotile	$Mg_3Si_2O_5(OH)_4$	4.72	—
Sepiolite	$Mg_4Si_6O_{15}(OH)_2 \cdot 6H_2O$	3.93	—
Anthophyllite	$Mg_7Si_8O_{22}(OH)_2$	3.48	—
Dolomite	$CaMg(CO_3)_2$	3.46	50
Dolomite-ord	$CaMg(CO_3)_2$	3.46	—
Huntite	$CaMg_3(CO_3)_4$	2.13	—
Dolomite-dis	$CaMg(CO_3)_2$	1.91	—
Magnesite	$MgCO_3$	1.02	—
Calcite	$CaCO_3$	.81	—
Aragonite	$CaCO_3$	.64	—
Quartz	SiO_2	−.02	1

equilibrium with dolomite. The saturation index predicted by our model, therefore, may be unrealistically high.

Mass Balance and Mass Action

Because we formulated the governing equations from the principles of mass balance and mass action, we should now be able to show that our calculation results honor these principles. Demonstrating that the computer performed the calculations correctly is an important step for a geochemical modeler. Since no programmer is incapable of erring, no storage device is incorruptible, and no computer is infallible, the responsibility of showing correctness ultimately lies with the modeler. In geochemical modeling, fortunately, we can accomplish this relatively easily.

To show mass balance, we add the molalities of each species containing a component (but not species concentrations in mg/kg, since the mole weight of each species differs) to arrive at the input constraint. Taking component SO_4^{--} as an example, we find the total mole number (M_i) from the molalities (m_i and m_j) of the sulfur-bearing species

$$
\begin{array}{ll}
SO_4^{--} & .01607 \\
NaSO_4^- & .006386 \\
MgSO_4 & .005767 \\
CaSO_4 & .0008334 \\
KSO_4^- & .0001869 \\
HSO_4^- & 1.8\times10^{-9} \\
H_2SO_4 & 5.5\times10^{-21} \\
HS^- & 1.4\times10^{-142} \\
H_2S(aq) & 3.6\times10^{-144} \\
S^{--} & 1.2\times10^{-147} \\
\hline
& .02924\ molal
\end{array}
$$

Converting units,

$$(.02924\ molal)\times(1\ kg\ solv.)\times(96.058\ g\ SO_4/mol)\times$$
$$(1000\ mg/g)/(1.0364\ kg\ sol'n) = 2710\ mg\ SO_4/kg\ sol'n \qquad (6.7)$$

which is the input value from Table 6.2.

Repeating the procedure for component Ca^{++}

$$
\begin{array}{ll}
Ca^{++} & .005953 \\
CaCl^+ & .003780 \\
CaSO_4 & .0008334 \\
CaHCO_3^+ & .00003545 \\
CaCO_3 & .00002481 \\
CaH_3SiO_4^+ & 1.29\times10^{-7} \\
CaOH^+ & 9.64\times10^{-8} \\
CaH_2SiO_4 & 5.81\times10^{-9} \\
Ca(H_3SiO_4)_2 & 2.08\times10^{-10} \\
\hline
& .01063\ molal
\end{array}
$$

gives

$$(.01063\ molal)\times(1\ kg\ solv.)\times(40.080\ g\ Ca/mol)\times$$
$$(1000\ mg/g)/(1.0364\ kg\ sol'n) = 411\ mg\ Ca/kg\ sol'n \qquad (6.8)$$

Again, this is the input value.

To demonstrate mass action, we show that for any possible reaction the activity product Q matches the equilibrium constant K. This step is most easily accomplished by computing $\log Q$ as the sum of the products of the reaction coefficients and log activities of the corresponding species. The reaction for the sodium-sulfate ion pair, for example,

$$NaSO_4^- \rightleftarrows Na^+ + SO_4^{--} \qquad (6.9)$$

has a $\log K$ of $-.694$, according to the LLNL database. Calculating $\log Q$

$$
\begin{array}{ll}
\text{Na}^+ & 1 \times -0.4958 \\
\text{SO}_4^{--} & 1 \times -2.5657 \\
\text{NaSO}_4^- & -1 \times -2.3676 \\
\hline
& -.6939
\end{array}
$$

gives the value of log K, demonstrating that the reaction is indeed in equilibrium.

The procedure for verifying mineral saturation indices is similar. The program reports that log Q/K for the dolomite reaction

$$
\text{CaMg(CO}_3)_2 + 2\,\text{H}^+ \rightleftarrows \text{Ca}^{++} + \text{Mg}^{++} + 2\,\text{HCO}_3^- \tag{6.10}
$$
$$
\text{dolomite}
$$

is 3.457. The log K for the reaction is 2.5207. Determining log Q as before

$$
\begin{array}{ll}
\text{Mg}^{++} & 1 \times -1.9009 \\
\text{Ca}^{++} & 1 \times -2.8334 \\
\text{HCO}_3^- & 2 \times -2.9851 \\
\text{H}^+ & -2 \times -8.3411 \\
\hline
& 5.9777
\end{array}
$$

gives the expected result

$$
\log Q/K = \log Q - \log K = 3.457 \tag{6.11}
$$

Stable Phase Assemblage

After finding the equilibrium distribution of species in the initial fluid, the program sets about determining the system's theoretical state of true equilibrium. To do so, it searches for the stable mineral assemblage, following the procedure described in Section 5.4. A second block of results in the program output shows the equilibrium state corresponding to the predicted stable assemblage of fluid, dolomite, and quartz. These results are largely of academic interest: you could leave a bottle of seawater on a shelf for a very long time without fear that dolomite or quartz would form.

The search procedure, which we can trace through the flow chart shown in Fig. 5.6, is of interest. The steps in the procedure are:

1. Calculate initial species distribution; thirteen minerals are supersaturated.
2. Remove fugacity buffers.
    Swap MgCO_3 for $\text{CO}_2(\text{g})$, $\text{O}_2(\text{aq})$ for $\text{O}_2(\text{g})$.
3. Swap antigorite for MgCO_3.

       Iteration converges, six minerals are supersaturated.
4.   Swap dolomite for SiO_2(aq).
       Iteration converges, antigorite has dissolved away.
5.   Swap SiO_2(aq) for antigorite.
       Iteration converges, quartz is supersaturated.
6.   Swap quartz for SiO_2(aq).
       Iteration converges, there are no supersaturated minerals.

The program begins (step 1) with the distribution of species, as already described. Before beginning to search for the stable phase assemblage, it sets a closed system (step 2) by eliminating the fugacity buffers for CO_2(g) and O_2(g). It does so by swapping the aqueous species $MgCO_3$ and O_2(aq) into the basis. If these gases had been set as fixed buffers using the `fix` command, however, the program would skip this step.

Next (step 3), the program swaps antigorite, the most supersaturated mineral in the initial fluid, into the basis in place of species $MgCO_3$. With antigorite in the system, six minerals are supersaturated. In the next step (step 4), the program chooses to swap dolomite into the basis in place of SiO_2(aq). This swap seems strange until we write the reaction for dolomite in terms of the current basis:

$$\text{CaMg(CO}_3)_2 + 1.6\,H_2O + .7\,SiO_2(\text{aq}) \rightleftarrows$$
$$\textit{dolomite}$$

$$.04\,Mg_{24}Si_{17}O_{42.5}(OH)_{31} + Ca^{++} + 2\,HCO_3^- \qquad (6.12)$$
$$\textit{antigorite}$$

Since antigorite holds the place of magnesium in the basis, dolomite contains a negative amount of silica.

Dolomite precipitation consumes magnesium and produces H^+, causing the antigorite to dissolve completely. The resulting pH shift also causes quartz, which is most soluble under alkaline conditions, to become supersaturated. The program (step 5) replaces antigorite with SiO_2(aq), leading to a solution supersaturated with respect to only quartz. Including quartz in the mineral assemblage (step 6), the program converges to a saturated solution representing the system's theoretical equilibrium state.

Interpreting Saturation Indices

It is tempting to place significance on the relative magnitudes of the saturation indices calculated for various minerals and then to relate these values to the amounts of minerals likely to precipitate from solution. The data in Table 6.6, however, suggest no such relationship. Thirteen minerals are supersaturated in the initial fluid, but the phase rule limits to ten the number of minerals that can form; only two (dolomite and quartz) appear in the final phase assemblage.

For a number of reasons, using saturation indices as measures of the mineral masses to be formed as a fluid approaches equilibrium is a futile (if commonly undertaken) exercise. First, a mineral's saturation index depends on the choice of its formula unit. If we were to write the formula for quartz as Si_2O_4 instead of SiO_2, we would double its saturation index. Large formula units have been chosen for many of the clay and zeolite minerals listed in the LLNL database, and this explains why these minerals appear frequently at the top of the supersaturation list.

Second, at a given saturation index, supersaturated minerals with high solubilities have the potential to precipitate in greater mass than do less soluble ones. Consider a solution equally supersaturated with respect to halite (NaCl) and gypsum ($CaSO_4 \cdot 2H_2O$). Of the two minerals, halite is the more soluble and hence more of it must precipitate for the fluid to approach equilibrium.

Third, for minerals with binary or higher order reactions, there is no assurance that the reactants are available in stoichiometric proportions. We could prepare solutions equally supersaturated with respect to gypsum by using differing Ca^{++} to SO_4^{--} ratios. A solution containing these components in equal amounts would precipitate the most gypsum. Solutions rich in Ca^{++} but depleted in SO_4^{--}, or rich in SO_4^{--} but depleted in Ca^{++}, would produce lesser amounts of gypsum.

Finally, common ion effects link many mineral precipitation reactions, so the reactions do not operate independently. In the seawater example, dolomite precipitation consumed magnesium and produced hydrogen ions, significantly altering the saturation states of the other supersaturated minerals.

6.2 Amazon River Water

We turn our attention to developing a chemical model of water from the Amazon River, using a chemical analysis reported by Hem (1985, p. 9). The procedure is

```
pH = 6.5
SiO2(aq)  =   7,   mg/kg
Al+++     =    .07 mg/kg
Fe++      =    .06 mg/kg
Ca++      =   4.3  mg/kg
Mg++      =   1.1  mg/kg
Na+       =   1.8  mg/kg
HCO3-     = 19.    mg/kg
SO4--     =  3.    mg/kg
Cl-       =   1.9  mg/kg
O2(aq)    =   5.8  mg/kg

precip = off
go
```

TABLE 6.7 Calculated molalities (m), activity coefficients (γ), and log activities (a) of the most abundant species in Amazon River water

Species	m	γ	log a
HCO_3^-	$.1813 \times 10^{-3}$	.9735	-3.753
$O_2(aq)$	$.1810 \times 10^{-3}$	1.0001	-3.742
Cl^-	$.1382 \times 10^{-3}$	.9732	-3.871
$CO_2(aq)$	$.1297 \times 10^{-3}$	1.0	-3.887
$SiO_2(aq)$	$.1164 \times 10^{-3}$	1.0001	-3.934
Ca^{++}	$.1064 \times 10^{-3}$	.8991	-4.019
Na^+	$.7827 \times 10^{-4}$	.9734	-4.118
Mg^{++}	$.4499 \times 10^{-4}$	.9005	-4.392
SO_4^{--}	$.3049 \times 10^{-4}$	.8976	-4.563

The command `precip = off` tells the program to not let supersaturated minerals precipitate, since we are not especially interested in the fluid's true equilibrium state.

The resulting species distribution (Table 6.7), as would be expected, differs sharply from that in seawater (Table 6.4). Species approach $m$molal instead of molal concentrations and activity coefficients differ less from unity. In the Amazon River water, the most abundant cation and anion are Ca^{++} and HCO_3^-; in seawater, in contrast, Na^+ and Cl^- predominate. Seawater, clearly, is not simply concentrated river water.

In the river water, as opposed to seawater, the neutral species $O_2(aq)$, $CO_2(aq)$, and $SiO_2(aq)$ are among the species present in greatest concentration. Complexing among species is of little consequence in the river water, so the major cations and anions are present almost entirely as free ions.

The calculation predicts that a number of aluminum-bearing and iron-bearing minerals are supersaturated in the river water (Fig. 6.3). As discussed in the previous section, interpreting saturation indices calculated for natural waters can be problematic, since data errors can affect the predicted values so strongly.

In this case, a likely explanation for the apparent supersaturation is that the chemical analysis included not only dissolved aluminum and iron, but also a certain amount of aluminum and iron suspended in the water as colloids and fine sediments. Analytical error of this type occurs because the standard sampling procedure calls for passing the sample through a rather coarse filter of .45 μm pore size and then adding acid to "preserve" it during transport and storage.

Acidifying the sample causes colloids and fine sediments that passed through the filter to gradually dissolve, yielding abnormally high concentrations of elements such as aluminum, iron, silicon, and titanium when the fluid is analyzed. Figure 6.4, from a study of this problem by Kennedy et al. (1974),

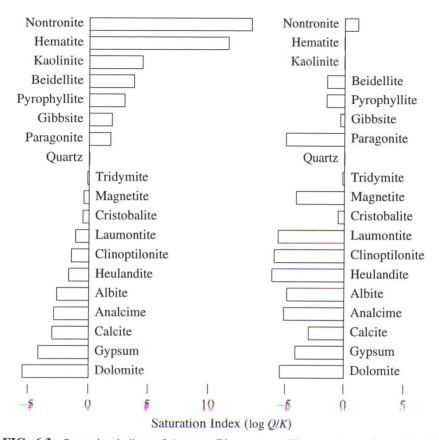

FIG. 6.3 Saturation indices of Amazon River water with respect to various minerals (left) calculated directly from a chemical analysis, and (right) computed assuming that equilibrium with kaolinite and hematite controls the fluid's aluminum and iron content.

shows how the pore size of the filter paper used during sample collection affects the concentrations determined for aluminum and iron.

To construct an alternative model of Amazon River water, we assume that equilibrium with kaolinite [a clay mineral, $Al_2Si_2O_5(OH)_4$] and hematite (ferric oxide, Fe_2O_3) controls the aluminum and iron concentrations:

```
(cont'd)
swap Kaolinite for Al+++
swap Hematite for Fe++
1 free cm3 Kaolinite
1 free cm3 Hematite
go
```

Here, we arbitrarily specify the amounts of the minerals that coexist with the

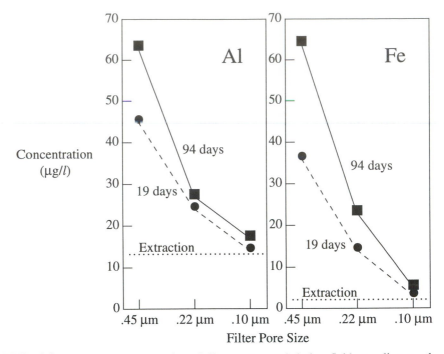

FIG. 6.4 Effects of the pore size of filter paper used during fluid sampling on the analytical concentrations reported for aluminum and iron (Kennedy et al., 1974). Samples were acidified, stored for 19 (●) or 94 (■) days, and analyzed by standard wet chemical methods. Dotted lines show dissolved concentrations determined by a solvent extraction technique.

fluid. The amount chosen (in this case, 1 cm^3), of course, has no effect on the predicted fluid composition. We set "free" constraints because we wish to specify amounts of actual minerals (n_k), rather than mineral components (M_k).

According to the second model (Fig. 6.3), only the nontronite clay minerals [e.g., $Ca_{.165}Fe_2Al_{.33}Si_{3.67}O_{10}(OH)_2$] are supersaturated by any significant amount. The nontronite minerals appear supersaturated so frequently in modeling natural waters that they are almost certainly erroneously stable in the database. In fact, the stabilities of iron-bearing alumino-silicate minerals are problematic in general because of the special difficulties entailed in controlling redox state during solubility experiments.

The second model is perhaps more attractive than the first because the predicted saturation states seem more reasonable. The assumption of equilibrium with kaolinite and hematite can be defended on the basis of known difficulties in analyzing for dissolved aluminum and iron. Nonetheless, on the basis of information available to us, neither model is correct or incorrect; they are simply founded on differing assumptions. The most that we can say is that one model may prove more useful for our purposes than the other.

6.3 Red Sea Brine

We turn our attention now to the hydrothermal brines of the Red Sea. An oceanic survey in 1963 discovered pools of hot, saline, and metal-rich brines along the axial rift of the Red Sea (Degens and Ross, 1969; Hoffmann, 1991). The dense brines pond in the rift's depressions, or deeps. The Atlantis II deep contains the largest pool, which measures 5×14 km and holds about 5 km^3 of supersaline brine. The deep holds two layers of brine. The lower brine contains about 25 wt.% dissolved salts and exists at temperatures up to 60°C. Table 6.8 shows the brine's average composition. A somewhat cooler, less saline water overlies the lower brine, separating it from normal seawater.

According to various lines of evidence (Shanks and Bischoff, 1977), subsea hot springs feed the brine pool by discharging each year about ten billion liters of brine, at temperatures of 150 to 250°C, upward across the sea floor. The springs have yet to be located.

The Red Sea deeps attracted the attention of geologists because of the metalliferous sediments that have accumulated as muds along the sea floor. The muds contain sulfide minerals including pyrite, chalcopyrite, and sphalerite, as well as hematite, magnetite, barite, and clay minerals. The minerals apparently precipitated from the brine pool over the past 25,000 years, accumulating into a metal-rich layer about 20–30 m thick. Geologists study the deeps to look at the process of ore deposition in a modern environment and thus gain insight to how ancient mineral deposits formed.

Modeling the chemistry of highly saline waters is perilous (as discussed in Chapter 7) because of the difficulty in computing activity coefficients at high ionic strengths. Here, we employ an extension of the Debye-Hückel method (the B-dot equation; see Section 7.1), using an ionic strength of 3 molal, the limiting value for the correlations, instead of the actual ionic strength, which is greater than 5 molal. In Chapter 7, we examine alternative methods for estimating activity coefficients in brines.

We attempt the calculation in the hope that error in estimating activity coefficients will not be so large as to render the results meaningless. In fact, the situation may be slightly better than might be feared; because the activity coefficients appear in the numerator and denominator of the mass action equations, the error tends to cancel itself.

To model the brine's chemistry, we need to estimate its oxidation state. We could use the ratio of sulfate to sulfide species to fix a_{O_2}, but chemical analysis has not detected reduced sulfur in the brine, which is dominated by sulfate species. A less direct approach is to assume equilibrium with a mineral containing reduced iron or sulfur, or with a pair of minerals that form a redox

TABLE 6.8 Chemical composition of
the lower brine, Atlantis II deep, Red
Sea (Shanks and Bischoff, 1977)

Na^+ (mg/kg)	92,600
K^+	1,870
Mg^{++}	764
Ca^{++}	5,150
Cl^-	156,030
SO_4^{--}	840
HCO_3^-	140
Ba^{++}	.9
Cu^+	.26
F^-	5
Fe^{++}	81
Pb^{++}	.63
Zn^{++}	5.4
T (°C)	60
pH (25°C, measured)	5.5
pH (60°C, estimated)	5.6

couple. Equilibrium with hematite and magnetite, for example,

$$3\ Fe_2O_3 \rightleftharpoons 2\ Fe_3O_4 + \tfrac{1}{2}\ O_2(aq) \qquad (6.13)$$
$$\text{hematite} \qquad \text{magnetite}$$

fixes a_{O_2}, as does the coexistence of sphalerite

$$Zn^{++} + SO_4^{--} \rightleftharpoons ZnS + 2\ O_2(aq) \qquad (6.14)$$
$$\text{sphalerite}$$

given constraints on the Zn^{++} and SO_4^{--} concentrations. We will assume the latter. We further constrain the Ba^{++} concentration by assuming equilibrium with barite

$$Ba^{++} + SO_4^{--} \rightleftharpoons BaSO_4 \qquad (6.15)$$
$$\text{barite}$$

This assumption provides a convenient check on the calculation's accuracy, since we already know the fluid's barium content.

The calculation procedure is to swap sphalerite and barite into the basis in place of $O_2(aq)$ and Ba^{++}

TABLE 6.9 Calculated molalities (m), activity coefficients (γ), and log activities (a) of the most abundant species in the Red Sea brine

Species	m	γ	log a
Cl^-	5.183	.6125	.5017
Na^+	4.861	.7036	.5341
$NaCl$	.5512	1.0	−.2587
$CaCl^+$	.1276	.7036	−1.047
K^+	$.5951\times10^{-1}$	.6125	−1.438
Ca^{++}	$.4445\times10^{-1}$	.1941	−2.064
$MgCl^+$	$.2496\times10^{-1}$	.7036	−1.756
Mg^{++}	$.1692\times10^{-1}$	.2895	−2.310
$NaSO_4^-$	$.7906\times10^{-2}$	.7036	−2.255
KCl	$.4736\times10^{-2}$	1.0	−2.325
SO_4^{--}	$.2675\times10^{-2}$	.0985	−3.579
$CO_2(aq)$	$.1809\times10^{-2}$	1.0	−2.743

```
swap Sphalerite for O2(aq)
swap Barite for Ba++
```

and then to set temperature and constrain each basis entry

```
(cont'd)
T = 60
pH = 5.6
TDS = 257000

Na+   =      92600   mg/kg
K+    =       1870   mg/kg
Mg++  =        764   mg/kg
Ca++  =       5150   mg/kg
Cl-   =     156030   mg/kg
SO4-- =        840   mg/kg
HCO3- =        140   mg/kg
Cu+   =        .26   mg/kg
F-    =          5   mg/kg
Fe++  =         81   mg/kg
Pb++  =        .63   mg/kg
Zn++  =        5.4   mg/kg

1.e-9 free grams Sphalerite
1.e-9 free grams Barite
print species=long
go
```

TABLE 6.10 Extent of complexing in three natural waters

	Amazon River	Seawater	Red Sea deep
I (molal)	.0006	.65	5.3
TDS (mg/kg)	45	35 000	260 000
Na^+ (% complexed)	.03	2	10
K^+	—	3	8
Mg^{++}	.5	28	60
Ca^{++}	.8	44	74
Cl^-	~ 0	3	12
SO_4^{--}	2.4	45	77
$HCO_3^- + CO_2(aq)$	.1	37	28

Here, we set a vanishingly small free mass for each mineral so that only negligible amounts can dissolve during the calculation's second step, when supersaturated minerals precipitate from the fluid.

The species distribution (Table 6.9) calculated for the brine differs from that of seawater and Amazon River water in the large molalities predicted and the predominance of ion pairs such as NaCl, $CaCl^+$, and $MgCl^+$. The complex species make up a considerable portion of the brine's dissolved load.

It is interesting to compare the effects of complexing in the three waters we have studied so far. As shown in Table 6.10, the complexed fraction of each of the major dissolved components increases with salinity. Whereas complex species are of minor importance in the Amazon River water, they are abundant in seawater and account for about three-fourths of the calcium and sulfate and more than half of the magnesium in the Red Sea brine.

The principle of mass action explains the relationship between concentration and complexation. The abundance of ion pairs in aqueous solution is controlled by reactions such as

$$Na^+ + Cl^- \rightleftarrows NaCl \tag{6.16}$$

and

$$Ca^{++} + Cl^- \rightleftarrows CaCl^+ \tag{6.17}$$

Starting with a dilute solution, for each doubling of the activities of the free ions on the left of these reactions, the activities of the ion pairs on the right sides must quadruple. As concentration increases, the ion pairs become progressively more important and eventually can come to overwhelm the free ions in solution. The higher temperature of the Red Sea brine also favors complexing because ion pairs gain stability relative to free ions as temperature increases.

We can check our results for the Red Sea brine against two independent pieces of information. In our results, sulfate species such as $NaSO_4^-$ dominate reduced sulfur species such as $H_2S(aq)$ and HS^-, in seeming accord with the failure of analysis to detect reduced sulfur in the brine. The predominance of sulfate over sulfide species in our calculation reflects the oxidation state resulting from our assumption of equilibrium with sphalerite.

To check that this oxidation state might be reasonable, we determine the total concentration of sulfide species to be

Species	Molality
$H_2S(aq)$	5.2×10^{-8}
HS^-	6.6×10^{-9}
S^{--}	1.4×10^{-15}
S_2^{--}	8.6×10^{-19}
S_3^{--}	7.4×10^{-23}
S_4^{--}	2.7×10^{-24}
S_5^{--}	1.4×10^{-28}
S_6^{--}	3.7×10^{-33}
	5.9×10^{-8}

or about .002 mg/kg as H_2S. The detection limit for reduced sulfur in a typical chemical analysis is about .01 mg/kg. The sulfide concentration and oxidation state, therefore, do not appear unreasonable.

Second, the total barium concentration in solution, which is constrained by equilibrium with barite, is 6.6 μmolal or .68 mg/kg. The concentration reported by chemical analysis (Table 6.8) is .9 mg/kg, in close agreement with the calculation. Considering the uncertainties in the calculation, these values are probably fortuitously similar.

The calculation predicts that the brine is supersaturated with respect to seven minerals: bornite (Cu_5FeS_4), chalcopyrite ($CuFeS_2$), chalcocite (Cu_2S), pyrite (FeS_2), fluorite (CaF_2), galena (PbS), and covellite (CuS). The saturation index for bornite, the most supersaturated mineral, is greater than 7, indicating significant supersaturation. The predicted saturation indices reflect in large part the calculated values for activity coefficients and, in the case of the sulfide minerals, the oxidation state. Unfortunately, the activity coefficient model and oxidation state represent two of the principal uncertainties in the calculation.

Uncertainties aside, it is interesting to pursue the question of whether the brine has the potential to precipitate sulfide minerals. In the second step of the calculation, where the model allows precipitation reactions to progress to the system's equilibrium state, three minerals form in small quantities:

Mineral	Mass (g)
Fluorite, CaF_2	7.3×10^{-3}
Chalcocite, Cu_2S	9.3×10^{-6}
Barite, $BaSO_4$	5.8×10^{-8}

Less than 10 μg of sulfide mineral formed as the fluid (which has a total mass of 1.25 kg) equilibrated. Bornite, the most supersaturated mineral, failed to form at all.

From the perspective of ore genesis, these results seem disappointing in light of the fluid's high degree of supersaturation. Shanks and Bischoff (1977), for example, estimate that about 60 mg of sphalerite alone precipitate from each kg of ore fluid feeding the Atlantis II deep. The reaction to form chalcocite

$$2\,Cu^+ + H_2S(aq) \;\rightarrow\; \underset{chalcocite}{Cu_2S} \; + 2\,H^+ \qquad (6.18)$$

explains why so little of that mineral formed. Since the fluid is nearly depleted in reduced sulfur species, the reaction can proceed to the right by only a minute increment. As noted in Section 6.1, a large saturation index does not guarantee that the reactants needed to form a mineral are available in suitable proportions for the reaction to proceed to any significant extent.

Because little mass can precipitate from it, the brine, if related to deposition of the metalliferous muds, is likely to be a residuum of the original ore fluid. As it discharged into the deep, the ore fluid was richer in metals than in reduced sulfur. Mineral precipitation depleted the fluid of nearly all of its reduced sulfur without exhausting the metals, leaving the metal-rich brine observed in the deep.

6.4 Redox Disequilibrium

The equilibrium model, despite its limitations, in many ways provides a useful if occasionally abstract description of the chemical states of natural waters. However, if used to predict the state of redox reactions, especially at low temperature, the model is likely to fail. This shortcoming does not result from any error in formulating the thermodynamic model. Instead, it arises from the fact that redox reactions in natural waters proceed at such slow rates that they commonly remain far from equilibrium.

Complicating matters even further is the fact that the platinum electrode, the standard tool for measuring Eh directly, does not respond to some of the most important redox couples in geochemical systems. The electrode, for example, responds incorrectly or not at all to the couples SO_4^{--}-HS^-, O_2-H_2O, CO_2-CH_4, NO_3^--N_2, and N_2-NH_4 (Stumm and Morgan, 1981; Hostettler, 1984). In a laboratory experiment, Runnells and Lindberg (1990) prepared solutions with differing ratios of selenium in the Se^{4+} and Se^{6+} oxidation states. They found that even under controlled conditions the platinum electrode was completely insensitive to the selenium composition. The meaning of an Eh measurement from a natural water, therefore, may be difficult or impossible to determine.

Geochemists (e.g., Thorstenson et al., 1979; Thorstenson, 1984) have long recognized that at low temperature many redox reactions are unlikely to achieve equilibrium, and that the meaning of Eh measurements is problematic. Lindberg and Runnells (1984) demonstrated the generality of the problem. They compiled from the WATSTORE database more than 600 water analyses that provide at least two measures of oxidation state. The measures included Eh, dissolved oxygen content, concentrations of dissolved sulfate and sulfide, ferric and ferrous iron, and so on.

They calculated species distributions for each sample and then computed the Nernst Eh for the redox couples in the analysis. Their results show that the redox couples generally failed to achieve equilibrium with each other, varying per sample by as much as 1000 mV. In addition, they could demonstrate little relationship between measured and Nernst Eh values. Similarly, Criaud et al. (1989) computed discordant Nernst Eh values for low-temperature geothermal fluids from the Paris basin.

There are, fortunately, some instances in which measured Eh values can be interpreted in a quantitative sense. Nordstrom et al. (1979), for example, showed that Eh measurements in acid mine drainage accurately reflect the $a_{Fe^{+++}}/a_{Fe^{++}}$ ratio. They further noted a number of other studies establishing agreement between measured and Nernst Eh values for various couples. Nonetheless, it is clearly dangerous for a geochemical modeler to assume *a priori* that a sample is in internal redox equilibrium or that an Eh measurement reflects a sample's redox state.

Redox Coupling

A flexible method for modeling redox disequilibrium is to divide the reaction database into two parts. The first part contains reactions between the basis species (e.g., Table 6.1) and a number of redox species, which represent the basis species in alternative oxidation states. For example, redox species Fe^{+++} forms a redox pair with basis species Fe^{++}, and HS^- forms a redox pair with SO_4^{--}. These coupling reactions are balanced in terms of an electron donor or acceptor, such as $O_2(aq)$. Table 6.11 shows coupling reactions from REACT's database.

The second part of the database contains reactions for the various secondary species, minerals, and gases. These reactions are balanced in terms of the basis and redox species, avoiding (to the extent practical) electron transfer. Species and minerals containing ferric iron, for example, are balanced in terms of the redox species Fe^{+++}

TABLE 6.11 Some of the redox couples in REACT's database

Redox pair	Coupling reaction
$AsH_3(aq)$ - $As(OH)_4^-$	$AsH_3(aq) + H_2O + 1.5\,O_2(aq) \rightleftarrows As(OH)_4^- + H^+$
AsO_4^{---} - $As(OH)_4^-$	$As(OH)_4^- + \frac{1}{2}\,O_2(aq) \rightleftarrows AsO_4^{---} + H_2O + 2\,H^+$
Au^{+++} - Au^+	$2\,H^+ + Au^+ + \frac{1}{2}\,O_2(aq) \rightleftarrows Au^{+++} + H_2O$
CH_3COO^- - HCO_3^-	$CH_3COO^- + 2\,O_2(aq) \rightleftarrows 2\,HCO_3^- + H^1$
$CH_4(aq)$ - HCO_3^-	$CH_4(aq) + 2\,O_2(aq) \rightleftarrows H_2O + H^+ + HCO_3^-$
ClO_4^- - Cl^-	$Cl^- + 2\,O_2(aq) \rightleftarrows ClO_4^-$
Co^{+++} - Co^{++}	$Co^{++} + H^+ + \frac{1}{4}\,O_2(aq) \rightleftarrows Co^{+++} + \frac{1}{2}\,H_2O$
Cr^{++} - Cr^{+++}	$Cr^{++} + H^+ + \frac{1}{4}\,O_2(aq) \rightleftarrows \frac{1}{2}\,H_2O + Cr^{+++}$
CrO_4^- - Cr^{+++}	$Cr^{+++} + 2.5\,H_2O + \frac{3}{4}\,O_2(aq) \rightleftarrows CrO_4^- + 5\,H^+$
CrO_4^{--} - Cr^{+++}	$Cr^{+++} + 3\,H_2O + \frac{1}{2}\,O_2(aq) \rightleftarrows CrO_4^{--} + 6\,H^+$
Cu^{++} - Cu^+	$Cu^+ + H^+ + \frac{1}{4}\,O_2(aq) \rightleftarrows Cu^{++} + \frac{1}{2}\,H_2O$
Eu^{++} - Eu^{+++}	$Eu^{++} + H^+ + \frac{1}{4}\,O_2(aq) \rightleftarrows \frac{1}{2}\,H_2O + Eu^{+++}$
Fe^{+++} - Fe^{++}	$Fe^{++} + H^+ + \frac{1}{4}\,O_2(aq) \rightleftarrows Fe^{+++} + \frac{1}{2}\,H_2O$
$H_2(aq)$ - $O_2(aq)$	$H_2(aq) + \frac{1}{2}\,O_2(aq) \rightleftarrows H_2O$
HS^- - SO_4^{--}	$HS^- + 2\,O_2(aq) \rightleftarrows SO_4^{--} + H^+$
Hg_2^{++} - Hg^{++}	$Hg_2^{++} + 2\,H^+ + \frac{1}{2}\,O_2(aq) \rightleftarrows 2\,Hg^{++} + H_2O$
MnO_4^- - Mn^{++}	$Mn^{++} + 1.5\,H_2O + 1.25\,O_2(aq) \rightleftarrows MnO_4^- + 3\,H^+$
MnO_4^{--} - Mn^{++}	$Mn^{++} + 2\,H_2O + O_2(aq) \rightleftarrows MnO_4^{--} + 4\,H^+$
$N_2(aq)$ - NO_3^-	$N_2(aq) + H_2O + 2.5\,O_2(aq) \rightleftarrows 2\,H^+ + 2\,NO_3^-$
NH_4^+ - NO_3^-	$NH_4^+ + 2\,O_2(aq) \rightleftarrows NO_3^- + 2\,H^+ + H_2O$
NO_2^- - NO_3^-	$NO_2^- + \frac{1}{2}\,O_2(aq) \rightleftarrows NO_3^-$
Se^{--} - SeO_3	$Se^{--} + 1.5\,O_2(aq) \rightleftarrows SeO_3^{--}$
SeO_4 - SeO_3	$SeO_3^{--} + \frac{1}{2}\,O_2(aq) \rightleftarrows SeO_4^{--}$
Sn^{++} - Sn^{++++}	$Sn^{++} + 2\,H^+ + \frac{1}{2}\,O_2(aq) \rightleftarrows Sn^{++++} + H_2O$
U^{+++} - U^{++++}	$U^{+++} + H^+ + \frac{1}{4}\,O_2(aq) \rightleftarrows U^{++++} + \frac{1}{2}\,H_2O$
UO_2^+ - U^{++++}	$U^{++++} + 1.5\,H_2O + \frac{1}{4}\,O_2(aq) \rightleftarrows UO_2^+ + 3\,H^+$
UO_2^{++} - U^{++++}	$U^{++++} + H_2O + \frac{1}{2}\,O_2(aq) \rightleftarrows UO_2^{++} + 2\,H^+$
VO^{++} - V^{+++}	$V^{+++} + \frac{1}{2}\,H_2O + \frac{1}{4}\,O_2(aq) \rightleftarrows VO^{++} + H^+$
VO_4^{---} - V^{+++}	$V^{+++} + 3\,H_2O + \frac{1}{2}\,O_2(aq) \rightleftarrows VO_4^{---} + 6\,H^+$

$$Fe_2O_3 + 6\,H^+ \rightleftarrows 3\,H_2O + 2\,Fe^{+++} \tag{6.19}$$
hematite

whereas those containing ferrous iron are balanced with basis species Fe^{++}

$$FeSO_4 \rightleftarrows Fe^{++} + SO_4^{--} \tag{6.20}$$

The mineral magnetite (Fe_3O_4) contains oxidized and reduced iron, so its reaction

$$Fe_3O_4 + 8\,H^+ \rightleftarrows 2\,Fe^{+++} + Fe^{++} + 4\,H_2O \tag{6.21}$$
magnetite

contains both the basis and redox species.

The modeler controls which redox reactions should be in equilibrium by interactively coupling or decoupling the redox pairs. For each coupled pair, the model uses the corresponding coupling reaction to eliminate redox species from the reactions in the database. For example, if the pair Fe^{+++}-Fe^{++} is coupled, the model adds the coupling reaction to the reaction for hematite

$$Fe_2O_3 + 6\,H^+ \rightleftarrows 3\,H_2O + 2\,Fe^{+++}$$

$$\frac{2 \times (Fe^{+++} + \tfrac{1}{2}\,H_2O \rightleftarrows Fe^{++} + H^+ + \tfrac{1}{4}\,O_2(aq)\,)}{Fe_2O_3 + 4\,H^+ \rightleftarrows 2\,H_2O + 2\,Fe^{++} + \tfrac{1}{2}\,O_2(aq) \tag{6.22}}$$
hematite

to eliminate the redox species Fe^{+++}. The same procedure is applied to the reactions for the other species and minerals that contain ferric iron.

Models calculated assuming redox disequilibrium generally require more input data than equilibrium models, in which a single variable constrains the system's oxidation state. The modeler can decouple as many redox pairs as can be independently constrained. A completely decoupled model, therefore, would require analytical data for each element in each of its redox states. Unfortunately, analytical data of this completeness are seldom collected

Example of a Disequilibrium Model

As an example of modeling a fluid in redox disequilibrium, we use an analysis, slightly simplified from Nordstrom et al. (1992), of a groundwater sampled near the Morro do Ferro ore district in Brazil (Table 6.12). There are three measures of oxidation state in the analysis: the Eh value determined by platinum electrode, the dissolved oxygen content, and the distribution of iron between ferrous and ferric species.

To calculate an equilibrium model, the procedure is

```
T  = 22
pH = 6.05

O2(aq)  = 4.3    mg/l
HCO3-   = 1.8    mg/l
Ca++    = 0.238  mg/l
Mg++    = 0.352  mg/l
Na+     = 0.043  mg/l
K+      = 0.20   mg/l
Fe++    = 0.73   mg/l
```

TABLE 6.12 Chemical analysis of a
groundwater from near the Morro do Ferro
deposits, Brazil (Nordstrom et al., 1992)

T (°C)	22
pH	6.05
Eh (mV)	504
HCO_3 (mg/l)	1.8
Ca^{++}	0.238
Mg^{++}	0.352
Na^+	0.043
K^+	0.20
Fe (II)	0.73
Fe (total)	0.76
Mn++	0.277
Zn^{++}	0.124
SO_4^{--}	1.5
Cl^-	< 2.0
Dissolved O_2	4.3

```
Mn++    = 0.277 mg/l
Zn++    = 0.124 mg/l
SO4     = 1.5   mg/l
balance on Cl-

precip = off
print species = long
go
```

Here, we set oxidation state in the model using the dissolved oxygen content.

To account for the possibility of redox disequilibrium among iron species, we use the analysis for ferrous as well as total iron:

```
(cont'd)
decouple Fe+++
Fe+++   = 0.03  mg/l
Fe++    = 0.73  mg/l
go
```

By decoupling the ferric-ferrous reaction with the decouple command, we add Fe^{+++} as a new basis entry in the calculation. We constrain the entry using the difference between the total and ferrous iron contents.

As shown in Table 6.13, the two calculations predict broadly differing species distributions for iron. In the first calculation, the fluid is almost devoid of

TABLE 6.13 Concentrations (molal) of predominant iron species in Morro do Ferro groundwater, calculated assuming equilibrium and redox disequilibrium

Species	Equilibrium	Disequilibrium
Ferrous		
Fe^{++}	$.11 \times 10^{-12}$	$.13 \times 10^{-4}$
$FeSO_4$	$.24 \times 10^{-15}$	$.29 \times 10^{-7}$
$FeHCO_3^+$	$.20 \times 10^{-16}$	$.25 \times 10^{-8}$
$FeCl^+$	$.68 \times 10^{-17}$	$.13 \times 10^{-8}$
$FeOH^+$	$.65 \times 10^{-17}$	$.78 \times 10^{-9}$
Ferric		
$Fe(OH)_2^+$	$.92 \times 10^{-5}$	$.37 \times 10^{-6}$
$Fe(OH)_3$	$.39 \times 10^{-5}$	$.16 \times 10^{-6}$
$FeOH^{++}$	$.29 \times 10^{-7}$	$.12 \times 10^{-8}$
$Fe(OH)_4^-$	$.92 \times 10^{-9}$	$.38 \times 10^{-10}$

ferrous iron species, reflecting the high concentration of dissolved oxygen. This result contradicts the dominance of ferrous over ferric species reported in the chemical analysis. The disequilibrium calculation, in which we separately constrain the fluid's ferrous and ferric iron contents, provides a species distribution in which ferrous iron species predominate, in accord with the analytical data.

We can compare the Eh measured for the Morro do Ferro groundwater (Table 6.12) with the Nernst Eh values (Eqn. 3.42) given by the reactions for dissolved oxygen and iron oxidation, as reported in the program output:

	Eh (mV)
Measured by electrode	504
$\frac{1}{4} O_2(aq) + H^+ + e^- \rightleftarrows \frac{1}{2} H_2O$	861
$Fe^{+++} + e^- \rightleftarrows Fe^{++}$	306

The ratio of ferrous to ferric species represents a redox state considerably less oxidizing than suggested by the dissolved oxygen content. The measured Eh falls between these values. Since the values vary over a range of more than 500 mV, this water clearly is not in redox equilibrium; assuming that it is gives an incorrect distribution of iron species.

7

Activity Coefficients

Among the most vexing tasks for geochemical modelers, especially when they work with concentrated solutions, is estimating values for the activity coefficients of electrolyte species. To understand in a qualitative sense why activity coefficients in electrolyte solutions vary, we can imagine how solution concentration affects species activities. In the solution, electrical attraction draws anions around cations and cations around anions. We might think of a dilute solution as an imperfect crystal of loosely packed, hydrated ions that, within a matrix of solvent water, is constantly rearranging itself by Brownian motion. A solution of uncharged, nonpolar species, by contrast, would be nearly random in structure.

The electrolyte solution is lower in free energy G than it would be if the species did not interact electrically because of the energy liberated by moving ions of opposite charge together while separating those of like charge. The chemical potentials μ_i of the species, for the same reason, are smaller than they would be in the absence of electrostatic forces. By the equation

$$\mu_i \equiv \frac{\partial G_i}{\partial n_i} = \mu_i^\circ + RT_K \ln \gamma_i m_i \qquad (7.1)$$

(Eqn. 3.5, taking $a_i = \gamma_i m_i$), the reduction in a species' chemical potential is reflected by a decreased value (relative to one in an ideal solution) for its activity coefficient. By Coulomb's law, electrostatic forces vary inversely with the square of the distance of ion separation. For this reason, activity coefficients in dilute fluids decrease as concentration increases because the coulombic forces become stronger as ions pack together more closely.

When electrical attraction and repulsion operate over distances considerably larger than the hydrated sizes of the ions, we can compute species' activities

107

quite well from electrostatic theory, as demonstrated in the 1920s by the celebrated physical chemists Debye and Hückel. At moderate concentrations, however, the ions pack together rather tightly. In a one molal solution, for example, just a few molecules of solvent separate ions. Because electrical interactions at short range are complicated in nature, it is no longer sufficient to treat ions as points of electrical charge, as Debye and Hückel did in their analysis. Rather, we need to account for the distribution of electrical charge throughout the hydration sphere of each ion. Since this distribution is complex, simple analysis cannot account for the energetics of attraction and repulsion among closely packed ions.

In concentrated solutions, the energetic effect of ion repulsion seems to dominate. This effect is sometimes termed "hard core repulsion." With increasing concentration, furthermore, much and then most of the water in the solution is taken up in the species' hydration spheres, reducing the amount of water acting as solvent. These two effects cause the activity coefficients to increase. Increasing concentration, on the other hand, leads to greater degrees of ion association, which serves to decrease a species' free molality and, hence, activity relative to that expected if the ion were fully dissociated. Depending on the activity model employed, the effects of ion association may be accounted for directly by mass action (thereby decreasing activity by lowering species concentration) or indirectly, by adjusting activity coefficients while assuming complete dissociation.

The modeler should bear in mind a further complication: the requirement of electroneutrality (as discussed in Chapter 3) precludes the possibility of observing either the chemical potential of a charged species or its activity coefficient. We can measure only the mean activity coefficient of the ions comprising a salt. This value can be separated into the coefficients for single ions by following any of a number of conventions. The MacInnes convention, which is commonly but not universally employed in geochemical calculations, holds that the coefficients for K^+ and Cl^- are equal in a solution of a given ionic strength. The modeler must guard against mixing activity coefficients determined using differing conventions.

Geochemical modelers currently employ two types of methods to estimate activity coefficients (Plummer, 1992; Wolery, 1992b). The first type consists of applying variants of the Debye-Hückel equation, a simple relationship that treats a species' activity coefficient as a function of the species' size and the solution's ionic strength. Methods of this type take into account the distribution of species in solution and are easy to use, but can be applied with accuracy to modeling only relatively dilute fluids.

Virial methods, the second type, employ coefficients that account for interactions among the individual components (rather than species) in solution. The virial methods are less general, rather complicated to apply, require considerable amounts of data, and allow little insight into the distribution of species in solution. They can, however, reliably predict mineral solubilities even in concentrated brines.

The following sections describe the two estimation techniques. The discussion here leans toward the practical aspects of estimating activity coefficients. For an understanding of the theoretical basis of the activity models, the reader may refer to a thermodynamics text (such as Robinson and Stokes, 1968, or Anderson and Crerar, 1993) and the papers referenced herein.

7.1 Debye-Hückel Methods

In 1923, Debye and Hückel published their famous papers describing a method for calculating activity coefficients in electrolyte solutions. They assumed that ions behave as spheres with charges located at their center points. The ions interact with each other by coulombic forces. Robinson and Stokes (1968) present their derivation, and the papers are available (Interscience Publishers, 1954) in English translation.

The result of their analysis, known as the Debye-Hückel equation

$$\log \gamma_i = - \frac{A z_i^2 \sqrt{I}}{1 + \mathring{a}_i B \sqrt{I}} \tag{7.2}$$

gives the activity coefficient γ_i of an ion with electrical charge z_i. In this chapter we need not differentiate between the basis and secondary species, A_i and A_j. Hence, we will let γ_i, z_i, and so on represent the properties of the aqueous species in general.

Variable $\mathring{a}_i$ in Eqn. 7.2 is the ion size parameter. In practice, this value is determined by fitting the Debye-Hückel equation to experimental data. Variables A and B are functions of temperature, and I is the solution ionic strength. At 25°C, given I in molal units and taking $\mathring{a}_i$ in Å, the value of A is .5092, and B is .3283.

The ionic strength

$$I = \frac{1}{2} \sum_i m_i z_i^2 \tag{7.3}$$

is half the sum of the product of each species molality m_i and the square of its charge. In the literature, ionic strength may be reported in molar or molal units, and may be calculated either by accounting for the effect of ion pairing or assuming that the electrolyte dissociates completely. We use molal units here and follow Helgeson's (1969) terminology regarding the question of ion pairing. Ionic strength I refers to the "true ionic strength" calculated by Eqn. 7.3, accounting for the role of complexing in reducing the number of free ions in solution. We refer to the value calculated assuming complete dissociation (i.e., neglecting ion pairs) as the "stoichiometric ionic strength," which we label I_S.

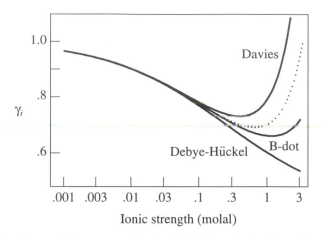

FIG. 7.1   Activity coefficients γ_i predicted at 25°C for a singly charged ion with size $\overset{\circ}{a}$ of 4 Å, according to the Debye-Hückel (Eqn. 7.2), Davies (Eqn. 7.4), and B-dot (Eqn. 7.5) equations. Dotted line shows the Davies equation evaluated with a coefficient of 0.2 instead of 0.3.

Equation 7.2 is notable in that it predicts a species' activity coefficient using only two numbers (z_i and $\overset{\circ}{a}_i$) to account for the species' properties and a single value I to represent the solution. As such, it can be applied readily to study a variety of geochemical systems, simple and complex.

Unfortunately, the equation becomes inaccurate at moderate ionic strength, above about 0.1 molal (e.g., Stumm and Morgan, 1981). As can be seen from Eqn. 7.2, Debye-Hückel activity coefficients approach unity when ionic strength nears zero. With increasing ionic strength, the coefficient decreases monotonically (Fig. 7.1). This decrease reflects the increasing strength of the long-range coulombic forces in solution, when short-range forces and hydration effects are ignored.

Davies Equation

The Davies (1962) equation is a variant of the Debye-Hückel equation (Eqn. 7.2) that can be carried to somewhat higher ionic strengths. The equation follows from Eqn. 7.2 by noting that at 25°C the product $\overset{\circ}{a}_i \cdot B$ is about one. Including an empirical term $0.3\,I$ to the correlation gives

$$\log \gamma_i = -Az_i^2 \left[\frac{\sqrt{I}}{1 + \sqrt{I}} - 0.3\,I \right] \tag{7.4}$$

A coefficient of 0.2 is used sometimes instead of 0.3. The only variable specific to the species in question is the charge z_i, which of course is known. For this reason, the Davies equation is especially easy to apply within geochemical models designed for work at 25°C, such as WATEQ2 (Ball et al., 1979) and its successors, and PHREEQE (Parkhurst et al., 1980).

As can be seen in Fig. 7.1, the Davies equation does not decrease monotonically with ionic strength, as the Debye-Hückel equation does. Beginning at ionic strengths of about 0.1 molal, it deviates above the Debye-Hückel function and at about 0.5 molal starts to increase in value. The Davies equation is reasonably accurate to an ionic strength of about 0.3 or 0.5 molal.

B-Dot Model

Helgeson (1969; see also Helgeson and Kirkham, 1974) presented an activity model based on an equation similar in form to the Davies equation. The model, adapted from earlier work (see Pitzer and Brewer, 1961, p. 326, p. 578, and Appendix 4, and references therein), is parameterized from 0°C to 300°C for solutions of up to 3 molal ionic strength in which NaCl is the dominant solute. The model takes it name from the "B-dot" equation

$$\log \, \gamma_i = - \frac{A z_i^2 \, \sqrt{I}}{1 + \mathring{a}_i \, B \, \sqrt{I}} + \dot{B} \, I \tag{7.5}$$

which is an extension of the Debye-Hückel equation (Eqn. 7.2). Coefficients A, B, and $\dot{B}$ vary with temperature, as shown in Fig. 7.2, whereas the ion size parameter $\mathring{a}_i$ for each species remains constant.

The B-dot equation is widely applied in geochemical models designed to operate over a range of temperatures, such as EQ3/EQ6 (Wolery, 1979, 1992b), CHILLER (Reed, 1982), SOLMNEQ (Kharaka et al., 1988), and REACT (Bethke, 1994). The equation is considered reasonably accurate in predicting the activities of Na^+ and Cl^- ions to concentrations as large as several molal, and of other species to ionic strengths up to about 0.3 to 1 molal. Figure 7.3 shows the activity coefficients predicted at 25°C for species of differing charge and ion size.

In the B-dot model, as currently applied (Wolery, 1992b), the activity coefficients of electrically neutral, nonpolar species [$B(OH)_3$, $O_2(aq)$, $SiO_2(aq)$, $CH_4(aq)$, and $H_2(aq)$] are calculated from ionic strength using an empirical relationship

$$\log \, \gamma_o = aI + bI^2 + cI^3 \tag{7.6}$$

Here, a, b, and c are polynomial coefficients that vary with temperature:

	a	b	c
25°C	.1127	−.01049	1.545×10^{-3}
100°C	.08018	−.001503	$.5009\times10^{-3}$
200°C	.09892	−.01040	1.386×10^{-3}
300°C	.1967	−.01809	-2.497×10^{-3}

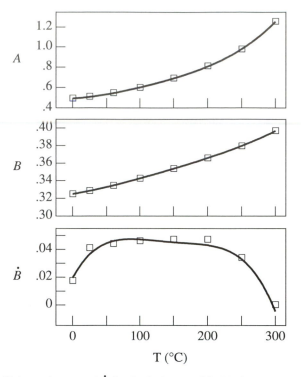

FIG. 7.2 Values of A, B, and $\dot{B}$ for the B-dot (modified Debye-Hückel) equation at 0°C, 25°C, 60°C, 100°C, 150°C, 200°C, 250°C, and 300°C (squares) and interpolation functions (lines). Values correspond to I taken in molal and $\overset{\circ}{a}$ in Å. Data from the LLNL database, after Helgeson (1969) and Helgeson and Kirkham (1974).

Figure 7.4 shows the function plotted against ionic strength at 25°C, 100°C, and 300°C. The rapid increase in value at high ionic strength represents the "salting out" effect by which gas solubility decreases with increasing salinity. In the model, polar neutral species are simply assigned activity coefficients of unity.

The ideality of the solvent in aqueous electrolyte solutions is commonly tabulated in terms of the osmotic coefficient ϕ (e.g., Pitzer and Brewer, 1961, p. 321; Denbigh, 1971, p. 288), which assumes a value of unity in an ideal dilute solution under standard conditions. By analogy to a solution of a single salt, the water activity can be determined from the osmotic coefficient and the stoichiometric ionic strength I_S according to

$$\ln a_w = -\frac{2 I_S \phi}{55.5} \tag{7.7}$$

In the B-dot model, the osmotic coefficient is taken to be described by a power series

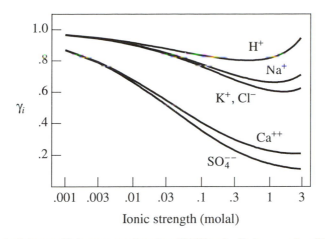

FIG. 7.3 Activity coefficients γ_i predicted at 25°C by the B-dot equation (Eqn. 7.5) for some singly and doubly charged ions, as a function of ionic strength. Corresponding $\mathring{a}$ values are 3 Å (K^+ and Cl^-), 4 Å (Na^+ and SO_4^{--}), 6 Å (Ca^{++}), and 9 Å (H^+).

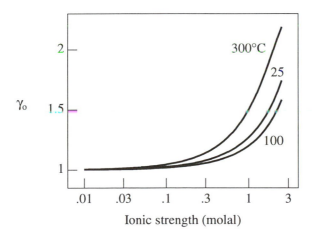

FIG. 7.4 Activity coefficients γ_o for neutral, nonpolar species as a function of ionic strength (molal) at 25°C, 100°C, and 300°C, according to the activity model of Helgeson (1969).

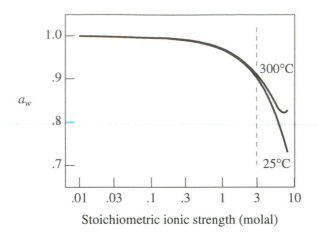

FIG. 7.5   Water activity a_w versus stoichiometric ionic strength I_S of NaCl solutions at 25°C and 300°C, according to the activity model of Helgeson (1969). Dashed line shows 3 molal limit to the model parameterization; values to right of this line are extrapolations of the original data.

$$\phi = 1 - \frac{2.303\,A}{a^3\,I_S}\left[\hat{b} - 2\ln\hat{b} - \frac{1}{\hat{b}}\right] + \frac{b\,I_S}{2} + \frac{2c\,I_S^2}{3} + \frac{3d\,I_S^3}{4} \qquad (7.8)$$

in terms of regression coefficients a, b, c, and d, which differ from those in Eqn. 7.6. Term $\hat{b}$ in this equation is given as

$$\hat{b} = 1 + a\,\sqrt{I_S} \qquad (7.9)$$

and representative values of the regression coefficients, which vary with temperature, are

	a	b	c	d
25°C	1.454	.02236	9.380×10^{-3}	-5.362×10^{-4}
100°C	1.555	.03648	6.437×10^{-3}	-7.132×10^{-4}
200°C	1.623	.04589	4.522×10^{-3}	-8.312×10^{-4}

Figure 7.5 shows the predicted water activity at 25°C and 300°C, plotted against I_S.

Program REACT differs somewhat from other models in that it sets limits to the values for I and I_S used to evaluate the B-dot model (Eqns. 7.5–7.7). Reflecting the fact that there is no basis for extrapolating the B-dot model to high ionic strength, REACT calculates activity coefficients using the lesser value of the actual ionic strength (I or I_S) and the limiting value. The limiting values are carried internally as variables timax and simax, which, by default, are set to 3 molal, but may be reset by the user.

7.2 Virial Methods

Virial equations offer a conceptual alternative to the Debye-Hückel methods for calculating electrolyte activities. The semi-empirical equations are sometimes called specific interaction equations, phenomenological equations, or simply the Pitzer equations, after Kenneth Pitzer, who has been largely responsible for their development since the 1970s (among his and his co-workers' many papers on the subject, see reviews by Pitzer, 1979, 1987). Virial methods are frequently employed in geochemical modeling because they can be applied with accuracy at high ionic strength. Programs SOLMNEQ.88 (Kharaka et al., 1988), PHRQPITZ (Plummer et al., 1988), EQ3/EQ6 (Wolery, 1992a), and REACT (Bethke, 1994), for example, include provision for using the virial methods.

The virial methods differ conceptually from other techniques in that they take little or no explicit account of the distribution of species in solution. In their simplest form, the equations recognize only free ions, as if each salt has fully dissociated in solution. The molality m_i of the Na^+ ion, then, is taken to be the analytical concentration of sodium. All of the calcium in solution is represented by Ca^{++}, the chlorine by Cl^-, the sulfate by SO_4^{--}, and so on. In many chemical systems, however, it is desirable to include some complex species in the virial formulation. Species that protonate and deprotonate with pH, such as those in the series CO_3^{--}-HCO_3^--CO_2(aq) and Al^{+++}-$AlOH^{++}$-$Al(OH)_2^+$, typically need to be included, and incorporating strong ion pairs such as $CaSO_4$(aq) may improve the model's accuracy at high temperatures. Weare (1987, pp. 148–153) discusses the criteria for selecting complex species to include in a virial formulation.

In the virial methods, therefore, the activity coefficients account implicitly for the reduction in the free ion's activity due to the formation of whatever ion pairs and complex species are not included in the formulation. As such, they describe not only the factors traditionally accounted for by activity coefficient models, such as the effects of electrostatic interaction and ion hydration, but also the distribution of species in solution. There is no provision in the method for separating the traditional part of the coefficients from the portion attributable to speciation. For this reason, the coefficients differ (even in the absence of error) in meaning and value from activity coefficients given by other methods. It might be more accurate and less confusing to refer to the virial methods as activity models rather than as activity coefficient models.

The virial methods work by assuming that the solution's excess free energy G_{ex} (i.e., the free energy in excess of that in an ideal solution) can be described by a function of the form

$$\frac{G^{ex}}{n_w R T_K} = f^{dh}(I) + \sum_i \sum_j \lambda_{ij}(I) \, m_i m_j + \sum_i \sum_j \sum_k \mu_{ijk} \, m_i m_j m_k \qquad (7.10)$$

Here, i, j, and k are subscripts representing the various species in solution and

f^{dh} is a function of ionic strength similar in form to the Debye-Hückel equation. The terms λ_{ij} and μ_{ijk} are second and third virial coefficients, which are intended to account for short-range interactions among ions; the second virial coefficients vary with ionic strength, whereas the third virial coefficients do not.

Equation 7.10 is notable in that it ascribes specific energetic effects to the interactions of the aqueous species taken in pairs (the first summation) and triplets (second summation). The equation's general form is not *ad hoc* but suggested by statistical mechanics (Anderson and Crerar, 1993, pp. 446–451). The values of the virial coefficients, however, are largely empirical, being deduced from chemical potentials determined from solutions of just one or two salts.

An expression for the ion activity coefficients γ_i follows from differentiating Eqn. 7.10 with respect to m_i. The result in general form is

$$\ln \gamma_i = \ln \gamma_i^{dh} + \sum_j D_{ij}(I) \, m_j + \sum_j \sum_k E_{ijk} \, m_j m_k \tag{7.11}$$

Here, γ_i^{dh} is a Debye-Hückel term, and D_{ij} and E_{ijk} are second and third virial coefficients, defined for each pair and triplet of ions in solution. As before, the values of D_{ij} vary with ionic strength, whereas the terms E_{ijk} are constant at a given temperature.

One of the most useful implementations for geochemical modeling, published by Harvie et al. (1984), is known as the Harvie-Møller-Weare (or HMW) method. The method treats solutions in the Na-K-Mg-Ca-H-Cl-SO$_4$-OH-HCO$_3$-CO$_3$-CO$_2$-H$_2$O system at 25°C. Notably absent from the method are the components SiO$_2$ and Al^{+++}, which are certainly important in many geochemical studies. There is, in fact, no published dataset regressed to include virial coefficients for both silica and aluminum.

Tables 7.1–7.3 show the calculation procedure for the HMW method, and Tables 7.4–7.7 list the required coefficients. In these tables, subscripts c, c', and M refer to cations, a, a', and X to anions, and n to species of neutral charge; subscripts i, j, and k refer to species in general. In Appendix 2, we carry through an example calculation by hand to provide a clearer idea of how the method is implemented. The reader should work through the example calculation before attempting to program the method.

Considering the rather large amount of data required to implement virial methods even at 25°C (e.g., Tables 7.4–7.7), it is tempting to dismiss the methods as no more than statistical fits to experimental data. In fact, however, virial methods take chemical potentials measured from simple solutions containing just one or two salts to provide an activity model capable of accurately predicting species activities in complex fluids. Eugster et al. (1980), for example, used the virial method of Harvie and Weare (1980) to accurately trace the evaporation of seawater almost to the point of desiccation. Using any other activity model, such a calculation could not even be contemplated. Other

TABLE 7.1 Procedure (Part 1) for evaluating the HMW activity model

GIVEN DATA. The following data are known at the onset of the calculation:

For each cation-anion pair: model parameters $\beta_{MX}^{(0)}$, $\beta_{MX}^{(1)}$, $\beta_{MX}^{(2)}$, C_{MX}^{ϕ}, and α_{MX}.

For each cation-cation and anion-anion pair: model parameters θ_{ij}.

For each pairing of an ion with a neutral species: model parameters λ_{ni}.

For each cation-cation-anion and anion-anion-cation triplet: model parameters ψ_{ijk}.

For each species considered: the molality m_i and charge z_i.

STEP 1. Calculate solution ionic strength I and the total molal charge Z

$$I = \frac{1}{2} \sum_i m_i z_i^2$$

$$Z = \sum_i m_i \, |z_i|$$

STEP 2. Determine for each possible pairing of like-signed charges the values of the functions ${}^E\theta_{ij}(I)$ and ${}^E\theta'_{ij}(I)$ of I by numerical integration or approximation (for details, see Pitzer, 1987, p. 130-132; Harvie and Weare, 1980). A computer program for this purpose is listed in Appendix 2. The functions are zero for like charges and symmetrical about zero, so only the positive unlike pairings (e.g., 2-1, 3-1, 3-2) need be evaluated.

STEP 3. For each cation-anion pair MX, evaluate functions $g(x)$ and $g'(x)$ for $x = \alpha_{MX} \sqrt{I}$

$$g(x) = \frac{2\,[1 - (1 + x)\,e^{-x}]}{x^2}$$

$$g'(x) = -\frac{2\,[1 - (1 + x + x^2/2)\,e^{-x}]}{x^2}$$

STEP 4. Compute for each cation-anion pair the second virial coefficients B_{MX}, B'_{MX}, B_{MX}^{ϕ}

$$B_{MX} = \beta_{MX}^{(0)} + \beta_{MX}^{(1)}\, g(\alpha_{MX} \sqrt{I}) + \beta_{MX}^{(2)}\, g(12\sqrt{I})$$

$$B'_{MX} = \beta_{MX}^{(1)}\, g'(\alpha_{MX} \sqrt{I})/I + \beta_{MX}^{(2)}\, g'(12\sqrt{I})/I$$

$$B_{MX}^{\phi} = B_{MX} + I\, B'_{MX}$$

TABLE 7.2 Procedure (Part 2) for evaluating the HMW activity model

STEP 5. Calculate for each cation-anion pair the third virial coefficient C_{MX}

$$C_{MX} = \frac{C_{MX}^{\phi}}{2\sqrt{|z_M \, z_X|}}$$

STEP 6. Compute for each cation-cation and anion-anion pair the second virial coefficients Φ_{ij}, Φ'_{ij}, Φ^{ϕ}_{ij}

$$\Phi_{ij} = \theta_{ij} + {}^E\theta_{ij}(I)$$

$$\Phi'_{ij} = {}^E\theta'_{ij}(I)$$

$$\Phi^{\phi}_{ij} = \Phi_{ij} + I \, \Phi'_{ij}$$

STEP 7. Determine the intermediate value F

$$F = -A^{\phi}\left(\frac{\sqrt{I}}{1 + 1.2\sqrt{I}} + \frac{2}{1.2} \ln{(1 + 1.2\sqrt{I})}\right)$$

$$+ \sum_{c=1}^{N_c}\sum_{a=1}^{N_a} m_c m_a B'_{ca} + \sum_{c=1}^{N_c-1}\sum_{c'=c+1}^{N_c} m_c m_{c'}\Phi'_{cc'} + \sum_{a=1}^{N_a-1}\sum_{a'=a+1}^{N_a} m_a m_{a'}\Phi'_{aa'}$$

Here $A^{\phi} = 2.303\,A/3$, where A is the Debye-Hückel parameter.

STEP 8. Calculate activity coefficients γ_M, γ_X, and γ_N for cations, anions, and neutral species

$$\ln \gamma_M = z_M^2 F + \sum_{a=1}^{N_a} m_a \, (2B_{Ma} + ZC_{Ma}) + \sum_{c=1}^{N_c} m_c \left[2\Phi_{Mc} + \sum_{a=1}^{N_a} m_a \Psi_{Mca}\right]$$

$$+ \sum_{a=1}^{N_a-1}\sum_{a'=a+1}^{N_a} m_a m_{a'}\Psi_{aa'M} + |z_M| \sum_{c=1}^{N_c}\sum_{a=1}^{N_a} m_c m_a C_{ca} + \sum_{n=1}^{N_n} m_n \, (2\lambda_{nM})$$

$$\ln \gamma_X = z_X^2 F + \sum_{c=1}^{N_c} m_c \, (2B_{cX} + ZC_{cX}) + \sum_{a=1}^{N_a} m_a \left[2\Phi_{Xa} + \sum_{c=1}^{N_c} m_c \Psi_{Xac}\right]$$

$$+ \sum_{c=1}^{N_c-1}\sum_{c'=c+1}^{N_c} m_c m_{c'}\Psi_{cc'X} + |z_X| \sum_{c=1}^{N_c}\sum_{a=1}^{N_a} m_c m_a C_{ca} + \sum_{n=1}^{N_n} m_n \, (2\lambda_{nX})$$

$$\ln \gamma_N = \sum_{c=1}^{N_c} m_c \, (2\lambda_{nc}) + \sum_{a=1}^{N_a} m_a \, (2\lambda_{na})$$

TABLE 7.3 Procedure (Part 3) for evaluating the HMW activity model

STEP 9. Calculate the osmotic coefficient ϕ according to

$$\sum_i m_i\,(\phi-1) = 2\Big[\frac{-A^\phi I^{3/2}}{1+1.2\sqrt{I}} + \sum_{c=1}^{N_c}\sum_{a=1}^{N_a} m_c m_a\,(B_{ca}^\phi + ZC_{ca})$$

$$+ \sum_{c=1}^{N_c-1}\sum_{c'=c+1}^{N_c} m_c m_{c'} \left(\Phi_{cc'}^\phi + \sum_{a=1}^{N_a} m_a \Psi_{cc'a}\right)$$

$$+ \sum_{a=1}^{N_a-1}\sum_{a'=a+1}^{N_a} m_a m_{a'} \left(\Phi_{aa'}^\phi + \sum_{c=1}^{N_c} m_a \Psi_{aa'c}\right)$$

$$+ \sum_{n=1}^{N_n}\sum_{a=1}^{N_a} m_n m_a \lambda_{na} + \sum_{n=1}^{N_n}\sum_{c=1}^{N_c} m_n m_c \lambda_{nc} \Big]$$

STEP 10. Calculate the activity a_w of water from the relation

$$\ln a_w = -\frac{W}{1000}\,\Big(\sum_i m_i\Big)\,\phi$$

where W is the mole weight of water (18.016 g/mol).

accomplishments in geochemistry (Weare, 1987) include prediction of mineral precipitation in alkaline lakes and in fluid inclusions within evaporite minerals.

7.3 Comparison of the Methods

It is interesting to compare the Debye-Hückel and virial methods, since each has its own advantages and limitations. The Debye-Hückel equations are simple to apply and readily extensible to include new species in solution, since they require few coefficients specific to either species or solution. The method can be applied as well over the range of temperatures most important to an aqueous geochemist. There is an extensive literature on ion association reactions, so there are few limits to the complexity of the solutions that can be modeled.

The Debye-Hückel methods work poorly, however, when carried to moderate or high ionic strength, especially when salts other than NaCl dominate the solute. In the theory, the ionic strength represents all the properties of a solution. For this reason, a Debye-Hückel method applied to any solution of a certain ionic strength (whether dominated by NaCl, KCl, HCl, H_2SO_4, or any salt or salt mixture) gives the same set of activity coefficients, regardless of the solution's composition. This result, except for dilute solutions, is, of course, incorrect.

TABLE 7.4 HMW model parameters for cation-anion pairs

c	a	$\beta_{ca}^{(0)}$	$\beta_{ca}^{(1)}$	$\beta_{ca}^{(2)}$	C_{ca}^{ϕ}	α_{ca}
Na^+	Cl^-	.0765	.2664*	—	.00127	2
Na^+	SO_4^{--}	.01958	1.113	—	.00497	2
Na^+	HSO_4^-	.0454	.398	—	—	2
Na^+	OH^-	.0864	.253	—	.0044	2
Na^+	HCO_3^-	.0277	.0411	—	—	2
Na^+	CO_3^{--}	.0399	1.389	—	.0044	2
K^+	Cl^-	.04835	.2122	—	−.00084	2
K^+	SO_4^{--}	.04995	.7793	—	—	2
K^+	HSO_4^-	−.0003	.1735	—	—	2
K^+	OH^-	.1298	.320	—	.0041	2
K^+	HCO_3^-	.0296	−.013	—	−.008	2
K^+	CO_3^{--}	.1488	1.43	—	−.0015	2
Ca^{++}	Cl^-	.3159	1.614	—	−.00034	2
Ca^{++}	SO_4^{--}	.20	3.1973	−54.24	—	1.4
Ca^{++}	HSO_4^-	.2145	2.53	—	—	2
Ca^{++}	OH^-	−.1747	−.2303	−5.72	—	2
Ca^{++}	HCO_3^-	.4	2.977	—	—	2
Ca^{++}	CO_3^{--}	—	—	—	—	—
Mg^{++}	Cl^-	.35235	1.6815	—	.00519	2
Mg^{++}	SO_4^{--}	.2210	3.343	−37.23	.025	1.4
Mg^{++}	HSO_4^-	.4746	1.729	—	—	2
Mg^{++}	OH^-	—	—	—	—	—
Mg^{++}	HCO_3^-	.329	.6072	—	—	2
Mg^{++}	CO_3^{--}	—	—	—	—	—
$MgOH$	Cl^-	−.10	1.658	—	—	2
$MgOH$	SO_4^{--}	—	—	—	—	—
$MgOH$	HSO_4^-	—	—	—	—	—
$MgOH$	OH^-	—	—	—	—	—
$MgOH$	HCO_3^-	—	—	—	—	—
$MgOH$	CO_3^{--}	—	—	—	—	—
H^+	Cl^-	.1775	.2945	—	.0008	2
H^+	SO_4^{--}	.0298	—	—	.0438	—
H^+	HSO_4^-	.2065	.5556	—	—	2
H^+	OH^-	—	—	—	—	—
H^+	HCO_3^-	—	—	—	—	—
H^+	CO_3^{--}	—	—	—	—	—

*Corrected value.

TABLE 7.5 HMW model parameters for cation-cation pairs and triplets

c	c'	$\theta_{cc'}$	$\Psi_{cc'Cl}$	$\Psi_{cc'SO_4}$	$\Psi_{cc'HSO_4}$	$\Psi_{cc'OH}$	$\Psi_{cc'HCO_3}$	$\Psi_{cc'CO_3}$
Na^+	K^+	-.012	-.0018	.010	—	—	-.003	.003
Na^+	Ca^{++}	.07	-.007	-.055	—	—	—	—
Na^+	Mg^{++}	.07	-.012	-.015	—	—	—	—
Na^+	$MgOH$	—	—	—	—	—	—	—
Na^+	H^+	.036	-.004	—	-.0129	—	—	—
K^+	Ca^{++}	.032	-.025	—	—	—	—	—
K^+	Mg^{++}	0.	-.022	-.048	—	—	—	—
K^+	$MgOH$	—	—	—	—	—	—	—
K^+	H^+	.005	-.011	.197	-.0265	—	—	—
Ca^{++}	Mg^{++}	.007	-.012	.024	—	—	—	—
Ca^{++}	$MgOH$	—	—	—	—	—	—	—
Ca^{++}	H^+	.092	-.015	—	—	—	—	—
Mg^{++}	$MgOH$	—	.028	—	—	—	—	—
Mg^{++}	H^+	.10	-.011	—	-.0178	—	—	—
$MgOH$	H^+	—	—	—	—	—	—	—

TABLE 7.6 HMW model parameters for anion-anion pairs and triplets

a	a'	$\theta_{aa'}$	$\Psi_{aa'Na}$	$\Psi_{aa'K}$	$\Psi_{aa'Ca}$	$\Psi_{aa'Mg}$	$\Psi_{aa'MgOH}$	$\Psi_{aa'H}$
Cl^-	SO_4^{--}	.02	.0014	—	-.018	-.004	—	—
Cl^-	HSO_4^-	-.006	-.006	—	—	—	—	.013
Cl^-	OH^-	-.050	-.006	-.006	-.025	—	—	—
Cl^-	HCO_3^-	.03	-.015	—	—	-.096	—	—
Cl^-	CO_3^{--}	-.02	.0085	.004	—	—	—	—
SO_4^{--}	HSO_4^-	—	-.0094	-.0677	—	-.0425	—	—
SO_4^{--}	OH^-	-.013	-.009	-.050	—	—	—	—
SO_4^{--}	HCO_3^-	.01	-.005	—	—	-.161	—	—
SO_4^{--}	CO_3^{--}	.02	-.005	-.009	—	—	—	—
HSO_4^-	OH^-	—	—	—	—	—	—	—
HSO_4^-	HCO_3^-	—	—	—	—	—	—	—
HSO_4^-	CO_3^{--}	—	—	—	—	—	—	—
OH^-	HCO_3^-	—	—	—	—	—	—	—
OH^-	CO_3^{--}	.10	-.017	-.01	—	—	—	—
HCO_3^-	CO_3^{--}	-.04	.002	.012	—	—	—	—

TABLE 7.7 HMW model parameters for
neutral species-ion pairs

i	$\lambda_{CO_2\ i}$	$\lambda_{CaCO_3\ i}$	$\lambda_{MgCO_3\ i}$
H^+	0.	—	—
Na^+	.100	—	—
K^+	.051	—	—
Ca^{++}	.183	—	—
Mg^{++}	.183	—	—
$MgOH$	—	—	—
Cl^-	−.005	—	—
SO_4^{--}	.097	—	—
HSO_4^-	−.003	—	—
OH^-	—	—	—
HCO_3^-	—	—	—
CO_3^{--}	—	—	—

Clearly, we cannot rely on a single value to describe how the properties of a concentrated solution depend on its solute content.

The virial methods, on the other hand, provide remarkably accurate results over a broad range of solution concentrations and with a variety of dominant solutes. The methods, however, are limited in breadth. Notably lacking at present are data for redox reactions and for components such as aluminum and silica with low solubilities. Data for extending the methods to apply beyond room temperature (e.g., Møller, 1988; Greenberg and Møller, 1989), furthermore, are limited currently to relatively simple chemical systems.

Unlike the Debye-Hückel equations, the virial methods provide little or no information about the distribution of species in solution. Geochemists like to identify the dominant species in solution in order to write the reactions that control a system's behavior. In the virial methods, this information is hidden within the complexities of the virial equations and coefficients. Many geochemists, therefore, find the virial methods to be less satisfying than methods that predict the species distribution. The information given by Debye-Hückel methods about species distributions in concentrated solutions, however, is not necessarily reliable and should be used with caution.

To explore the differences between the methods, we use REACT to calculate at 25°C the solubility of gypsum ($CaSO_4 \cdot 2H_2O$) as a function of NaCl concentration. We use two datasets: thermo.data, which invokes the B-dot equation, and thermo.hmw, based on the HMW model. The log K values for the gypsum dissolution reaction vary slightly between the datasets. To limit our

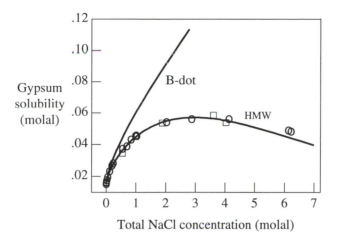

Total NaCl concentration (molal)

FIG. 7.6 Solubility of gypsum ($CaSO_4 \cdot 2\,H_2O$) at 25°C as a function of NaCl concentration, calculated according to the Harvie-Møller-Weare and B-dot (modified Debye-Hückel) activity models. Circles and squares, respectively, show experimental determinations by Marshall and Slusher (1966) and Block and Waters (1968).

comparison to the activity models, we use the `alter` command to set log K in each case to the value of -4.5805 used by Harvie et al. (1984).

To perform the calculation for a one molal NaCl solution, for example, we use the `data` command to set the appropriate dataset (`thermo.data` or `thermo.hmw`) and then enter

```
swap Gypsum for Ca++
100 free grams Gypsum
balance on SO4--

1 molal Na+
1 molal Cl-

precip = off
alter Gypsum 0 -4.5805
go
```

We then repeat the calculation over a range of NaCl concentrations. (To save effort, we can perform the calculation in one step by titrating NaCl into an initially dilute solution, as we discuss in Chapter 11, and then plotting the results with GTPLOT.)

Figure 7.6 shows the calculation results plotted against measured solubilities from laboratory experiments. The HMW calculations closely coincide with the experimental data, reflecting the fact that these same data were used in parameterizing the model (Harvie and Weare, 1980). The B-dot results coincide

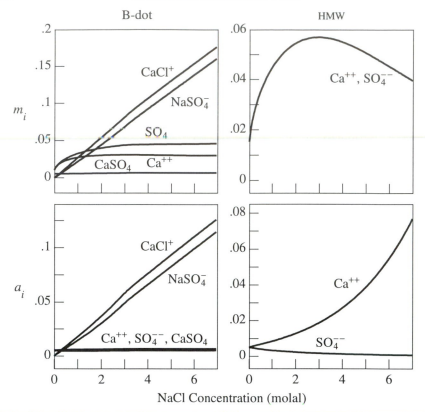

FIG. 7.7 Molal concentrations m_i and activities a_i of calcium and sulfate species in equilibrium with gypsum at 25°C as functions of NaCl concentration, calculated using the B-dot equation (left) and the HMW activity model (right).

closely with the data only at NaCl concentrations less than about 0.5 molal. At 3 molal concentration, the predicted solubility is about double the measured value.

Whether error of this extent is acceptable depends primarily on the modeler's goals (e.g., Weare, 1987, pp. 160–162; Brantley et al., 1984; Felmy and Weare, 1986). If a model is designed to accurately predict mineral solubility in concentrated solutions, clearly the B-dot equation would be considered incorrect. A model created to explore the behavior of a fluid in equilibrium with gypsum might be useful, however, even though the modeler recognizes that the predicted gypsum solubility is inaccurate by a factor of two or more. Indeed, as discussed in Chapter 2, errors of this magnitude are not uncommon in geochemical modeling.

Figure 7.7 shows how concentrations and activities of the calcium and sulfate species vary with NaCl concentration. In the B-dot model, there are three ion pairs ($CaCl^+$, $NaSO_4^-$, and $CaSO_4$) in addition to the free ions Ca^{++} and SO_4^{--}. The activities of the free ions remain roughly constant with NaCl concentration,

TABLE 7.8 Chemical compositions (g/l) of brines from the Sebkhat El Melah brine deposit, Zarzis, Tunisia (Jarraya and El Mansar, 1987)

Well	K^+	Mg^{++}	Na^+	Ca^{++}	Br^-	Cl^-	SO_4^{--}	HCO_3^-	pH
RZ-2	6.90	52.1	43.1	.2	2.25	195	27.4	.14	7.15
RZ-7	7.65	47.2	50.0	.30	2.11	195	26.9	.19	7.0
RZ-8	7.46	52.3	37.5	.20	2.60	192	27.9	.20	6.8
RZ-9	6.70	47.3	50.8	.20	2.35	188	30.2	.17	6.8
RZ-10	6.6	34.4	51.3	.4	1.36	177	21.7	.16	6.85
RZ-11	7.85	54.7	32.	.2	2.32	206	19.2	.17	6.
RZ-16	6.80	46.2	47.	.20	3.29	195	18.7	.20	7.1
RZ-17	8.85	55.9	39.3	.20	3.2	202	28.2	.18	7.1
RZ-19	7.23	50.0	44.2	.5	2.	200	28.8	.24	6.9

and their concentrations increase only moderately, reflecting the decrease in the B-dot activity coefficients with increasing ionic strength (Fig. 7.3). Formation of the complex species $CaCl^+$ and $NaSO_4^-$ drives the general increase in gypsum solubility with NaCl concentration predicted by the B-dot model.

In the HMW model, in contrast, Ca^{++} and SO_4^{--} are the only calcium or sulfate-bearing species considered. The species maintain equal concentration, as required by electroneutrality, and mirror the solubility curve in Fig. 7.6. Unlike the B-dot model, the species' activities follow trends dissimilar to their concentrations. The Ca^{++} activity rises sharply while that of SO_4^{--} decreases. In this case, variation in gypsum solubility arises not from the formation of ion pairs, but from changes in the activity coefficients for Ca^{++} and SO_4^{--} as well as in the water activity. The latter value, according to the model, decreases with NaCl concentration from one to about 0.7.

7.4 Brine Deposit at Sebkhat El Melah

As a test of our ability to calculate activity coefficients in natural brines, we consider groundwater from the Sebkhat El Melah brine deposit near Zarzis, Tunisia (Perthuisot, 1980). The deposit occurs in a buried evaporite basin composed of halite (NaCl), anhydrite ($CaSO_4$), and dolomite [$CaMg(CO_3)_2$]. The Tunisian government would like to exploit the brines for their chemical content, especially for the potassium, which is needed to make fertilizer.

Since the deposit contains halite and anhydrite, the brines should be saturated with respect to these minerals and hence provide a good test of the activity models. Table 7.8 shows analyses of brine samples from the deposit. Note that the reported pH values are almost certainly incorrect because pH electrodes do

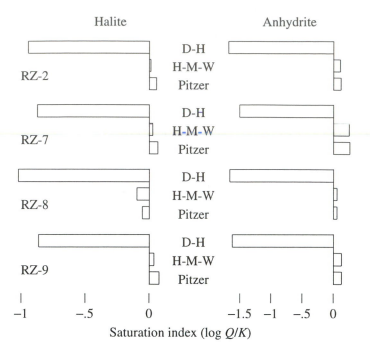

FIG. 7.8 Saturation indices of Sebkhat El Melah brine samples with respect to halite (left) and anhydrite (right), calculated using the B-dot (modified Debye-Hückel), Harvie-Møller-Weare, and Pitzer activity models.

not respond accurately in concentrated solutions. Hence, there is little to be gained by calculating dolomite saturation.

To model the brine, we enter the chemical composition and set the activity model. The REACT commands debye-huckel, hmw, and pitzer, respectively, set the Debye-Hückel (B-dot), Harvie-Møller-Weare, and Pitzer activity models. Here, the Pitzer model refers to the method of Pitzer (1979), as adapted at LLNL by Jackson and Wolery (1985). The HMW model does not account for bromine, so we must type remove Br- before invoking it. Similarly, the Pitzer model does not contain HCO_3^-.

A complication is that the analyses in Table 7.8 are reported in units of grams per liter, which REACT must convert to its internal units of molality. To do so, the program requires values for the solution's density and total dissolved solid content. A quick way to solve this problem is to iterate by running the program, then using the calculated density and TDS as constraints on the next run, and so on until the values converge. Taking the analysis for well RZ-2, for example,

```
K+ = 6.9 g/l
Mg++ = 52.1 g/l
(etc., from Table 7.8)
```

```
precip = off
debye-huckel
go

pickup density
pickup TDS
go
```

(repeat last three commands until density and TDS converge)

To verify convergence, we can check the current density and TDS by typing `show variables`, or by searching the output dataset with the commands `grep density` and `grep solids`.

Figure 7.8 shows the resulting saturation indices for halite and anhydrite, calculated for the first four samples in Table 7.8. The Debye-Hückel (B-dot) method, which of course is not intended to be used to model saline fluids, predicts that the minerals are significantly undersaturated in the brine samples. The Harvie-Møller-Weare and Pitzer models, on the other hand, predict that halite and anhydrite are near equilibrium with the brine, as we would expect. As usual, we cannot determine whether the remaining discrepancies result from the analytical error, error in the activity model, or error from other sources.

8

Surface Complexation

An important consideration in constructing certain types of geochemical models, especially those applied to environmental problems, is to account for the sorption of ions from solution onto mineral surfaces. Metal oxides and aluminosilicate minerals, as well as other phases, can sorb electrolytes strongly because of their high reactivities and large surface areas (e.g., Davis and Kent, 1990). When a fluid comes in contact with minerals such as iron or aluminum oxides and zeolites, sorption may significantly diminish the mobility of dissolved components in solution, especially those present in minor amounts.

Sorption, for example, may retard the spread of radionuclides near a radioactive waste repository or the migration of contaminants away from a polluting landfill. In acid mine drainages, ferric oxide sorbs heavy metals from surface water, helping limit their downstream movement (see Chapter 23). A geochemical model useful in investigating such cases must provide an accurate assessment of the effects of surface reactions.

Many of the sorption theories now in use are too simplistic to be incorporated into a geochemical model intended for general use. To be useful in modeling electrolyte sorption, a theory must account for the electrical charge on the mineral surface and provide for mass balance on the sorbing sites. In addition, an internally consistent and sufficiently broad database of sorption reactions must accompany the theory.

The Freundlich and Langmuir theories, which use distribution coefficients K_d to set the ratios of sorbed to dissolved ions, are applied widely in groundwater studies (Domenico and Schwartz, 1990) and used with considerable success to describe sorption of uncharged organic molecules (Adamson, 1976). The models, however, do not account for the electrical state of the surface, which varies sharply with pH, ionic strength, and solution composition. Freundlich theory prescribes no concept of mass balance, so that a surface might be predicted to sorb from solution without limit. Both theories require that distribution coefficients be determined experimentally for individual fluid and rock compositions, and hence both theories lack generality. Ion exchange theory (Stumm and Morgan, 1981; Sposito, 1989) suffers from similar limitations.

Surface complexation models, on the other hand, account explicitly for the electrical state of the sorbing surface (e.g., Adamson, 1976; Stumm, 1992). This class of models includes the constant capacitance, double layer, and triple layer theories (e.g., Westall and Hohl, 1980; Sverjensky, 1993). Of these, double layer theory (also known as diffuse layer theory) is most fully developed in the literature and probably the most useful in geochemical modeling (e.g., Dzombak and Morel, 1987).

In this chapter, we will discuss how double layer theory can be incorporated into a geochemical model. We will consider hydrous ferric oxide (FeOOH·nH$_2$O), which is one of the most important sorbing minerals at low temperature under oxidizing conditions. Sorption by hydrous ferric oxide has been widely studied and Dzombak and Morel (1990) have compiled an internally consistent database of its complexation reactions. The model we develop, however, is general and can be applied equally well to surface complexation with other metal oxides for which a reaction database is available.

8.1 Complexation Reactions

Surface complexation theory is well described in a number of texts on surface chemistry, including Adamson (1976), Stumm and Morgan (1981), Sposito (1989), Dzombak and Morel (1990), and Stumm (1992); therefore, we merely summarize it in this section. According to the theory, the sorbing surface is composed of metal-hydroxyl sites that can react with ions in solution.

In Dzombak and Morel's (1990) development, hydrous ferric oxide holds two site types, one weakly and the other strongly binding. In their uncomplexed forms, the sites are labeled >(w)FeOH and >(s)FeOH; the notation ">" represents bonding to the mineral structure, and "(w)" and "(s)" signify the weak and strong sites.

Each site can protonate or deprotonate to form surface species such as >(w)FeOH$_2^+$ and >(w)FeO$^-$. The corresponding reactions are

$$>(w)FeOH_2^+ \rightleftarrows >(w)FeOH + H^+ \tag{8.1a}$$

$$>(w)FeO^- + H^+ \rightleftarrows >(w)FeOH \tag{8.1b}$$

As well, the sites can react with cations and anions from solution to form complexes such as >(w)FeOCa$^+$ and >(w)FeSO$_4^-$

$$>(w)FeOCa^+ + H^+ \rightleftarrows >(w)FeOH + Ca^{++} \tag{8.1c}$$

$$>(w)FeSO_4^- + H_2O \rightleftarrows >(w)FeOH + SO_4^{--} + H^+ \tag{8.1d}$$

Following our convention, we place surface complexes to the left of the reactions and the uncomplexed sites to the right. For use with program REACT, file "FeOH.data" contains the database of surface complexation reactions prepared by Dzombak and Morel (1990).

The surface sites and complexes lie in a layer on the mineral surface which, because of the charged complexes, has a net electrical charge that can be either positive or negative. A second layer, the diffuse layer, separates the surface layer from the bulk fluid. The role of the diffuse layer is to achieve local charge balance with the surface; hence, its net charge is opposite that of the sorbing surface. Double layer theory, applied to a mixed ionic solution, does not specify which ions make up the diffuse layer.

To cast the equations in general terms, we use the label A_p to represent each type of surface site. In the case of hydrous ferric oxide, there are two such entries, $>(w)FeOH$ and $>(s)FeOH$. There are M_p total moles of each site type in the system, divided between uncomplexed and complexed sites. This value is the product of the mass (in moles) of the sorbing mineral and the site density (moles of sites per mole of mineral) for each site type.

Label A_q represents each possible surface complex, including protonated and deprotonated sites (e.g., $>(w)FeO^-$ and $>(w)FeOH_2^+$) and complexes with cations and anions ($>(w)FeOZn^+$ and $>(w)FePO_4^{--}$, for example). The molalities of the uncomplexed and complexed sites, respectively, are m_p and m_q.

As can be seen by Reactions 8.1a–d, the state of the Stern layer depends on the chemistry of the solution it contacts. As *p*H decreases, the numbers of protonated sites (e.g., $>(w)FeOH_2^+$) and sites complexed with bivalent anions (e.g., $>(w)FeSO_4^-$) increase. If protonated sites dominate, as is likely under acidic conditions, the surface has a net positive charge.

Increasing *p*H causes sites to deprotonate and complex with bivalent cations, forming species such as $>(w)FeO^-$ and $>(w)FeOCa^+$. In contact with an alkaline fluid, deprotonated sites are likely to dominate the surface, and the net charge will be negative. The point of zero charge (PZC) is the *p*H at which positive and negative complexes balance. The *p*H at which protonated and deprotonated sites achieve charge balance is the pristine point of zero charge, or PPZC. When there are no sorbing cations or anions, the PZC and PPZC are equivalent.

We can determine the surface charge directly from the molalities m_q of the surface complexes. The surface charge density σ (the charge per unit area of the sorbing surface, in C/m^2) is given as

$$\sigma = \frac{Fn_w}{A_{sf}} \sum_q z_q m_q \tag{8.2}$$

where F is the Faraday constant (96,485 C/mol), n_w is the mass of solvent water (kg), A_{sf} is the sorbing surface area (m^2), and z_q is the electrical charge on each complex.

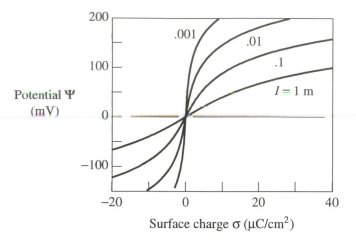

FIG. 8.1   Relationship (Eqn. 8.3) between surface charge density σ and surface potential Ψ for a sorbing surface in contact with solutions of differing ionic strengths I (molal).

Since the sorbing surface holds a charge, its electrical potential differs from that of the solution. The potential difference between surface and fluid is known as the surface potential Ψ and can be expressed in volts. The product $e \cdot \Psi$ is the work required to bring an elementary charge e from the bulk solution to the sorbing surface. According to one of the main results of double layer theory, the surface potential is related to the surface charge density by

$$\sigma = (8\, RT_K\, \varepsilon\varepsilon_o\, I \times 10^3)^{\frac{1}{2}} \sinh\left[\frac{z_{\pm}\Psi F}{2RT_K}\right] \qquad (8.3)$$

Here, R is the gas constant (8.3143 J/mol·K or V·C/mol·K), T_K is absolute temperature (K), ε is the dielectric constant (78.5 at 25°C), ε_o is the permittivity of free space (8.854×10^{-12}), I is ionic strength (molal), and $z_{\pm}$ is the charge on the background solute (assumed here to be unity). Ionic strength in this equation serves as a proxy for solute concentration, as carried in the original derivation, since the derivation formally considers a solution of a single symmetrical electrolyte, rather than a mixed solution. Figure 8.1 shows the relationship graphically.

In order for an ion to sorb from solution, it must first move through the electrical potential field and then react chemically at the surface. To write a mass balance equation for sorption reactions (such as Reactions 8.1a–d), therefore, we must account for both the electrostatic and chemical contributions to the free energy change of reaction.

Moving a positively charged ion toward a positively charged surface, for example, requires that work be done on the ion, increasing the system's free energy. An ion of the same charge escaping the surface would have the opposite energetic effect. Electrostatic effects, therefore, are important only for reactions that affect a net change in surface charge.

As before (see Chapter 3), an equilibrium constant K represents the chemical effects on free energy. Multiplying K by the Boltzman factor

$$\exp\left[\frac{-\Delta z \, \Psi \, F}{RT_K}\right] \tag{8.4}$$

gives a complete account of the reaction's free energy, including electrostatic effects. Here, Δz is the change over the reaction in the charge on surface species. The mass action equation for Reaction 8.1a, then, is

$$K \exp\left[\frac{\Psi \, F}{RT_K}\right] = \frac{m_{>(w)FeOH} \, a_{H^+}}{m_{>(w)FeOH_2^+}} \tag{8.5}$$

since Δz for the reaction is -1. Note that activity coefficients (i.e., $\gamma_{>(w)FeOH}$ and $\gamma_{>(w)FeOH_2^+}$) are not defined for the surface species.

Despite the seeming exactitude of the mathematical development, the modeler should bear in mind that the double layer model involves uncertainties and data limitations in addition to those already described (Chapter 2). Perhaps foremost is the nature of sorbing material itself. The complexation reactions are studied in laboratory experiments performed using synthetically precipitated ferric oxide. This material ripens with time, changing in water content and extent of polymerization. It eventually begins to crystallize to form goethite (FeOOH).

Since laboratories follow different aging procedures, results of their studies can be difficult to compare. Values reported for surface area and site densities vary over a relatively broad range (Dzombak and Morel, 1990). It is not clear, furthermore, how closely the synthetic material resembles sorbing ferric oxides (e.g., ferrihydrite) encountered in nature.

Nearly all of the data are collected at room temperature, and there is no accepted method for correcting them to other temperatures. Far fewer data have been collected for sorption of anions than for cations. The theory does not account for the kinetics of sorption reactions nor the hysteresis commonly observed between the adsorption and desorption of a strongly bound ion. Finally, much work remains to be done before the results of laboratory experiments performed on simple mineral-water systems can be applied to the study of complex soils.

8.2 Governing Equations

Because it is based on chemical reactions, the double layer model can be integrated into the equations describing the equilibrium state of a multicomponent system, as developed in Chapter 3. The basis appears as before (Table 3.1)

$$\mathbf{B} = \left[A_w, A_i, A_k, A_m, A_p\right] \qquad (8.6)$$

with the addition of an entry A_p for each type of surface site considered.

Parallel to the reactions to form the secondary species (Eqn. 3.22)

$$A_j - v_{wj}A_w + \sum_i v_{ij}A_i + \sum_k v_{kj}A_k + \sum_m v_{mj}A_m \qquad (8.7)$$

there is a reaction to form each surface complex

$$A_q = v_{wq}A_w + \sum_i v_{iq}A_i + \sum_k v_{kq}A_k + \sum_m v_{mq}A_m + \sum_p v_{pq}A_p \qquad (8.8)$$

Here, v_{wq}, v_{iq}, v_{kq}, v_{mq}, and v_{pq}, are coefficients in the reaction, written in terms of the basis $\mathbf{B}$, for surface complex A_q. We have already shown (Eqn. 3.27) that the molality of each secondary species is given by a mass action equation

$$m_j = \frac{1}{K_j \gamma_j}\left[a_w^{v_{wj}} \cdot \prod_i^{i} (\gamma_i m_i)^{v_{ij}} \cdot \prod_k^{k} a_k^{v_{kj}} \cdot \prod_m^{m} f_m^{v_{mj}}\right] \qquad (8.9)$$

In Reaction 8.8 the change Δz in surface charge is $-z_q$ because the uncomplexed sites A_p carry no charge. With this in mind, we can write a generalized mass action equation cast in the form of Eqn. 8.5,

$$m_q = \frac{1}{K_q \, e^{z_q F \Psi / RT_K}}\left[a_w^{v_{wq}} \cdot \prod_i^{i} (\gamma_i m_i)^{v_{iq}} \cdot \prod_k^{k} a_k^{v_{kq}} \cdot \prod_m^{m} f_m^{v_{mq}} \cdot \prod_p^{p} m_p^{v_{pq}}\right] \quad (8.10)$$

that sets the molality of each surface complex.

As before, we write mass balance equations for each basis entry. The equations

$$M_w = n_w \left[55.5 + \sum_j v_{wj}m_j + \sum_q v_{wq}m_q\right] \qquad (8.11a)$$

$$M_i = n_w \left[m_i + \sum_j v_{ij}m_j + \sum_q v_{iq}m_q\right] \qquad (8.11b)$$

$$M_k = n_k + n_w \left[\sum_j v_{kj}m_j + \sum_q v_{kq}m_q\right] \qquad (8.11c)$$

$$M_m = n_w \left[\sum_j v_{mj}m_j + \sum_q v_{mq}m_q\right] \qquad (8.11d)$$

differ from Eqns. 3.28a–d by the inclusion in each of a summation over the surface complexes. The summations account for the sorbed mass of each component. We write an additional mass balance equation

$$M_p = n_w \left[m_p + \sum_q \nu_{pq} m_q \right] \tag{8.11e}$$

for each entry A_p in the basis. This equation constrains, for each site type, the number of uncomplexed sites and surface complexes to the total number of sites M_p. Together, these relationships form a set of governing equations describing multicomponent equilibrium in the presence of a sorbing mineral surface.

8.3 Numerical Solution

The procedure for solving the governing equations parallels the technique described in Chapter 5, with the added complication of accounting for electrostatic effects. We begin as before by identifying the nonlinear portion of the problem to form the reduced basis

$$\mathbf{B}_r = \left[A_w, A_i, A_p \right]_r \tag{8.12}$$

For each basis entry we cast a residual function, which is the difference between the right and left sides of the mass balance equations (Eqns. 8.11a, b, and e)

$$R_w = n_w \left[55.5 + \sum_j \nu_{wj} m_j + \sum_q \nu_{wq} m_q \right] - M_w \tag{8.13a}$$

$$R_i = n_w \left[m_i + \sum_j \nu_{ij} m_j + \sum_q \nu_{iq} m_q \right] - M_i \tag{8.13b}$$

$$R_p = n_w \left[m_p + \sum_q \nu_{pq} m_q \right] - M_p \tag{8.13c}$$

We employ the Newton-Raphson method to iterate toward a set of values for the unknown variables $(n_w, m_i, m_p)_r$ for which the residual functions become vanishingly small.

To do so, we calculate the Jacobian matrix, which is composed of the partial derivatives of the residual functions with respect to the unknown variables. Noting the results of differentiating the mass action equations (Eqns. 8.9 and 8.10)

$$\frac{\partial m_j}{\partial n_w} = 0 \qquad \frac{\partial m_j}{\partial m_i} = \nu_{ij} \frac{m_j}{m_i} \qquad \frac{\partial m_j}{\partial m_p} = 0$$

$$\frac{\partial m_q}{\partial m_w} = 0 \qquad \frac{\partial m_q}{\partial m_i} = \nu_{iq} \frac{m_q}{m_i} \qquad \frac{\partial m_q}{\partial m_p} = \nu_{pq} \frac{m_q}{m_p} \tag{8.14}$$

simplifies the derivation. The entries in the Jacobian matrix are

$$J_{ww} = \frac{\partial R_w}{\partial n_w} = 55.5 + \sum_j \nu_{wj} m_j + \sum_q \nu_{wq} m_q \tag{8.15a}$$

$$J_{wi} = \frac{\partial R_w}{\partial m_i} = \frac{n_w}{m_i} \left[\sum_j \nu_{wj} \nu_{ij} m_j + \sum_q \nu_{wq} \nu_{iq} m_q \right] \tag{8.15b}$$

$$J_{wp} = \frac{\partial R_w}{\partial m_p} = \frac{n_w}{m_p} \sum_q \nu_{wq} \nu_{pq} m_q \tag{8.15c}$$

$$J_{iw} = \frac{\partial R_i}{\partial n_w} = m_i + \sum_j \nu_{ij} m_j + \sum_q \nu_{iq} m_q \tag{8.15d}$$

$$J_{ii'} = \frac{\partial R_i}{\partial m_{i'}} = n_w \delta_{ii'} + \frac{n_w}{m_{i'}} \left[\sum_j \nu_{ij} \nu_{i'j} m_j + \sum_q \nu_{iq} \nu_{i'q} m_q \right] \tag{8.15e}$$

$$J_{ip} = \frac{\partial R_i}{\partial m_p} = \frac{n_w}{m_p} \sum_q \nu_{iq} \nu_{pq} m_q \tag{8.15f}$$

$$J_{pw} = \frac{\partial R_p}{\partial n_w} = m_p + \sum_q \nu_{pq} m_q \tag{8.15g}$$

$$J_{pi} = \frac{\partial R_p}{\partial m_i} = \frac{n_w}{m_i} \left[\sum_q \nu_{pq} \nu_{iq} m_q \right] \tag{8.15h}$$

$$J_{pp'} = \frac{\partial R_p}{\partial m_{p'}} = n_w \delta_{pp'} + \frac{n_w}{m_{p'}} \left[\sum_q \nu_{pq} \nu_{p'q} m_q \right] \tag{8.15i}$$

Here, the Kronecker delta function is defined as

$$\delta_{ii'} = \begin{cases} 1 & \text{if } i=i' \\ 0 & \text{if } i \neq i' \end{cases} \quad \text{and} \quad \delta_{pp'} = \begin{cases} 1 & \text{if } p=p' \\ 0 & \text{if } p \neq p' \end{cases} \tag{8.16}$$

At each step in the iteration, we evaluate the residual functions and Jacobian matrix. We then calculate a correction vector as the solution to the matrix equation

$$\begin{bmatrix} J_{ww} & J_{wi} & J_{wp} \\ J_{iw} & J_{ii'} & J_{ip} \\ J_{pw} & J_{pi} & J_{pp'} \end{bmatrix}_r \begin{bmatrix} \Delta_w \\ \Delta_i \\ \Delta_p \end{bmatrix}_r = \begin{bmatrix} -R_w \\ -R_i \\ -R_p \end{bmatrix}_r \qquad (8.17)$$

To assure non-negativity of the unknown variables, we determine an underrelaxation factor δ_{UR} according to

$$\frac{1}{\delta_{UR}} = \max\left[1, -\frac{\Delta_w}{\frac{1}{2} n_w^{(q)}}, -\frac{\Delta_i}{\frac{1}{2} m_i^{(q)}}, -\frac{\Delta_p}{\frac{1}{2} m_p^{(q)}} \right]_r \qquad (8.18)$$

and then update values from the current (q) iteration level

$$\begin{bmatrix} n_w \\ m_i \\ m_p \end{bmatrix}_r^{(q+1)} = \begin{bmatrix} n_w \\ m_i \\ m_p \end{bmatrix}_r^{(q)} + \delta_{UR} \begin{bmatrix} \Delta_w \\ \Delta_i \\ \Delta_p \end{bmatrix}_r \qquad (8.19)$$

to give those at the new $(q+1)$ level.

The iteration step, however, is complicated by the need to account for the electrostatic state of the sorbing surface when setting values for m_q. The surface potential Ψ affects the sorption reactions, according to the mass action equation (Eqn. 8.10). In turn, according to Eqn. 8.2, the concentrations m_q of the sorbed species control the surface charge and hence (by Eqn. 8.3) potential. Since the relationships are nonlinear, we must solve numerically (e.g., Westall, 1980) for a consistent set of values for the potential and species concentrations.

The solution, performed at each step in the Newton-Raphson iteration, is accomplished by setting Eqn. 8.2 equal to Eqn. 8.3. We write a residual function for the surface potential

$$R_\Psi = \frac{A_{sf}}{F n_w} (8 \, RT_K \, \varepsilon\varepsilon_o \, I \times 10^3)^{\frac{1}{2}} \sinh\left[\frac{z_\pm \Psi F}{2RT_K} \right] - \sum_q z_q m_q \qquad (8.20)$$

which we wish to minimize. The function's derivative with respect to Ψ is

$$\frac{dR_\Psi}{d\Psi} = \frac{z_\pm A_{sf}}{2n_w RT_K} (8 \, RT_K \, \varepsilon\varepsilon_o \, I \times 10^3)^{\frac{1}{2}} \cosh\left[\frac{z_\pm \Psi F}{2RT_K} \right] + \frac{F}{RT_K} \sum_q z_q^2 m_q \qquad (8.21)$$

Using Newton's method (described in Chapter 5), we can quickly locate the appropriate surface potential by decreasing the residual function until it approaches zero.

8.4 Example Calculation

As an example of an equilibrium calculation accounting for surface complexation, we consider the sorption of mercury, lead, and sulfate onto hydrous ferric oxide at pH 4 and 8. We use ferric hydroxide [Fe(OH)$_3$] precipitate from the LLNL database to represent in the calculation hydrous ferric oxide (FeOOH·nH$_2$O). Following Dzombak and Morel (1990), we assume a sorbing surface area of 600 m^2/g and site densities for the weakly and strongly binding sites, respectively, of 0.2 and 0.005 mol/mol FeOOH. We choose a system containing 1 kg of solvent water (the default) in contact with 1 g of ferric hydroxide.

To set up the calculation, we enter the commands

```
surface_data = FeOH.data
sorbate include

decouple Fe+++
swap Fe(OH)3(ppd) for Fe+++
1 free gram Fe(OH)3(ppd)
suppress Hematite, Goethite
```

First, we read in the dataset of complexation reactions and specify that the initial mass balance calculations should include the sorbed as well as aqueous species. We disable the ferric-ferrous redox couple (since we are not interested in ferrous iron), specify that the system contains 1 g of sorbing mineral, and suppress the ferric oxides hematite (Fe$_2$O$_3$) and goethite (FeOOH), which are more stable thermodynamically than ferric hydroxide.

We set a dilute NaCl solution containing small concentrations of Hg^{++}, Pb^{++}, and SO$_4^{--}$. For the first calculation, we set pH to 4

```
(cont'd)
Na+ = 10 mmolal
Cl- = 10 mmolal
Hg++ = 100 umolal
Pb++ = 100 umolal
SO4-- = 200 umolal

pH = 4
go
```

and run the model. We repeat the calculation assuming a pH of 8

```
(cont'd)
pH = 8
go
```

Table 8.1 summarizes the calculation results.

The predicted state of the sorbing surface in the two calculations differs considerably. At pH 4, the surface carries a positive surface charge and

TABLE 8.1 Calculated examples of surface complexation

	pH = 4		pH = 8	
Surface charge ($\mu C/cm^2$)	16.0		.3	
Surface potential (mV)	168.		13.9	
SITE OCCUPANCY				
Weak sites	*mmolal*	*% of sites*	*mmolal*	*% of sites*
>(w)FeOH$_2^+$	1.23	65.7	.152	8.12
>(w)FeOH	.434	23.2	1.35	72.2
>(w)FeO$^-$	$.350\times10^{-2}$	.19	.271	14.5
>(w)FeOHg$^+$	$.415\times10^{-6}$	.000	.0986	5.27
>(w)FeSO$_4^-$	.117	6.25	$.174\times10^{-3}$	.009
>(w)FeOHSO$_4^{-}$	.0825	4.41	$.307\times10^{-2}$	.16
	1.871	100.	1.871	100.
Strong sites				
>(s)FeOH$_2^+$	.00557	11.9	$.930\times10^{-10}$	.000
>(s)FeOH	.00196	4.19	$.643\times10^{-9}$	.000
>(s)FeO$^-$	$.158\times10^{-4}$	.03	$.130\times10^{-9}$	.000
>(s)FeOHg$^+$	$.383\times10^{-7}$	.000	$.963\times10^{-9}$	.000
>(s)FeOPb$^+$	.0392	83.8	.0468	100.
	.0468	100.	.0468	100.
SORBED FRACTIONS				
		% sorbed		*% sorbed*
Hg^{++}		.000		98.7
Pb^{++}		39.2		46.8
SO$_4^{--}$		100.		.002

potential. The electrical charge arises largely from the predominance of the protonated surface species >(w)FeOH$_2^+$, which occupies about two thirds of the weakly binding sites. At pH 8, however, the surface charge and potential nearly vanish because of the predominance of the uncomplexed species >(w)FeOH, which is electrically neutral.

We can quickly verify the calculated charge density using Eqn. 8.2 and the data in Table 8.1. At pH 4, we add the products of each species charge and concentration

		mmolal
>(w)FeOH$_2^+$	+1 ×	1.23
>(w)FeO$^-$	−1 ×	.0035
>(w)FeSO$_4^-$	−1 ×	.117
>(w)FeOHSO$_4^{--}$	−2 ×	.0825
>(s)FeOH$_2^+$	+1 ×	.0056
>(s)FeOPb$^+$	+1 ×	.0392

0.989

and then calculate the charge density as

$$(96,485 \text{ C/mol}) \times (1 \text{ kg H}_2\text{O}) \times (0.989 \times 10^{-3} \text{ molal}) / (600 \text{ m}^2)$$
$$= .16 \text{ C/m}^2 \qquad (8.22)$$

or 16 μC/cm^2. Taking ionic strength as .01 molal, we can read the corresponding surface potential from Fig. 8.1. We can verify the results at pH 8 in a similar fashion.

According to the calculations, the surface's ability to sorb cations and anions differs markedly between pH 4 and 8, reflecting both electrostatic influences and mass action. Nearly all of the sulfate is sorbed at pH 4, whereas most of the mercury remains in solution; the opposite holds at pH 8. At pH 4, the surface carries a positive charge that attracts sulfate ions but repels mercury ions. The electrostatic effect is almost nil at pH 8, however, where the surface charge approaches zero. As well, the complexation reactions

$$>(w)FeSO_4^- + H_2O \rightleftharpoons >(w)FeOH + H^+ + SO_4^{--} \qquad (8.23)$$

$$>(w)FeOHg^+ + H^+ \rightleftharpoons >(w)FeOH + Hg^{++} \qquad (8.24)$$

conspire by mass action to favor complexing of sulfate at low pH and mercury at high pH. The reaction to form the surface complex >(w)FeOHSO$_4^{--}$

$$>(w)FeOHSO_4^{--} \rightleftharpoons >(w)FeOH + SO_4^{--} \qquad (8.25)$$

has no dependence on pH. Electrostatic forces and the variation in the amount of SO_4^{--} in solution [depending on the amount sorbed as >(w)FeSO$_4^-$] explain the variation in abundance of this species.

The sorption of lead

$$>(s)FeOPb^+ + H^+ \rightleftharpoons >(s)FeOH + Pb^{++} \qquad (8.26)$$

differs from that of sulfate and mercury in that it occurs only at strongly binding sites, which are far less abundant than weak sites. At pH 8, about 40% of the lead remains in solution even though lead complexes occupy nearly all of the strong sites. In this case, the sorption of lead is limited by the availability of the sorbing surface.

9

Automatic Reaction Balancing

Conveniently, perhaps even miraculously, the equations developed in Chapter 4 to accomplish basis swaps can be used to balance chemical reactions automatically. Once the equations have been coded into a computer program, there is no need to balance reactions, compute equilibrium constants, or even determine equilibrium equations by hand. Instead, these procedures can be performed quickly and reliably on a small computer.

To balance a reaction, we first choose a species to appear on the reaction's left side, and express that species' composition in terms of a basis **B**. The basis might be a list of the elements in the species' stoichiometry, or an arbitrary list of species that combine to form the left-side species. Then we form a second basis **B′** composed of species that we want to appear on the reaction's right side. To balance the reaction, we calculate the transformation matrix relating basis **B′** to **B**, following the procedures in Chapter 4. The transformation matrix, in turn, gives the balanced reaction and its equilibrium constant.

9.1 Calculation Procedure

Two methods of balancing reactions are of interest. We can balance reactions in terms of the stoichiometries of the species considered. In this case, the existing basis **B** is a list of elements and, if charged species are involved, the electron e^-. Alternatively, we may use a dataset of balanced reactions, such as the LLNL database. Basis **B**, in this case, is the one used in the database to write reactions. We will consider each possibility in turn.

Using Species' Stoichiometries

A straightforward way to balance reactions is to use as the initial basis the stoichiometries of the species involved. If the species' free energies of formation are known, the reaction's equilibrium constant can be determined as well. In the stoichiometric approach, basis $\mathbf{B}$ is the list of elements that will appear in the reaction, plus the electron if needed. We write swap reactions and calculate a transformation matrix as described in Section 3.1. The equations in Sections 3.2 and 3.3 give the balanced reaction and associated equilibrium constant.

The process is best shown by example. Suppose that we wish to balance the reaction by which calcium clinoptilolite $(CaAl_2Si_{10}O_{24}\cdot8H_2O)$, a zeolite mineral, reacts to form muscovite $[KAl_3Si_3O_{10}(OH)_2]$ and quartz (SiO_2). We choose to write the reaction in terms of the aqueous species Ca^{++}, K^+, and OH^-.

Reserving clinoptilolite for the reaction's left side, we write the stoichiometry of each remaining species in matrix form

$$
\begin{bmatrix} H_2O \\ Ca^{++} \\ K^+ \\ Muscovite \\ Quartz \\ OH^- \\ e^- \end{bmatrix}
=
\begin{bmatrix}
2 & 1 & 0 & 0 & 0 & 0 & 0 \\
0 & 0 & 1 & 0 & 0 & 0 & -2 \\
0 & 0 & 0 & 1 & 0 & 0 & -1 \\
2 & 12 & 0 & 1 & 3 & 3 & 0 \\
0 & 2 & 0 & 0 & 0 & 1 & 0 \\
1 & 1 & 0 & 0 & 0 & 0 & 1 \\
0 & 0 & 0 & 0 & 0 & 0 & 1
\end{bmatrix}
\begin{bmatrix} H \\ O \\ Ca \\ K \\ Al \\ Si \\ e^- \end{bmatrix}
\tag{9.1}
$$

Notice that we have added the electron to $\mathbf{B}$ and $\mathbf{B}'$ in order to account for the electrical charge on the aqueous species. This incorporation provides a convenient check: the electron's reaction coefficient must work out to zero in order for the reaction to be charge balanced.

We reverse the equation by computing the inverse to the coefficient matrix, giving

$$
\begin{bmatrix} H \\ O \\ Ca \\ K \\ Al \\ Si \\ e^- \end{bmatrix}
=
\begin{bmatrix}
1 & 0 & 0 & 0 & 0 & -1 & 1 \\
-1 & 0 & 0 & 0 & 0 & 2 & -2 \\
0 & 1 & 0 & 0 & 0 & 0 & 2 \\
0 & 0 & 1 & 0 & 0 & 0 & 1 \\
\frac{4}{3} & 0 & -\frac{1}{3} & \frac{1}{3} & -1 & -\frac{10}{3} & 3 \\
2 & 0 & 0 & 0 & 1 & -4 & 4 \\
0 & 0 & 0 & 0 & 0 & 0 & 1
\end{bmatrix}
\begin{bmatrix} H_2O \\ Ca^{++} \\ K^+ \\ Muscovite \\ Quartz \\ OH^- \\ e^- \end{bmatrix}
\tag{9.2}
$$

The inverted matrix is the transformation matrix $(\beta)^{-1}$.

Now, we write the stoichiometry of the clinoptilolite and substitute the result above, giving the balanced reaction:

$$\text{Ca--clinoptilolite} = \begin{bmatrix} 16 & 32 & 1 & 0 & 2 & 10 & 0 \end{bmatrix} \begin{bmatrix} \text{H} \\ \text{O} \\ \text{Ca} \\ \text{K} \\ \text{Al} \\ \text{Si} \\ \text{e}^- \end{bmatrix}$$

$$= \begin{bmatrix} 16 & 32 & 1 & 0 & 2 & 10 & 0 \end{bmatrix} \begin{bmatrix} 1 & 0 & 0 & 0 & 0 & -1 & 1 \\ -1 & 0 & 0 & 0 & 0 & 2 & -2 \\ 0 & 1 & 0 & 0 & 0 & 0 & 2 \\ 0 & 0 & 1 & 0 & 0 & 0 & 1 \\ \frac{4}{3} & 0 & -\frac{1}{3} & \frac{1}{3} & -1 & -\frac{10}{3} & 3 \\ 2 & 0 & 0 & 0 & 1 & -4 & 4 \\ 0 & 0 & 0 & 0 & 0 & 0 & 1 \end{bmatrix} \begin{bmatrix} \text{H}_2\text{O} \\ \text{Ca}^{++} \\ \text{K}^+ \\ \text{Musc.} \\ \text{Quartz} \\ \text{OH}^- \\ \text{e}^- \end{bmatrix}$$

$$= \begin{bmatrix} \frac{20}{3} & 1 & -\frac{2}{3} & \frac{2}{3} & 8 & \frac{4}{3} & 0 \end{bmatrix} \begin{bmatrix} \text{H}_2\text{O} \\ \text{Ca}^{++} \\ \text{K}^+ \\ \text{Muscovite} \\ \text{Quartz} \\ \text{OH}^- \\ \text{e}^- \end{bmatrix} \qquad (9.3)$$

More simply

$$\text{CaAl}_2\text{Si}_{10}\text{O}_{24}\cdot 8\,\text{H}_2\text{O} + \tfrac{2}{3}\,\text{K}^+ \rightleftarrows \tfrac{20}{3}\,\text{H}_2\text{O} + \text{Ca}^{++}$$
$$\text{Ca--clinoptilolite}$$

$$+ \tfrac{2}{3}\,\text{KAl}_3\text{Si}_3\text{O}_{10}(\text{OH})_2 + 8\,\text{SiO}_2 + \tfrac{4}{3}\,\text{OH}^- \quad (9.4)$$
$$\textit{muscovite} \qquad\qquad \textit{quartz}$$

which is the result we seek.

To calculate the reaction's equilibrium constant, we note that the free energy change ΔG_{sw} of each the swap reactions in Eqn. 9.1 is the negative free energy of formation from the elements of the corresponding species

$$\Delta G^o_{sw} = -\Delta G^o_f \qquad (9.5)$$

The sign on the reaction free energies is reversed because the species appear on the left side of Eqn. 9.1. In other words, we are decomposing the species rather than forming them from the elements. We determine $\Delta G°$ for the reaction by adding the values for $\Delta G°_{sw}$, just as we added log K's in Section 2.1:

$$\Delta G° = -\Delta G°_f \text{(Ca-clinopt)} + \begin{bmatrix} \dfrac{20}{3} & 1 & -\dfrac{2}{3} & \dfrac{2}{3} & 8 & \dfrac{4}{3} & 0 \end{bmatrix} \begin{bmatrix} \Delta G°_f \text{(H}_2\text{O)} \\ \Delta G°_f \text{(Ca}^{++}) \\ \Delta G°_f \text{(K}^{+}) \\ \Delta G°_f \text{(Musc)} \\ \Delta G°_f \text{(Qtz)} \\ \Delta G°_f \text{(OH}^{-}) \\ \Delta G°_f \text{(e}^{-}) \end{bmatrix} \quad (9.6)$$

Substituting values of $\Delta G°_f$ taken (in kJ/mol) from Robie et al. (1979) and the LLNL database, the equation

$$\Delta G° = 12\,764.67 + \begin{bmatrix} \dfrac{20}{3} & 1 & -\dfrac{2}{3} & \dfrac{2}{3} & 8 & \dfrac{4}{3} & 0 \end{bmatrix} \begin{bmatrix} -237.14 \\ -553.54 \\ -282.49 \\ -5\,590.76 \\ -856.29 \\ -157.29 \\ 0 \end{bmatrix} \quad (9.7)$$

predicts a free energy change of 31.32 kJ/mol. The value can be expressed as an equilibrium constant using the relation

$$\log K = -\frac{\Delta G°}{2.303\,RT_K} \quad (9.8)$$

which gives a value for log K of −5.48.

Using a Reaction Database

A reaction dataset, such as the LLNL database, provides an alternative method for balancing reactions. Such a database contains reactions to form a number of aqueous species, minerals, and gases, together with the corresponding equilibrium constants. Reactions are written in terms of a generic basis set **B**, which probably does not correspond to set **B′**, our choice of species to appear in the reaction.

To balance a reaction, we write swap reactions relating **B′** to **B**. Returning to the previous example, we wish to compute the reaction by which Ca-clinoptilolite transforms to muscovite and quartz. Reserving the clinoptilolite for the reaction's left side, we write the swap reactions for the basis transformation in matrix form. The reactions and associated equilibrium constants at 25°C are:

$$
\begin{bmatrix} H_2O \\ Ca^{++} \\ K^+ \\ Muscovite \\ Quartz \\ OH^- \end{bmatrix}
=
\begin{bmatrix} 1 & 0 & 0 & 0 & 0 & 0 \\ 0 & 1 & 0 & 0 & 0 & 0 \\ 0 & 0 & 1 & 0 & 0 & 0 \\ 6 & 0 & 1 & 3 & 3 & -10 \\ 0 & 0 & 0 & 0 & 1 & 0 \\ 1 & 0 & 0 & 0 & 0 & -1 \end{bmatrix}
\begin{bmatrix} H_2O \\ Ca^{++} \\ K^+ \\ Al^{+++} \\ SiO_2(aq) \\ H^+ \end{bmatrix}
\qquad \log K^{sw} = \begin{bmatrix} 0 \\ 0 \\ 0 \\ 14.56 \\ -4.00 \\ 13.99 \end{bmatrix} \qquad (9.9)
$$

We reverse the equation by inverting the coefficient matrix to give

$$
\begin{bmatrix} H_2O \\ Ca^{++} \\ K^+ \\ Al^{+++} \\ SiO_2(aq) \\ H^+ \end{bmatrix}
=
\begin{bmatrix} 1 & 0 & 0 & 0 & 0 & 0 \\ 0 & 1 & 0 & 0 & 0 & 0 \\ 0 & 0 & 1 & 0 & 0 & 0 \\ \frac{4}{3} & 0 & -\frac{1}{3} & \frac{1}{3} & -1 & -\frac{10}{3} \\ 0 & 0 & 0 & 0 & 1 & 0 \\ 1 & 0 & 0 & 0 & 0 & -1 \end{bmatrix}
\begin{bmatrix} H_2O \\ Ca^{++} \\ K^+ \\ Muscovite \\ Quartz \\ OH^- \end{bmatrix}
\qquad (9.10)
$$

The inverted matrix is the transformation matrix for the basis we have chosen.
  The reaction in the LLNL database for Ca-clinoptilolite

$$
Ca\text{--clinoptilolite} = \begin{bmatrix} 12 & 1 & 0 & 2 & 10 & -8 \end{bmatrix}
\begin{bmatrix} H_2O \\ Ca^{++} \\ K^+ \\ Al^{+++} \\ SiO_2(aq) \\ H^+ \end{bmatrix}
\qquad (9.11)
$$

has a $\log K$ at 25°C of -9.12. Substituting the transformation matrix and multiplying through gives

$$
Ca\text{--clinoptilolite} = \begin{bmatrix} 12 & 1 & 0 & 2 & 10 & -8 \end{bmatrix}
\begin{bmatrix} 1 & 0 & 0 & 0 & 0 & 0 \\ 0 & 1 & 0 & 0 & 0 & 0 \\ 0 & 0 & 1 & 0 & 0 & 0 \\ \frac{4}{3} & 0 & -\frac{1}{3} & \frac{1}{3} & -1 & -\frac{10}{3} \\ 0 & 0 & 0 & 0 & 1 & 0 \\ 1 & 0 & 0 & 0 & 0 & -1 \end{bmatrix}
\begin{bmatrix} H_2O \\ Ca^{++} \\ K^+ \\ Muscovite \\ Quartz \\ OH^- \end{bmatrix}
$$

$$= \begin{bmatrix} \frac{20}{3} & 1 & -\frac{2}{3} & \frac{2}{3} & 8 & \frac{4}{3} \end{bmatrix} \begin{bmatrix} H_2O \\ Ca^{++} \\ K^+ \\ Muscovite \\ Quartz \\ OH^- \end{bmatrix} \qquad (9.12)$$

or, as before,

$$CaAl_2Si_{10}O_{24} \cdot 8\,H_2O + \frac{2}{3}\,K^+ \rightleftarrows \frac{20}{3}\,H_2O + Ca^{++}$$
Ca−clinoptilolite

$$+ \frac{2}{3}\,KAl_3Si_3O_{10}(OH)_2 + 8\;SiO_2 + \frac{4}{3}\,OH^- \quad (9.13)$$
$$\qquad\qquad\quad muscovite \qquad\quad quartz$$

The formula in Section 3.3 gives the reaction's equilibrium constant:

$$\log K = -9.12 - \begin{bmatrix} \frac{20}{3} & 1 & -\frac{2}{3} & \frac{2}{3} & 8 & \frac{4}{3} \end{bmatrix} \begin{bmatrix} 0 \\ 0 \\ 0 \\ 14.56 \\ -4.00 \\ 13.99 \end{bmatrix} \qquad (9.14)$$

or $\log K = -5.48$, as determined in the previous section.

The program RXN performs such calculations automatically. To follow the procedure above, we enter the commands

```
react Clinoptil Ca
swap Muscovite for Al+++
swap Quartz for SiO2(aq)
swap OH- for H+
T = 25 C
go
```

giving the same result without hand calculation.

9.2 Dissolution of Pyrite

To further illustrate how the basis-swapping algorithm can be used to balance reactions, we consider several ways to represent the dissolution reaction of pyrite, FeS_2. Using the program RXN, we retrieve the reaction for pyrite as written in the LLNL database

```
react Pyrite
go
```

producing the result

$$\text{FeS}_2 + \text{H}_2\text{O} + \frac{7}{2}\,\text{O}_2(\text{aq}) \rightleftarrows \text{Fe}^{++} + 2\,\text{SO}_4^{--} + 2\,\text{H}^+ \qquad (9.15)$$
pyrite

By this reaction, sulfur from the pyrite oxidizes to form sulfate ions, liberating protons that acidify the solution.

The above reaction represents, in a simplified way, the origin of acid mine drainage. Streambeds in areas of acid drainage characteristically become coated with a orange layer of ferric precipitate. We can write a reaction representing the overall process by swapping ferric hydroxide in place of the ferrous ion:

```
(cont'd)
swap Fe(OH)3(ppd) for Fe++
go
```

The resulting reaction,

$$\text{FeS}_2 + \frac{7}{2}\,\text{H}_2\text{O} + \frac{15}{4}\,\text{O}_2(\text{aq}) \rightleftarrows \text{Fe(OH)}_3(\text{ppd}) + 2\,\text{SO}_4^{-} + 4\,\text{H}^+ \quad (9.16)$$
pyrite $\qquad\qquad\qquad\qquad$ *ferric hydroxide*

consumes more oxygen than the previous result, because not only sulfur but iron oxidizes.

Pyrite can dissolve into reducing as well as oxidizing solutions. To find the reaction by which the mineral dissolves to form H_2S, we swap this species into the basis in place of the sulfate ion

```
(cont'd)
unswap Fe++
swap H2S(aq) for SO4--
go
```

The result

$$\text{FeS}_2 + \text{H}_2\text{O} + 2\,\text{H}^+ \rightleftarrows \text{Fe}^{++} + 2\,\text{H}_2\text{S}(\text{aq}) + \frac{1}{2}\,\text{O}_2(\text{aq}) \qquad (9.17)$$
pyrite

differs from the previous reactions in that it consumes protons and produces oxygen.

Is there a reaction by which pyrite can dissolve without changing the overall oxidation state of its sulfur? To see, we return the sulfate ion to the basis and swap H_2S for dissolved oxygen:

```
(cont'd)
unswap SO4--
swap H2S(aq) for O2(aq)
go
```

The reaction written in terms of the new basis is

$$FeS_2 + H_2O + \frac{3}{2} H^+ \rightleftarrows Fe^{++} + \frac{1}{4} SO_4^{--} + \frac{7}{4} H_2S(aq) \qquad (9.18)$$
pyrite

which neither consumes nor produces oxygen. Calculating the transformation matrices in these examples provides an interesting exercise, which is left to the reader.

9.3 Equilibrium Equations

A reaction's equilibrium equation is given directly from the form of the reaction and the value of the equilibrium constant. Hence, its an easy matter to extend a reaction balancing program to report equilibrium lines. For example, the reaction

$$Fe_2O_3 + 4 H^+ \rightleftarrows 2 Fe^{++} + 2 H_2O + \frac{1}{2} O_2(aq) \qquad (9.19)$$
hematite

has an equilibrium constant at 25°C of $10^{-16.9}$. From the mass action equation and definition of pH, the general equilibrium line is

$$\log K = 4\,pH + 2 \log a_{Fe^{++}} + 2 \log a_{H_2O} + \frac{1}{2} \log a_{O_2(aq)} \qquad (9.20)$$

since the activity of hematite is one. The specific equilibrium line at 25°C for $a_{Fe^{++}} = 10^{-10}$ and $a_{H_2O} = 1$ is

$$\log a_{O_2(aq)} = 6.17 - 8\,pH \qquad (9.21)$$

To use RXN to calculate the equilibrium lines above in general and specific forms, we type

```
react Hematite
pH = ?
long
go


T = 25
log a Fe++ = -10
a H2O = 1
go
```

The `long` command tells the program to show the reaction's equilibrium constant versus temperature and calculate its equilibrium equation; `pH = ?` causes the program to render the equation in terms of pH instead of $\log a_{H^+}$. To find the equilibrium lines written in terms of pe and Eh, we type

(cont'd)
```
swap e- for O2(aq)
```

```
pe = ?
go

Eh = ?
go
```

giving the results

$$pe = 23.03 - 3\,pH \qquad (9.22)$$

and

$$Eh = 1.363 - .178\,pH \qquad (9.23)$$

Equilibrium Activity Ratio

Many mineralogic reactions involve exchange of cations or anions. Hence, geochemists commonly need to determine equilibrium lines in terms of activity ratios. Consider, for example, the reaction at 25°C between the clay kaolinite $[Al_2Si_2O_5(OH)_4]$ and the mica muscovite. The RXN commands

```
react Muscovite
swap Kaolinite for Al+++
T = 25
go
```

give the reaction

$$\underset{muscovite}{KAl_3Si_3O_{10}(OH)_2} + 1.5\,H_2O + H^+ \rightleftarrows K^+ + 1.5\,\underset{kaolinite}{Al_2Si_2O_5(OH)_4} \qquad (9.24)$$

whose $\log K$ is 3.42. To calculate the equilibrium ratio a_{K^+}/a_{H^+}, assuming unit water activity, we type

```
(cont'd)
swap K+/H+ for H+
a H2O = 1
long
go
```

giving the result $a_{K^+}/a_{H^+} = 10^{3.42}$.

In our second example, we calculate the same ratio for the reaction between muscovite and potassium feldspar ($KAlSi_3O_8$; "maximum microcline" in the database) in the presence of quartz:

$$K^+ + \frac{1}{2}\underset{muscovite}{KAl_3Si_3O_{10}(OH)_2} + 3\,\underset{quartz}{SiO_2} \rightleftarrows \frac{3}{2}\underset{microcline}{KAlSi_3O_8} + H^+ \qquad (9.25)$$

The commands

```
T = 25
react "Maximum Microcline"
swap Muscovite for Al+++
swap Quartz for SiO2(aq)
swap K+/H+ for H+
long
factor 3/2
go
```

give the result $a_{K^+}/a_{H^+} = 10^{4.84}$. For convenience, the `factor` command above applies a multiplier of 1.5 to the reaction coefficients. We can quickly recalculate the equilibrium activity ratio for reaction in the presence of the silica polymorphs tridymite and amorphous silica:

```
(cont'd)
swap Tridymite for SiO2(aq)
go

swap Amrph^silica for SiO2(aq)
go
```

The resulting a_{K^+}/a_{H^+} values, respectively, are $10^{4.34}$ and $10^{0.98}$.

As a final example, we balance the reaction between the zeolite calcium clinoptilolite and the mica prehnite [$Ca_2Al_2Si_3O_{10}(OH)_2$] in the presence of quartz, and calculate at 200°C the equilibrium activity ratio $a_{Ca^{++}}/a_{H^+}^2$. The commands

```
react Clinoptil-Ca
swap Prehnite for Al+++
swap Quartz for SiO2(aq)
T = 200
go
```

give the reaction

$$CaAl_2Si_{10}O_{24} \cdot 8H_2O + Ca^{++} \rightleftarrows 6\,H_2O + Ca_2Al_2Si_3O_{10}(OH)_2 + 7\,SiO_2 + 2\,H^+$$

$$\underset{\textit{Ca–clinoptilolite}}{\qquad} \qquad \underset{\textit{prehnite}}{\qquad} \quad \underset{\textit{quartz}}{\qquad} \tag{9.26}$$

which has a log K of -10.23. To calculate the activity ratio, we type

```
swap Ca++/H+^2 for Ca++
a H2O = 1
long
go
```

which gives, as expected, the value calculated for log K.

Equilibrium Temperature

When the activity of each species in a reaction is known, we can determine the temperature (or temperatures) at which the reaction is in equilibrium. As an example, we calculate the temperature at which gypsum ($CaSO_4 \cdot 2\,H_2O$) dehydrates to form anhydrite ($CaSO_4$). The RXN commands

```
react Gypsum
swap Anhydrite for Ca++
long
go
```

give the reaction

$$CaSO_4 \cdot 2\,H_2O \rightleftarrows CaSO_4 + 2\,H_2O \qquad (9.27)$$
$$\textit{gypsum} \qquad\quad \textit{anhydrite}$$

and equilibrium equation

$$\log K = 2 \log a_{H_2O} \qquad (9.28)$$

The equilibrium temperature for any water activity is the temperature at which $\log K$ satisfies this equality. To find this value when the activity of water is one, we type

```
(cont'd)
a H2O = 1
go
```

The resulting equilibrium temperature, 43.7°C, is the temperature at which the $\log K$ is zero. For a water activity of 0.7, the equilibrium temperature drops to 11.8°C. Typing

```
(cont'd)
a H2O = ?
T = 25 C
```

we find that the two minerals are in equilibrium at 25°C when the water activity is 0.815.

10

Uniqueness

A practical question that arises in quantitative modeling is whether the results of a modeling study are unique. In other words, is it possible to arrive at results that differ, at least slightly, from the original ones but nonetheless satisfy the governing equations and honor the input constraints?

In the broadest sense, of course, no model is unique (see, for example, Oreskes et al., 1994). A geochemical modeler could conceptualize the problem differently, choose a different compilation of thermodynamic data, include more or fewer species and minerals in the calculation, or employ a different method of estimating activity coefficients. The modeler might allow a mineral to form at equilibrium with the fluid or require it to precipitate according to any of a number of published kinetic rate laws and rate constants, and so on. Since a model is a simplified version of reality that is useful as a tool (Chapter 2), it follows that there is no "correct" model, only a model that is most useful for a given purpose.

A more precise question (Bethke, 1992) is the subject of this chapter: in geochemical modeling is there but a single root to the set of governing equations that honors a given set of input constraints? We might call such a property mathematical uniqueness, to differentiate it from the broader aspects of uniqueness. The property of mathematical uniqueness is important because once the software has discovered a root to a problem, the modeler may abandon any search for further solutions. There is no concern that the choice of a starting point for iteration has affected the answer. In the absence of a demonstration of uniqueness, on the other hand, the modeler cannot be completely certain that another solution, perhaps a more realistic or useful one, remains undiscovered.

Geochemists, following early theoretical work in other fields, have long considered the multicomponent equilibrium problem (as defined in Chapter 3) to be mathematically unique. In fact, however, this assumption is not correct. Although relatively uncommon, there are examples of geochemical models in which more than one root of the governing equations satisfy the modeling constraints equally well. In this chapter, we consider the question of uniqueness and pose three simple problems in geochemical modeling that have nonunique solutions.

10.1 The Question of Uniqueness

As noted in Chapter 1, chemical modeling developed from efforts to calculate the thrust of a rocket fuel from its bulk composition. Mathematicians of the era (for example, Brinkley, 1947; White et al., 1958; Boynton, 1960) analyzed the multicomponent equilibrium problem, including the uniqueness of its roots, in detail. Warga (1963) considered the problem of a thermodynamically ideal solution of known bulk composition. He showed that the free energy surface representing the sum of the free energies of individual species, when traced along trajectories satisfying mass balance, is concave upward. The surface, hence, has a single minimum point and therefore a unique equilibrium state. Van Zeggeren and Storey (1970, pp. 31–32) cite several other studies that offer mathematical proofs of uniqueness.

Geochemists have generally believed that the uniqueness proofs hold in their field. Their proofs, however, are limited in two regards: they consider thermodynamically ideal solutions and assume that, as in the rocket fuel problem, the calculation is posed in terms of mass balance constraints. The first limitation is not known to be a serious problem in modeling speciation in aqueous fluids. The second limitation is important, however, because in geochemical modeling the bulk composition of the modeled system is seldom known completely. For this reason, geochemists generally construct models using a combination of mass balance and mass action constraints (see Chapter 5).

Mass balance constraints are specifications, when solving the governing equations (Eqns. 3.29a–d), of the mole numbers M_w, M_i, and M_k of the components in the basis. Setting the bulk sodium content of a fluid, for example, represents a mass balance constraint. Mass action constraints include setting a species' molality m_i or activity a_i, the free mass n_k of a mineral, or the fugacity f_m of a gas. In setting pH, the volume of quartz in a system, or the CO_2 fugacity, the modeler poses a mass action constraint. By doing so, he violates an underlying assumption of the uniqueness proofs and opens the possibility of mathematical nonuniqueness.

10.2 Examples of Nonunique Solutions

To demonstrate nonuniqueness, we pose here three problems in geochemical modeling that each have two physically realistic solutions. In the first example, based on data from an aluminum solubility experiment, we assume equilibrium with an alumina mineral to fix the *p*H of a fluid of otherwise known composition. Setting *p*H by mineral equilibrium is a widespread practice in modeling the chemistry of deep groundwaters, and of fluids sampled from hydrothermal experiments, because it is difficult to directly measure the *in situ p*H of hot fluids. In this case, however, there are two possible solutions because many aluminous minerals, including hydroxides, clays, and micas, are amphoteric and hence equally soluble at low and high *p*H.

In REACT, we prepare our calculation by swapping boehmite (AlOOH) for H^+,

```
swap Boehmite for H+
1 free cm3 Boehmite
```

so the equilibrium with this mineral controls *p*H according to reactions such as

$$\underset{boehmite}{AlOOH} + 2\,H^+ \rightleftarrows H_2O + AlOH^{++} \qquad (10.1)$$

$$\underset{boehmite}{AlOOH} + H^+ \rightleftarrows Al(OH)_2^+ \qquad (10.2)$$

or

$$\underset{boehmite}{AlOOH} + 2\,H_2O \rightleftarrows Al(OH)_4^- + H^+ \qquad (10.3)$$

depending on the predominant aluminum species in solution. In general, species $AlOH^{++}$ and $Al(OH)_2^+$ predominate under acidic conditions, whereas the hydroxy species $Al(OH)_4^-$ dominates under alkaline conditions (Fig. 10.1).

We assume a 0.1 molal KCl solution containing hypothetical amounts of silica, aluminum, and carbonate. We set temperature to 200°C and run the calculation

```
(cont'd)
K+        = 100 mmolal
Cl-       = 100 mmolal
SiO2(aq) =   3 mmolal
Al+++     =  10 umolal
HCO3-     =  60 umolal
```

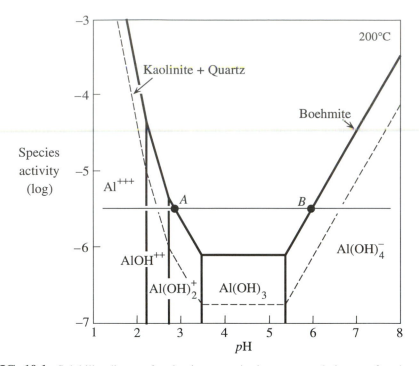

FIG. 10.1 Solubility diagram for aluminum species in aqueous solution as a function of pH at 200°C in the presence of boehmite (solid lines) and kaolinite plus quartz (dashed lines). Aluminum is soluble at a specific activity (horizontal line) either under acidic conditions as species $Al(OH)_2^+$, $AlOH^{++}$, or Al^{+++} (e.g., point A), or under alkaline conditions as $Al(OH)_4^-$ (point B).

```
T = 200
precip = off
print species = long
go
```

For simplicity, we do not allow supersaturated minerals to precipitate. The result is an acidic fluid in which $AlOH^{++}$ and $Al(OH)_2^+$ predominate among aluminum species.

We repeat the calculation, this time swapping $Al(OH)_4^-$ into the basis to represent dissolved aluminum.

```
(cont'd)
swap Al(OH)4- for Al+++
go
```

This swap favors iteration toward a second root at alkaline pH where $Al(OH)_4^-$ is the predominant aluminum species in solution.

The chemistries corresponding to the two roots can be summarized as

	Root A	Root B
pH	3.1	6.3
total Al^{+++} (molal)	$10.\times10^{-6}$	$10.\times10^{-6}$
free $AlOH^{++}$	4.3×10^{-6}	1.5×10^{-12}
free $Al(OH)_2^+$	3.0×10^{-6}	1.8×10^{-9}
free $Al(OH)_4^-$	5.5×10^{-9}	9.2×10^{-6}

Repeating the calculations using minerals such as kaolinite $[Al_2Si_2O_5(OH)_4]$ or muscovite $[KAl_3Si_3O_{10}(OH)_2]$ to fix pH also produces nonunique results.

As a second example, we constrain a fluid's oxidation state by assuming equilibrium with pyrite (FeS_2). As before, direct information on this variable can be difficult to obtain, so it is not uncommon for modelers to use mineral equilibrium to fix a fluid's redox state. The choice of pyrite to buffer oxidation state, however, is perilous because pyrite sulfur, which is in the S^{1-} oxidation state, may dissolve by oxidation to sulfate (S^{6+})

$$FeS_2 + H_2O + \tfrac{7}{2}O_2(g) \rightleftarrows Fe^{++} + 2\,SO_4^- + 2\,H^+ \qquad (10.4)$$
pyrite

or by reduction to H_2S (S^{2-})

$$FeS_2 + H_2O + 2\,H^+ \rightleftarrows Fe^{++} + 2\,H_2S + \tfrac{1}{2}O_2(g) \qquad (10.5)$$
pyrite

As such, there are two redox states that satisfy the assumption of equilibrium with pyrite (Fig. 10.2).

In our example, we know the pH and iron and sulfur contents of a 1 molal NaCl solution at 100°C. We swap pyrite into the basis in place of $O_2(aq)$ and run the model

```
swap Pyrite for O2(aq)
1 free cm3 Pyrite

Na+   =  1 molal
Cl-   =  1 molal
Fe++  = 10 mmolal
SO4-- = 10 mmolal
```

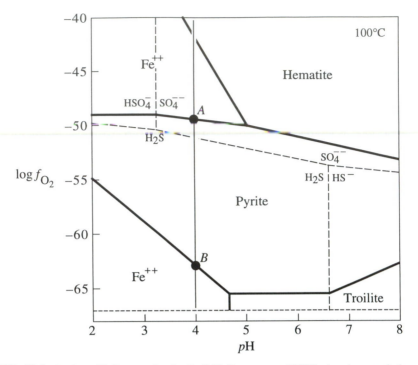

FIG. 10.2 Redox-pH diagram for the Fe-S-H_2O system at 100°C, showing speciation of sulfur (dashed line) and the stability fields of iron minerals (solid lines). Diagram is drawn assuming sulfur and iron species activities, respectively, of 10^{-3} and 10^{-4}. Broken line at bottom of diagram is the water stability limit. At pH 4, there are two oxidation states (points A and B) in equilibrium with pyrite under these conditions

```
pH = 4
T = 100
print species = long
go
```

In this case, the program converges to a relatively oxidized solution in which sulfur speciation is dominated by sulfate species.

Swapping $H_2S(aq)$ into the basis in the place of SO_4^{--}, on the other hand

```
(cont'd)
swap H2S(aq) for SO4--
go
```

changes the starting point of the iteration and causes the program to converge to a more reduced solution nearly devoid of sulfate species. The two roots are

	Root *A*	Root *B*
pH	4.0	4.0
$\log f_{O_2}$	−50	−67
total Fe^{++} (molal)	.010	.010
total S	.010	.010
$\Sigma\ SO_4^{--}$	.010	3.3×10^{-7}
$\Sigma\ H_2S(aq)$	5.0×10^{-34}	.010

Here, the notations $\Sigma\ SO_4^{--}$ and $\Sigma\ H_2S(aq)$ refer, respectively, to the sum of the molalities of the sulfate and sulfide sulfur species in solution.

Analogous examples of nonuniqueness can be constructed using any mineral or gas of intermediate oxidation state. Buffering the fugacity of $N_2(g)$ or $SO_2(g)$, for example, would be a poor choice for constraining oxidation state, since the gases can either oxidize to NO_3^- and SO_4^{--}, respectively, or reduce to NH_4 and $H_2S(aq)$ species.

As a final example, we consider a fluid of known fluoride concentration whose calcium content is set by equilibrium with fluorite (CaF_2). The speciation of fluorine provides for two solutions to this problem. In dilute solutions, in which the free ion F^- dominates, the reaction

$$CaF_2 \rightleftarrows Ca^{++} + 2\ F^- \qquad (10.6)$$
$$\textit{fluorite}$$

requires that calcium content vary inversely with fluorine concentration. By this reaction, increasing the fluorine concentration leads to solutions that are less calcic, and vice versa.

As calcium content increases, especially at elevated temperature, the CaF^+ ion pair becomes predominant. The CaF^+ activity exceeds that of F^- at 200°C whenever the activity of Ca^{++} is greater than about 10^{-3}; at 300°C, the ion pair is favored at Ca^{++} activities as small as $10^{-4.8}$. Where CaF^+ dominates, the reaction

$$CaF_2 + Ca^{++} \rightleftarrows 2\ CaF^+ \qquad (10.7)$$
$$\textit{fluorite}$$

controls fluorite solubility. This reaction, in contrast to this previous one, requires that fluids become proportionally richer in calcium as their fluorine contents increase.

As shown in Fig. 10.3, fluids of identical fluorine content but two distinct Ca^{++} activities can exist in equilibrium with fluorite. At 200°C, setting the activity of dissolved fluorine to $10^{-3.5}$ allows two equilibrium activities of Ca^{++}: $10^{-4.3}$ and $10^{-1.6}$ (points *A* and *B*); the corresponding activities at 300°C are $10^{-6.5}$ and $10^{-3.1}$.

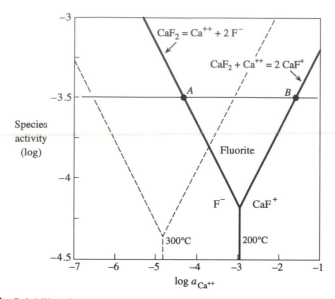

FIG. 10.3 Solubility diagram for fluorite as a function of calcium ion activity at 200°C (solid) and 300°C (dashed lines). Fluorite is soluble at a specific activity (horizontal line) either as F^- at small Ca^{++} activity (point A) or as CaF^+ at high Ca^{++} activity (point B).

To discover the root at low Ca^{++} concentration, we swap fluorite for Ca^{++} and arbitrarily set a 1 molal NaCl solution of pH 5 that contains slightly less than 1 *mmolal* fluorine. We set temperature to 200°C

```
swap Fluorite for Ca++
1 free cm3 Fluorite

F-  = 0.8 mmolal
Na+ = 1 molal
Cl- = 1 molal
pH = 5

T = 200
print species = long
go
```

and run the model.

It is difficult to persuade REACT to locate the high Ca^{++} root directly. By experimentation we learn that if we gradually add 1.0165 moles of $CaCl_2$ to the fluid, the fluid initially becomes supersaturated with respect to fluorite but then returns to equilibrium.

```
(cont'd)
react 1 mol of Ca++
react 2 mol of Cl-
```

```
reactants times 1.0165

dump
fix pH
precip = off
go
```

Here, we set the `dump` option to remove excess fluorite from the system before beginning the path and fix pH to a constant value to prevent it from wandering as activity coefficients change with ionic strength.

The end point of the reaction path is the second root to the problem. At 300°C, we find by trial and error that we need to add .08992 moles of $CaCl_2$ to reach the second root. The differences among the roots are summarized

	200°C		300°C	
	Root *A*	Root *B*	Root *A*	Root *B*
*p*H	5.0	5.0	5.0	5.0
Ca^{++} (molal)	.0026	1.0	.00013	.090
F^-	$.80\times10^{-3}$	$.80\times10^{-3}$	$.80\times10^{-3}$	$.80\times10^{-3}$
free F^-	$.34\times10^{-3}$	$.021\times10^{-3}$	$.73\times10^{-3}$	$.027\times10^{-3}$
free CaF^+	$.050\times10^{-3}$	$.75\times10^{-3}$	$.027\times10^{-3}$	$.77\times10^{-3}$
I (molal)	.91	2.7	1.0	1.1

where *I* is ionic strength.

10.3 Coping with Nonuniqueness

The examples in the previous section demonstrate that nonunique solutions to the equilibrium problem can occur when the modeler constrains the calculation by assuming equilibrium between the fluid and a mineral or gas phase. In each example, the nonuniqueness arises from the nature of the multicomponent equilibrium problem and the variety of species distributions that can exist in an aqueous fluid. When more than one root exists, the iteration method and its starting point control which root the software locates.

In each of the cases, the dual roots differ from each other in terms of *p*H, sulfide content, or ionic strength, so that in a modeling study the ''correct'' root could readily be selected. The danger of nonuniqueness is that a modeler, having reached an inappropriate root, might not realize that a separate, more meaningful root to the problem exists.

Unfortunately, no software techniques exist currently to automatically search for additional roots. Instead, modelers must rely on their understanding of geochemistry to demonstrate uniqueness to their satisfaction. Activity-activity diagrams such as those presented in Figs. 10.1–10.3 are the most useful tools for identifying additional roots.

11

Mass Transfer

In previous chapters we have discussed the nature of the equilibrium state in geochemical systems: how we can define it mathematically, what numerical methods we can use to solve for it, and what it means conceptually. With this chapter we begin to consider questions of process rather than state. How does a fluid respond to changes in composition as minerals dissolve into it, or as it mixes with other fluids? How does a fluid evolve in response to changing temperature or variations in the fugacity of a coexisting gas? In short, we begin to consider reaction modeling.

In this chapter we consider how to construct reactions paths that account for the effects of simple reactants, a name given to reactants that are added to or removed from a system at constant rates. We take on other types of mass transfer in later chapters. Chapter 12 treats the mass transfer implicit in setting a species' activity or gas' fugacity over a reaction path. In Chapter 14 we develop reaction models in which the rates of mineral precipitation and dissolution are governed by kinetic rate laws.

11.1 Simple Reactants

Simple reactants are those added to (or removed from) the system at constant rates over the reaction path. As noted in Chapter 2, we commonly refer to such a path as a titration model, because at each step in the process, much like in a laboratory titration, the model adds an aliquot of reactant mass to the system. Each reactant A_r is added at a rate n_r, expressed in moles per unit reaction progress, ξ. Negative values of n_r, of course, describe the removal rather than the addition of the reactant. Since ξ is unitless and varies from zero at the start of the path to one at the end, we can just as well think of n_r as the number of moles of the reactant to be added over the reaction path.

A simple reactant may be an aqueous species (including water), a mineral, a gas, or any entity of known composition. The only requirement is that we be able to form it by a reaction

$$A_r \rightleftharpoons v_{wr}A_w + \sum_i v_{ir}A_i + \sum_k v_{kr}A_k + \sum_m v_{mr}A_m \tag{11.1}$$

among the basis entries. Significantly, we need not know the $\log K$ for this reaction; only the reaction coefficients (v_{wr}, v_{ir}, and so on) come into play in the calculation. We can therefore employ a substance as a reactant, even if we do not know its stability.

The calculation procedure for tracing a titration path is straightforward. We begin by calculating the equilibrium state of the initial system, as described in Chapter 5. Once we know the initial equilibrium state, we substitute the resulting bulk compositions (M_w, M_i, M_k, and M_m) for any free constraints. If we specified pH to set up the calculation, for example, we now take the value of M_{H^+} to constrain the governing equation for the H^+ component, leaving m_{H^+} as an unknown variable. Similarly, we take as a constraint the mole number M_k for each mineral component, replacing the mineral's free mass n_k. The exceptions to this process are species set at fixed or sliding activity and gases of fixed and sliding fugacity, as described in the next chapter; for these exceptions, we retain the free constraints.

Once a substitution of constraints is accomplished, the calculation consists of incrementally changing the system's bulk composition as a function of reaction progress and, after each increment, recalculating the equilibrium state. The bulk composition is given from the component masses (M_w^0, M_i^0, and M_k^0) present at $\xi = 0$, the reaction rates n_r, and the reactants' stoichiometric coefficients (v_{wr}, and so on, from Reaction 11.1), according to

$$M_w(\xi) = M_w^0 + \xi \sum_r v_{wr} n_r \tag{11.2a}$$

$$M_i(\xi) = M_i^0 + \xi \sum_r v_{ir} n_r \tag{11.2b}$$

$$M_k(\xi) = M_k^0 + \xi \sum_r v_{kr} n_r \tag{11.2c}$$

Alternatively, we can compute the new composition from the composition at the end ξ' of the previous step as

$$M_w(\xi) = M_w(\xi') + (\xi - \xi') \sum_r v_{wr} n_r \tag{11.3a}$$

$$M_i(\xi) = M_i(\xi') + (\xi - \xi') \sum_r v_{ir} n_r \tag{11.3b}$$

$$M_k(\xi) = M_k(\xi') + (\xi - \xi') \sum_r \nu_{kr} \, n_r \qquad (11.3c)$$

The equations expressed in this stepwise manner are somewhat easier to integrate into certain reaction configurations, such as the flush or flow-through model described later in this chapter. We could also update M_m in this manner, but there is no need to do so. A gas species A_m appears in the basis only when its fugacity f_m is known, so the value of each M_m results from solving the governing equations, as described in Chapter 5.

To trace the reaction path, the model begins with the system at $\xi = 0$ and steps forward in reaction progress by setting ξ to $\Delta\xi$, where $\Delta\xi$ is the size of the reaction step. It then recomputes the bulk composition using Eqns. 11.3a–c and, honoring these values, iterates to the new equilibrium state. To start the iteration, the model takes the values of the unknown variables at the beginning of the step. The model then takes a further step forward in reaction progress, incrementing ξ to a value of $2 \cdot \Delta\xi$, and recomputes the equilibrium state. The calculation continues in this fashion until it reaches $\xi = 1$.

We consider as an example the hydrolysis of potassium feldspar ($KAlSi_3O_8$), the first reaction path traced using a computer (Helgeson et al., 1969). We specify the composition of a hypothetical water

```
Na+    =       5 mg/kg
K+    =        1 mg/kg
Ca++   =      15 mg/kg
Mg++   =       3 mg/kg
Al+++  =       1 ug/kg
SiO2(aq)  =    3 mg/kg
Cl-   =       30 mg/kg
SO4-- =        8 mg/kg
HCO3- =       50 mg/kg
pH  =  5
```

and then define a reaction path involving the addition of a small amount of feldspar

```
(cont'd)
react .15 cm3 of K-feldspar
```

Typing go triggers the calculation.

Figure 11.1 shows the mineralogical results of the calculation, plotted in log-log, semilog, and linear coordinates. Note that we have plotted each diagram in terms of the mass of feldspar reacted into the water. Common practice in reaction modeling is to present results plotted in terms of ξ, but this is an unfortunate convention. The reaction progress variable ξ has mathematical meaning within the modeling program, but its physical meaning vests only in terms of how the modeler sets the reaction rates n_r. By choosing a variable with

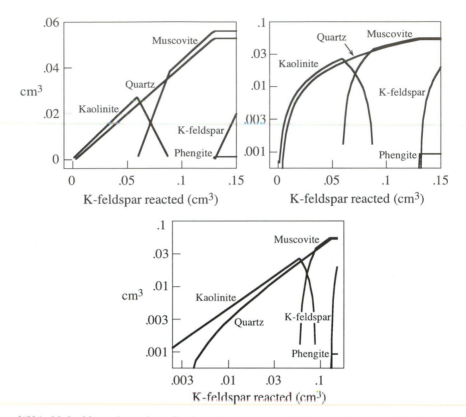

FIG. 11.1 Mineralogical results of reacting potassium feldspar into a hypothetical water at 25°C, plotted in linear, semilog (a "spaghetti diagram"), and log-log coordinates.

physical meaning (such as reacted mass) when plotting results, we can present our calculation results in a more direct manner.

In the first segment of the reaction path (Fig. 11.1), the feldspar dissolves into solution, producing kaolinite [$Al_2Si_2O_5(OH)_4$] and quartz (SiO_2). The reaction gradually increases the activity ratio a_{K^+}/a_{H^+} until the solution reaches equilibrium with muscovite [$KAl_3Si_3O_{10}(OH)_2$]. At this point, the kaolinite begins to dissolve, producing muscovite and quartz. The reaction continues after the kaolinite is consumed until the fluid reaches equilibrium with the feldspar [a small amount of phengite, $KAlMgSi_4O_{10}(OH)_2$, forms just before this point]. Once the fluid is in equilibrium with it, the feldspar simply accumulates as it is added to the system, and reaction with the fluid ceases. In the next section of this chapter, we consider how we can extract the chemical reactions occurring over each segment of the reaction path.

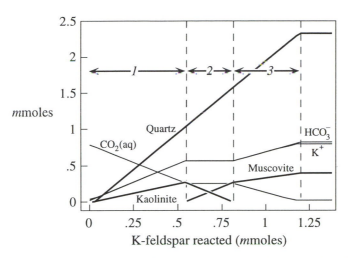

FIG. 11.2 Results of the reaction path shown in Fig. 11.1, plotted to allow the overall reaction to be extracted using the slopes-of-the-lines method.

11.2 Extracting the Overall Reaction

The ultimate goal in reaction modeling is to discover the overall reaction that occurs within a system. Strangely, whereas the results of nearly every published study involving reaction modeling are presented in a "spaghetti diagram" (see, e.g., Fig. 11.1), few papers report the overall reaction. For this reason, some of the most important information in the results is obscured. Who could blame the reader for thinking, "Enough pasta, let's get to the meatballs!"?

The procedure for determining the overall reaction, fortunately, is straightforward. The modeler plots the mole numbers of the species and minerals in the system against the mass of a reactant added to the system over the reaction path. The plot must be in linear coordinates with both axes in consistent units, such as moles or *m*moles.

The slopes of the lines in the plot give the reaction coefficients for each species and mineral in the overall reaction. Species with negative slopes appear to the left of the reaction (with their coefficients set positive), and those with positive slopes are placed to the right. The reactant plotted on the horizontal axis appears to the left of the reaction with a coefficient of one. If there are additional reactants, these also appear on the reaction's left with coefficients equal to the ratios of their reaction rates n_r to that of the first reactant.

As an example, we consider the reaction path traced in the previous section (Fig. 11.1). To extract the overall reaction for each segment of the path, we construct a plot as described above. The result is shown in Fig. 11.2. There are three segments in the reaction path: the precipitation of kaolinite, the transformation of kaolinite to muscovite, and the continued formation of muscovite once the kaolinite is exhausted. There is a distinct overall reaction for each segment.

From this plot (and one showing the mass of species H_2O, which does not fit on these axes), we can write down the slope of the line for each species and mineral. The results, compiled for each segment in the reaction, are

	Segment 1	Segment 2	Segment 3
$CO_2(aq)$	-1	0	$-.67$
HCO_3^-	$+1$	0	$+.67$
K^+	$+1$	0	$+.67$
H_2O	-1.5	$+1$	$-.67$
quartz	$+2$	$+2$	$+2$
kaolinite	$+.5$	-1	—
muscovite	—	$+1$	.33

The values in the first column give the overall reaction for the first segment of the reaction path

$$KAlSi_3O_8 + \frac{3}{2} H_2O + CO_2(aq) \rightleftarrows$$
K–feldspar

$$2 \ SiO_2 + \frac{1}{2} Al_2Si_2O_5(OH)_4 + HCO_3^- + K^+ \quad (11.4)$$
$$quartz \qquad\qquad kaolinite$$

Similarly, the overall reactions for the second segment

$$KAlSi_3O_8 + Al_2Si_2O_5(OH)_4 \rightleftarrows 2 \ SiO_2 + KAl_3Si_3O_{10}(OH)_2 + H_2O \quad (11.5)$$
$$K–feldspar \qquad kaolinite \qquad\qquad quartz \qquad muscovite$$

and third segment

$$KAlSi_3O_8 + \frac{2}{3} CO_2(aq) + \frac{2}{3} H_2O \rightleftarrows$$
K–feldspar

$$2 \ SiO_2 + \frac{1}{3} KAl_3Si_3O_{10}(OH)_2 + \frac{2}{3} HCO_3^- + \frac{2}{3} K^+ \quad (11.6)$$
$$quartz \qquad\qquad muscovite$$

are given by the coefficients in the second and third columns.

11.3 Special Configurations

In Chapter 2 we discussed three special configurations for tracing reaction paths: the dump, flow-through, and flush models. These models are special cases of mass transfer that can be implemented within the mathematical framework developed in this chapter.

In the dump configuration, the model discards the masses of any minerals present in the initial equilibrium system before beginning to trace the reaction path. To do so, the model updates the total composition to reflect the absence of minerals

$$M_k = M_k - n_k \tag{11.7}$$

and sets the mineral masses n_k to zero. Here, of course, we use the "=" to represent assignment of value, rather than algebraic equivalence. If the model considers surface complexation (Chapter 8), the mole numbers M_p of the surface components as well as the molalities m_p and m_q of the surface species must also be set to zero to reflect the disappearance of sorbing mineral surfaces.

The model then swaps aqueous species A_j into the basis locations A_k held by the mineral components. The technique for swapping the basis is explained in Chapter 4, and a method for selecting an appropriate species A_j to include in the basis is described in Chapter 5. When the procedure is complete, the equilibrium system contains only the original fluid.

In the flow-through model, any mineral mass present at the end of a reaction step is sequestered from the equilibrium system to avoid back-reaction. At the end of each step, the model eliminates the mineral mass (including any sorbed species) from the equilibrium system, keeping track of the total amount removed. To do so, it applies Eqn. 11.7 for each mineral component and sets each n_k to a vanishingly small number. It is best to avoid setting n_k to exactly zero in order to maintain the mineral entries A_k in the basis. The model then updates the system composition according to Eqn. 11.3a–c and takes another reaction step.

In a flush model, reactant fluid displaces existing fluid from the equilibrium system. It is simplest to implement this model by determining the mass of water entering the system over a step and eliminating an equal mass of water component and the solutes it contains from the system. In this case, we ignore any density differences between the fluids.

The model first determines (from the reaction rate n_r for water and the mass M_w of the water component) the fraction X_{disp} of fluid to be displaced from the system over a step. Typically, the model will limit the size $\Delta \xi$ of the reaction step to a value that will cause only a fraction (perhaps a tenth or a quarter) of the fluid present at the start of the step to be displaced, in the event that the modeler accidentally sets too large a step size. The formulae for determining the updated composition become

$$M_w(\xi) = (1 - X_{disp})\, M_w(\xi') + (\xi - \xi') \sum_r \nu_{wr}\, n_r \qquad (11.8a)$$

$$M_i(\xi) = (1 - X_{disp})\, M_i(\xi') + (\xi - \xi') \sum_r \nu_{ir}\, n_r \qquad (11.8b)$$

$$M_k(\xi) = (1 - X_{disp})\, (M_k(\xi') - n_k) + n_k + (\xi - \xi') \sum_r \nu_{kr}\, n_r \qquad (11.8c)$$

These are Eqns. 11.3a–c, modified to account for the loss of the displaced fluid.

12

Polythermal, Fixed, and Sliding Paths

In this chapter we consider how to construct reaction models that are somewhat more sophisticated than those discussed in the previous chapter, including reaction paths over which temperature varies and those in which species activities and gas fugacities are buffered. The latter cases involve the transfer of mass between the equilibrium system and an external buffer. Mass transfer in these cases occurs at rates implicit in solving the governing equations, rather than at rates set explicitly by the modeler. In Chapter 14 we consider the use of kinetic rate laws, a final method for defining mass transfer in reaction models.

12.1 Polythermal Reaction Paths

Polythermal reactions paths are those in which temperature varies as a function of reaction progress, ξ. In the simplest case, the modeler prescribes the temperatures T_o and T_f at the beginning and end of the reaction path. The model then varies temperature linearly with reaction progress. This type of model is sometimes called a "sliding temperature" path.

The calculation procedure for a sliding temperature path is straightforward. In taking a reaction step, the model evaluates the temperature to be attained at the step's end. Since ξ varies from zero to one, temperature at any point ξ in reaction progress is given

$$T(\xi) = T_o + \xi (T_f - T_o) \tag{12.1}$$

as a function of the initial and final temperatures, as set by the modeler. The model then reevaluates values for the reaction log Ks and the constants used to calculate activity coefficients (see Chapter 7). If reaction kinetics or isotopic

171

fractionation is considered (Chapters 14 and 15), the model recalculates the reaction rate constants and isotopic fractionation factors.

A second type of polythermal path traces temperature as reactants mix into the equilibrium system. This case differs from a sliding temperature path only in the manner in which temperature is determined. The modeler assigns a temperature T_o to the initial system, as before, and a distinct temperature T_r to the reactants. By assuming that the heat capacities C_{P_f}, C_{P_k}, and C_{P_r} of the fluid, minerals, and reactants are constant over the temperature range of interest, we can calculate temperature $T(\xi)$ from energy balance and the temperature $T(\xi')$ at the onset of the step according to

$$T(\xi) = \frac{\left[1000\,(1 - X_{disp})\,C_{P_f}n_f + \sum_k C_{P_k}M_{W_k}n_k\right] T(\xi') + (\xi - \xi')\,T_r\,\sum_r C_{P_r}M_{W_r}n_r}{1000\,(1 - X_{disp})\,C_{P_f}n_f + \sum_k C_{P_k}M_{W_k}n_k + (\xi - \xi')\sum_r C_{P_r}M_{W_r}n_r}$$

(12.2)

In this equation, ξ' and ξ are the values of reaction progress at the beginning and end of the step; n_f is the mass in kg of the fluid (equal to n_w, the water mass, plus the mass of the solutes); n_k is the mole number of each mineral; n_r is the reaction rate (moles) for each reactant; M_{W_k} is the mole weight (g/mol) of each mineral, and M_{W_r} is the mole weight for each reactant; and X_{disp} is the fraction of the fluid displaced over the reaction step in a flush model (X_{disp} is zero if a flush model is not invoked).

In an example of a sliding temperature path, we consider the effects of cooling from 300°C to 25°C a system in which a 1 molal NaCl solution is in equilibrium with the feldspars albite ($NaAlSi_3O_8$) and microcline ($KAlSi_3O_8$), quartz (SiO_2), and muscovite [$KAl_3Si_3O_{10}(OH)_2$]. To set up the calculation, we enter the commands

```
swap Albite for Al+++
swap "Maximum Microcline" for K+
swap Muscovite for H+
swap Quartz for SiO2(aq)

1 molal Na+
1 molal Cl-
20 free cm3 Albite
10 free cm3 "Maximum Microcline"
5 free cm3 Muscovite
2 free cm3 Quartz

suppress "Albite low"
```

to define the fluid in equilibrium with the minerals, which are present in excess amounts. For simplicity, we suppress the entry "albite low," which is almost identical in stability to the entry "albite" in the thermodynamic database. We

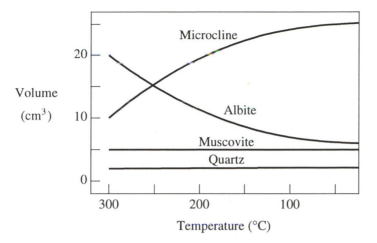

FIG. 12.1 Mineralogical results of tracing a polythermal reaction path. In the calculation, a 1 molal NaCl solution in equilibrium with albite, microcline, muscovite, and quartz cools from 300°C to 25°C.

then set a polythermal path with the command

> *(cont'd)*
> T initial = 300, final = 25

and type go to start the calculation.

In the calculation results (Fig. 12.1), albite reacts to form microcline

$$NaAlSi_3O_8 + K^+ \rightarrow KAlSi_3O_8 + Na^+ \qquad (12.3)$$
$$albite \qquad\qquad microcline$$

as the temperature decreases. The reaction is driven not by mass transfer but by variation in the stabilities of the minerals and the species in solution. The composition of the system (fluid plus minerals), in fact, remains constant over the reaction path. Chapters 16 and 17 give a number of further examples of the application of polythermal reaction paths.

12.2 Fixed Activity and Fugacity Paths

In a fixed activity path, the activity of an aqueous species (or those of several species) maintains a constant value over the course of the reaction path. A fixed fugacity path is similar, except that the model holds constant a gas fugacity instead of a species activity. Fixed activity paths are useful in modeling laboratory experiments in which an aspect of a fluid's chemistry is maintained mechanically. In studying reaction kinetics, for example, it is common practice to hold constant the pH of a solution with a pH-stat. Fixed activity paths are also convenient for calculating speciation diagrams, which by convention may be plotted at constant pH and oxygen activity. Fixed fugacity paths are useful for tracing reaction paths in which a fluid remains in contact with a gas phase, such as the atmosphere.

To calculate a fixed activity path, the model maintains within the basis each species A_i whose activity a_i is to be held constant. For each such species, the corresponding mass balance equation (Eqn. 5.3b) is reserved from the reduced basis, as described in Chapter 5, and the known value of a_i is used in evaluating the mass action equation (Eqn. 5.4). Similarly, the model retains within the basis each gas A_m whose fugacity is to be fixed. We reserve the corresponding mass balance equation (Eqn. 5.3d) from the reduced basis and use the corresponding fugacity f_m in evaluating the mass action equation.

A complication to the calculation procedure for holding an aqueous species at fixed activity is the necessity of maintaining ionic charge balance over the reaction path. If the species is charged, the model must enforce charge balance at each step in the calculation by adjusting the concentration of a specified component, as discussed in Section 5.3. For example, if the pH is fixed over a path and the charge balance component is Cl^-, then the model will behave as if HCl were added to or removed from the system in the quantities needed to maintain a constant H^+ activity.

In an example of a fixed fugacity path we model the dissolution of pyrite (FeS_2) at 25°C. We start in REACT with a hypothetical water in equilibrium with hematite (Fe_2O_3) and oxygen in the atmosphere

```
swap O2(g) for O2(aq)
swap Hematite for Fe++

f O2(g) = .2
1 free mg Hematite
pH = 6.5

 4 mg/kg Ca++
 1 mg/kg Mg++
 2 mg/kg Na+
18 mg/kg HCO3-
 3 mg/kg SO4--
 5 mg/kg Cl-
```

into which we react a small amount of pyrite

```
(cont'd)
react 10 mg Pyrite
go
```

In this initial calculation, we do not fix the oxygen fugacity.

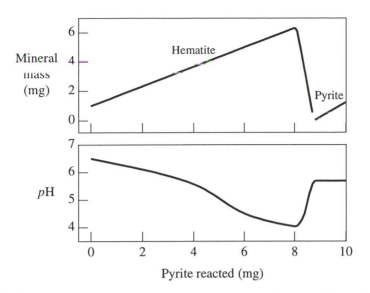

FIG. 12.2 Mineralogical results (top) of reacting pyrite at 25°C into a dilute water held closed to O_2, and variation in pH (bottom) over the reaction.

As shown in Fig. 12.2, about 8 mg of pyrite dissolves into the water, producing hematite. The reaction drives pH from the initial value of 6.5 to about 4 before the water becomes reducing. At this point, the hematite redissolves and the fluid reaches equilibrium with pyrite, bringing the reaction to an end.

We can write the overall reaction by which hematite forms, using the slopes-of-the-lines method discussed in Chapter 11. Initially, the reaction proceeds as

$$\underset{pyrite}{FeS_2} + \frac{15}{4} O_2(aq) + 4\,HCO_3^- \rightarrow \underset{hematite}{\frac{1}{2}\,Fe_2O_3} + 2\,SO_4^{--} + 4\,CO_2(aq) + 2\,H_2O \qquad (12.4)$$

as can be seen from Fig. 12.3. As the water becomes more acidic and the supply of HCO_3^- is depleted, a second reaction

$$\underset{pyrite}{FeS_2} + \frac{15}{4} O_2(aq) + 2\,H_2O \rightarrow \underset{hematite}{\frac{1}{2}\,Fe_2O_3} + 2\,SO_4^{--} + 4\,H^+ \qquad (12.5)$$

becomes dominant. Pyrite continues to dissolve until the available $O_2(aq)$ has been consumed.

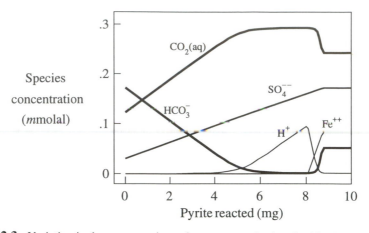

FIG. 12.3 Variation in the concentrations of aqueous species involved in the dissolution reaction of pyrite, for the reaction path shown in Fig. 12.2.

How would the reaction have proceeded if the oxygen fugacity had been fixed by equilibrium with the atmosphere? To find out, we repeat the calculation, this time holding the oxygen fugacity constant

```
(cont'd)
fix f O2(g)
react 1000 mg Pyrite
go
```

and specifying a hundred-fold increase in the supply of pyrite.

The fixed fugacity path (Fig. 12.4) differs from the previous calculation (in which the fluid was closed to the addition of oxygen) in that pyrite dissolution continues indefinitely, since there is an unlimited supply of oxygen gas. Initially, the reaction proceeds as

$$\text{FeS}_2 + \frac{15}{4} \text{O}_2(\text{g}) + 2\,\text{H}_2\text{O} \rightarrow \frac{1}{2}\,\text{Fe}_2\text{O}_3 + 2\,\text{SO}_4^{--} + 4\,\text{H}^+ \qquad (12.6)$$
$$\textit{pyrite} \qquad\qquad\qquad\qquad \textit{hematite}$$

as can be seen from Fig. 12.5. Later, a second reaction that produces HSO_4^- instead of SO_4^{--},

$$\text{FeS}_2 + \frac{15}{4} \text{O}_2(\text{aq}) + 2\,\text{H}_2\text{O} \rightarrow \frac{1}{2}\,\text{Fe}_2\text{O}_3 + 2\,\text{HSO}_4^- + 2\,\text{H}^+ \qquad (12.7)$$
$$\textit{pyrite} \qquad\qquad\qquad\qquad \textit{hematite}$$

becomes dominant. The H^+ produced by these reactions drives pH to values far more acidic than those in the closed-system case.

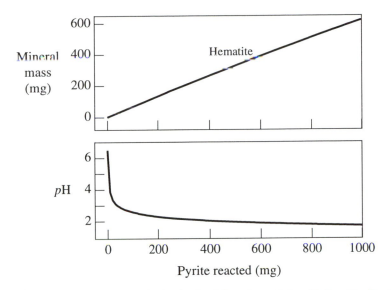

FIG. 12.4 Mineralogical results (top) of a fixed fugacity path in which pyrite dissolves at 25°C into water held in equilibrium with O_2 in the atmosphere, and the variation in pH (bottom) over the path.

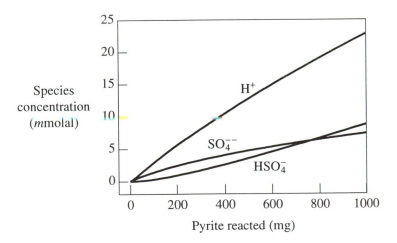

FIG. 12.5 Concentrations of species involved in the dissolution of pyrite, for the fixed fugacity path shown in Fig. 12.4.

12.3 Sliding Activity and Fugacity Paths

Sliding activity and sliding fugacity paths are similar to fixed activity and fixed fugacity paths, except that the model varies the buffered activity or fugacity over the reaction path rather than holding it constant. Once the equilibrium state of the initial system is known, the model stores the initial activity a_i^o or initial fugacity f_m^o of the buffered species or gas. (The modeler could set this value as a constraint on the initial system, but this is not necessary.)

The modeler supplies the final (or target) activity a_i^f or fugacity f_m^f, which will be achieved at the end of the reaction path, when $\xi = 1$. The modeler also specifies whether the activity or fugacity itself or the logarithm of activity or fugacity is to be varied. If the value is to be varied, the model determines it as a function of reaction progress according to

$$a_i(\xi) = a_i^o + \xi\,(a_i^f - a_i^o) \tag{12.8a}$$

or

$$f_m(\xi) = f_m^o + \xi\,(f_m^f - f_m^o) \tag{12.8b}$$

If, on the other hand, the value's logarithm is to be varied, the model calculates the value according to

$$\log a_i(\xi) = \log a_i^o + \xi\,(\log a_i^f - \log a_i^o) \tag{12.9a}$$

or

$$\log f_m(\xi) = \log f_m^o + \xi\,(\log f_m^f - \log f_m^o) \tag{12.9b}$$

To see the difference between sliding a value and sliding its logarithm, consider a path in which f_{O_2} varies from one (10^0) to 10^{-30}. If $\Delta\xi$ in this simple example is 0.2, the model will step along the reaction path in even steps of either f_{O_2} or $\log f_{O_2}$:

ξ	f_{O_2}	$\log f_{O_2}$
0	1	1
.2	.8	10^{-6}
.4	.6	10^{-12}
.6	.4	10^{-18}
.8	.2	10^{-24}
1	$10^{-30}\ (\approx 0)$	10^{-30}

If we select the linear option (the center column), the path will stop at a series of oxidizing points followed by a single reducing point. If we choose the logarithmic option (right column), however, the path will visit a range of oxidation states.

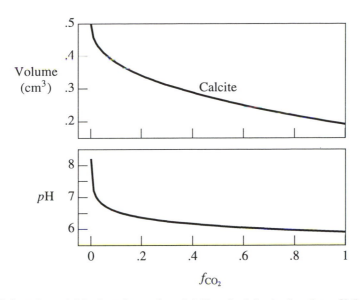

FIG. 12.6 Effect of CO_2 fugacity on the solubility of calcite (top) and on *pH* (bottom), calculated at 25°C using a sliding fugacity path.

In an example of a sliding fugacity path, we calculate how CO_2 fugacity affects the solubility of calcite ($CaCO_3$). We begin by defining a dilute solution in equilibrium with calcite and the CO_2 fugacity of the atmosphere

```
swap Calcite for Ca++
swap CO2(g) for H+

10 mmolal Na+
10 mmolal Cl-
1/2 free cm3 Calcite
log f CO2(g) = -3.5
balance on HCO3-
go
```

The resulting fluid has a *pH* of about 8.3. To vary the CO_2 fugacity, we set a path in which f_{CO_2} slides from the initial atmospheric value

```
(cont'd)
slide f CO2(g) to 1
go
```

to one.

In the calculation results (Fig. 12.6), increasing the CO_2 fugacity decreases the *pH* to about 6, causing calcite to dissolve into the fluid. The fugacity increase drives CO_2 from the buffer into the fluid, and most of the CO_2 (Fig. 12.7)

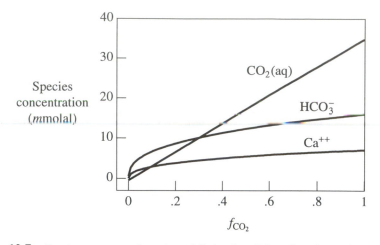

FIG. 12.7   Species concentrations (*m*molal) in the sliding fugacity path shown in Fig. 12.6.

becomes $CO_2(aq)$. The nearly linear relationship between the concentration of $CO_2(aq)$ and the fugacity of $CO_2(g)$ results from the reaction

$$CO_2(g) \rightleftarrows CO_2(aq) \qquad (12.10)$$

which holds a_{CO_2} proportional to f_{CO_2}. Some of the gas, however, dissociates to produce HCO_3^- and H^+, and the resulting acid is largely consumed by dissolving calcite. The overall reaction (derived using the slopes-of-the-lines method) is approximately

$$5\,CO_2(g) + \underset{calcite}{CaCO_3} + H_2O \rightarrow 4\,CO_2(aq) + Ca^{++} + 2\,HCO_3^- \qquad (12.11)$$

In a second example, we calculate how *p*H affects sorption onto hydrous ferric oxide, expanding on our discussion (Section 8.4) of Dzombak and Morel's (1990) surface complexation model. We start as before, setting the dataset of surface reactions, suppressing the ferric minerals hematite (Fe_2O_3) and goethite (FeOOH), and specifying the amount of ferric oxide [represented in the calculation by $Fe(OH)_3$ precipitate] in the system

```
surface_data = FeOH.data
sorbate include
decouple Fe+++
suppress Hematite, Goethite
swap Fe(OH)3(ppd) for Fe+++
1 free gram Fe(OH)3(ppd)
```

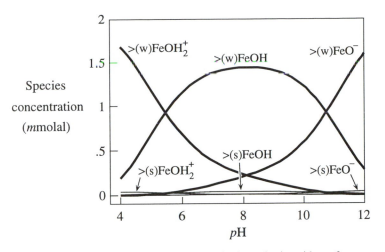

FIG. 12.8 Concentrations (*m*molal) of sites on a hydrous ferric oxide surface exposed at 25°C to a 0.1 molal NaCl solution, calculated using a sliding *p*H path.

We set a 0.1 molal NaCl solution and define a sliding activity path in which *p*H varies from 4 to 12

```
(cont'd)
0.1 molal Na+
0.1 molal Cl-
pH = 4

slide pH to 12
precip = off
go
```

Figure 12.8 shows how the concentrations of the surface species vary with *p*H in the calculation results.

According to the complexation model, neither Na^+ nor Cl^- reacts with the surface, so the species consist entirely of surface sites (>(w)FeOH and >(s)FeOH) in their uncomplexed, protonated

$$>(w)FeOH + H^+ \rightleftarrows >(w)FeOH_2^+ \tag{12.12}$$

and deprotonated forms

$$>(w)FeOH \rightleftarrows >(w)FeO^- + H^+ \tag{12.13}$$

depending on the H^+ activity. Protonated sites dominate at low *p*H, resulting in a positive surface potential (Fig. 12.9), whereas the predominance of deprotonated sites at high *p*H yields a negative potential.

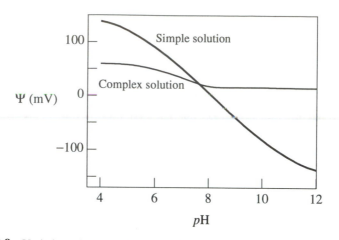

FIG. 12.9 Variation of surface potential Ψ (mV) with pH for a hydrous ferric oxide surface in contact at 25°C with a 0.1 molal NaCl solution (bold line) and a more complex solution (fine line) that also contains Ca, SO_4, Hg, Cr, As, and Zn.

To see how contact with a more complex solution affects the surface, we introduce to the fluid 10 $mmoles$ of $CaSO_4$ and 100 $\mu moles$ each of Hg^{++}, Cr^{+++}, $As(OH)_4^-$, and Zn^{++}

```
(cont'd)
10 mmolal Ca++
10 mmolal SO4--
100 umolal Hg++
100 umolal Cr+++
100 umolal As(OH)4-
100 umolal Zn++
```

Typing go triggers the model, which again slides pH from 2 to 12.

In this case (Fig. 12.10), we observe a more complicated distribution of species. At low pH, the H^+ activity and positive surface potential drive SO_4^{--} to sorb according to the reaction

$$SO_4^{--} + >(w)FeOH + H^+ \rightleftarrows >(w)FeSO_4^- + H_2O \qquad (12.14)$$

Under alkaline conditions, on the other hand, reactions such as

$$Zn^{++} + >(w)FeOH \rightleftarrows >(w)FeOZn^+ + H^+ \qquad (12.15)$$

$$Hg^{++} + >(w)FeOH \rightleftarrows >(w)FeOHg^+ + H^+ \qquad (12.16)$$

and

$$Ca^{++} + >(w)FeOH \rightleftarrows >(w)FeOCa^+ + H^+ \qquad (12.17)$$

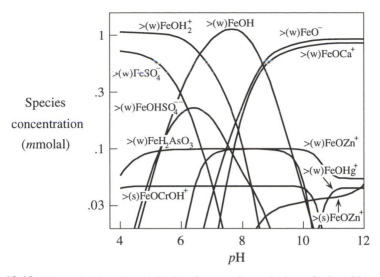

FIG. 12.10 Concentrations (*m*molal) of surface species on hydrous ferric oxide exposed at 25°C to a solution containing Ca, SO₄, Hg, Cr, As, and Zn, calculated using a sliding *p*H path.

promote the sorption of bivalent cations. These reactions produce negatively charged surface species at low *p*H and positively surface charged species at high *p*H, thereby reducing the magnitude of the surface charge under acidic as well as alkaline conditions. Hence, the surface is considerably less charged than it was in contact with the NaCl solution, as shown in Fig. 12.9.

The $As(OH)_4^-$ and Cr^{+++} components follow a pattern distinct from the other metals (Fig. 12.11), sorbing at only near-neutral *p*H. This pattern results from the manner in which the metals speciate in solution. As(III) appears as $As(OH)_3$ when *p*H is less than 9, and as $As(OH)_4^-$, or AsO_2OH^{--} at higher *p*H. The sorption reactions for these species are

$$As(OH)_3 + >(w)FeOH \rightleftarrows >(w)FeH_2AsO_3 + H_2O \qquad (12.18)$$

$$As(OH)_4^- + >(w)FeOH + H^+ \rightleftarrows >(w)FeH_2AsO_3 + 2\,H_2O \qquad (12.19)$$

$$AsO_2OH^{--} + >(w)FeOH + 2\,H^+ \rightleftarrows >(w)FeH_2AsO_3 + H_2O \qquad (12.20)$$

There is no *p*H dependence to the reaction for $As(OH)_3$, which sorbs strongly when *p*H is near neutral. Under acidic conditions, however, high H^+ activities drive the reactions for protonation and the sorption of SO_4^{--}, which displace arsenic from the mineral surface. Arsenic sorbs poorly under alkaline conditions because low H^+ activities work against the complexation of $As(OH)_4^-$, or AsO_2OH^{--}, as shown in the reactions for these species.

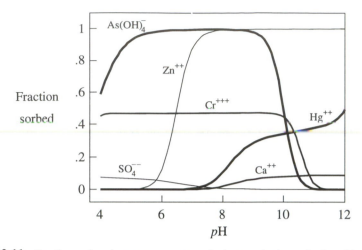

FIG. 12.11   Fractions of various components sorbed onto a hydrous ferric oxide surface at 25°C, as functions of pH, for the sliding activity path shown in Fig. 12.10.

The chromium follows a similar pattern. The component, present as Cr^{+++} at low pH, reacts successively to form $CrOH^{++}$, $Cr(OH)_2^+$, $Cr(OH)_3$, and $Cr(OH)_4^-$ as pH increases. The sorption reactions are

$$Cr^{+++} + {>}(s)FeOH + H_2O \rightleftarrows {>}(s)FeOCrOH^+ + 2\,H^+ \qquad (12.21)$$

$$CrOH^{++} + {>}(s)FeOH \rightleftarrows {>}(s)FeOCrOH^+ + H^+ \qquad (12.22)$$

$$Cr(OH)_2^+ + {>}(s)FeOH \rightleftarrows {>}(s)FeOCrOH^+ + H_2O \qquad (12.23)$$

$$Cr(OH)_3 + {>}(s)FeOH + H^+ \rightleftarrows {>}(s)FeOCrOH^+ + 2\,H_2O \qquad (12.24)$$

$$Cr(OH)_4^- + {>}(s)FeOH + 2\,H^+ \rightleftarrows {>}(s)FeOCrOH^+ + 3\,H_2O \qquad (12.25)$$

Significantly, the reactions for the species predominant at low pH favor the desorption of chromium when the H^+ activity is high, and those for the species predominant under alkaline conditions favor desorption when the H^+ activity is low. Hence Cr(III), like Ar(III), sorbs strongly only when pH is near neutral.

In a final example of the use of a sliding activity path, we calculate a speciation diagram, plotted versus pH, for hexavalent uranium in the presence of dissolved phosphate at 25°C. We take a 10 $m$molal NaCl solution containing 1 $m$molal each of UO_2^{++}, the basis species for U(VI), and HPO_4^-

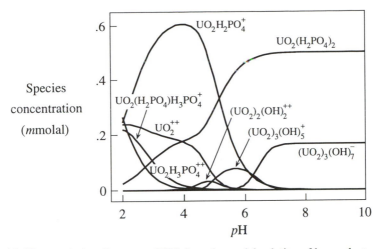

FIG. 12.12 Speciation diagram at 25°C for a 1 *m*molal solution of hexavalent uranium containing 1 *m*molal dissolved phosphate, calculated as a sliding activity path.

```
decouple UO2++

10 mmolal Na+
10 mmolal Cl-
1 mmolal UO2++
swap H3PO4 for HPO4--
1 mmolal H3PO4
```

Here we haved swapped H_3PO_4 for $HPO_4^=$ to help the program converge under acidic conditions. We specify the initial and final pH values for the calculation, set the program to avoid precipitating minerals (since we assume a fixed solution composition)

```
(cont'd)
pH = 2
slide pH to 10
precip = off
```

and type go to trigger the calculation. The resulting diagram (Fig. 12.12) shows the importance of complexing between U(VI) and phosphate. We can, of course, make many variations on the calculation, such as setting different concentrations, including other components, allowing minerals to precipitate, and so on.

13

Geochemical Buffers

Buffers are reactions that at least temporarily resist change to some aspect of a fluid's chemistry. A pH buffer, for example, holds pH to an approximately constant value, opposing processes that would otherwise drive the solution acid or alkaline. The bicarbonate-CO_2 buffer

$$HCO_3^- + H^+ \rightleftarrows CO_2(aq) + H_2O \qquad (13.1)$$

for example, consumes hydrogen ions when they are added to the system and produces them when they are consumed, thereby resisting variation in pH. The buffer operates until nearly all of the HCO_3^- is converted to $CO_2(aq)$, or vice-versa. The thirsty reader might be interested to know that the concept of buffering (as well as the notation pH) was introduced by the brewing industry in Europe (Sørensen, 1909; see Rosing 1993) as it sought to improve the flavor of beer.

Buffers such as the bicarbonate reaction are known as homogeneous buffers, because all of the constituents are found in the fluid phase. Many important buffering reactions in geochemical systems are termed heterogeneous (e.g., Rosing, 1993) because, in addition to the fluid, they involve minerals or a gas phase. Reducing minerals or oxygen in the atmosphere (examples of heterogeneous buffers) can control a fluid's oxidation state. Equilibrium with quartz fixes a fluid's silica content. Some buffers, such as those provided by assemblages of minerals, can be rather complex. Many reaction models, in fact, are designed to describe how buffers behave and how various buffering reactions interact.

In this chapter we construct models of buffering reactions, both homogeneous and heterogeneous. We concentrate on buffering reactions that are well known to geochemists, taking the opportunity to explore reaction modeling

187

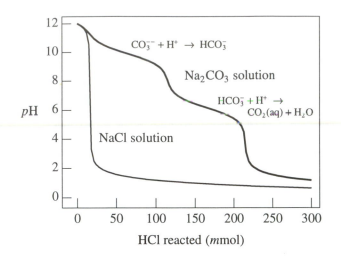

FIG. 13.1 Calculated effects on *p*H of reacting hydrochloric acid into a 0.2 molal NaCl solution and a 0.1 molal Na_2CO_3 solution, as functions of the amount of HCl added. The two plateaus on the second curve represent the buffering reactions between CO_3^{--} and HCO_3^-, and between HCO_3^- and $CO_2(aq)$.

on familiar geochemical terrain. The methods discussed here, however, can be readily applied to more complicated situations, such as those involving multiple buffers or buffers involving a larger number of phases.

13.1 Buffers in Solution

We begin by considering the well-known *p*H buffer provided by the aqueous carbonate system and its effects on the ease with which a fluid can be acidified. We start with an alkaline NaCl solution containing a small amount of carbonate, and add 300 *m*mol of hydrochloric acid to it. The procedure in REACT is

```
pH = 12
HCO3- =     1 mmolal
Cl-   = 200 mmolal
balance on Na+

react 300 mmol HCl
go
```

As can be seen in Fig. 13.1, the effect is to quickly drive the solution acidic. The only buffer is the presence of OH^- ions, which are quickly consumed by reaction with H^+ to produce water.

In a second experiment, we reverse the anion concentrations so that the fluid is dominantly a Na_2CO_3 solution:

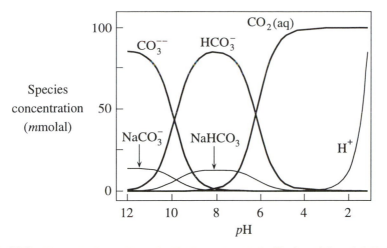

FIG. 13.2 Concentrations of species in the carbonate buffer in a 0.1 molal Na_2CO_3 solution, plotted against pH.

```
(cont'd)
HCO3- = 100 mmolal
Cl-   =   1 mmolal
go
```

Since the HCO_3^- component is present mostly as the doubly charged species CO_3^{--}, its molality is half that of the balancing cation, Na^+. In this case (Fig. 13.1), the fluid resists acidification until more than 200 *mmol* of HCl have been added.

The buffering occurs in two stages, as shown in Fig. 13.2: first the CO_3^{--} species in solution consume H^+ ions as they react to produce HCO_3^-, then the HCO_3^- reacts with H^+ to make $CO_2(aq)$. The two stages are represented in Fig. 13.1 by nearly horizontal portions of the pH curve. When all of the CO_3^{--} and HCO_3^- species have been consumed, the solution quickly becomes acidic.

We can readily derive the overall reactions in the buffer from Fig. 13.3 using the slopes-of-the-lines method, described in Chapter 11. In the first stage, the overall reaction is

$$.83 \, CO_3^{--} + .17 \, NaCO_3^- + H^+ \rightarrow .83 \, HCO_3^- + .17 \, NaHCO_3 \qquad (13.2)$$

The reaction for the second stage is

$$.83 \, HCO_3^- + .17 \, NaHCO_3 + H^+ \rightarrow CO_2(aq) + .17 \, Na^+ + H_2O \qquad (13.3)$$

It is common practice when writing overall reactions to omit mention of ion pairs whenever they are not considered important to the point being addressed. We could well write the reactions above as

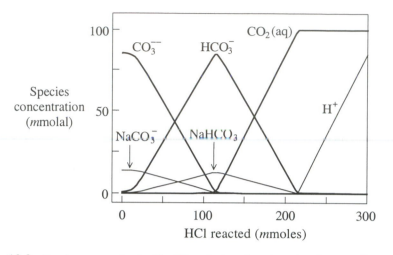

FIG. 13.3 Species masses in the Na_2CO_3 solution plotted so that the overall reaction can be determined using the slopes-of-the-lines method.

$$CO_3^{--} + H^+ \rightarrow HCO_3^- \qquad (13.4)$$

and

$$HCO_3^- + H^+ \rightarrow CO_2(aq) + H_2O \qquad (13.5)$$

The simplified form is not as exact but is less cluttered than the full form and shows more clearly the nature of the buffering reaction. In this book, we will often make simplifications of this sort.

In a practical example of the use of reaction modeling to trace buffering reactions, we consider the problem of interpreting the titration alkalinity of natural waters. Laboratories commonly report titration alkalinity rather than provide a direct analysis of a solution's carbonate content. Titration alkalinity is the solution's ability to neutralize strong acid (e.g., Snoeyink and Jenkins, 1980; Hem, 1985). The analyst titrates an acid such as H_2SO_4 into the solution until it reaches an endpoint pH, as indicated by the color change of an indicator such as methyl orange. The endpoint pH is generally in the range of 4.5 to 4.8.

The analyst reports the amount of acid required to reach the endpoint, generally expressed in terms of the number of mg of $CaCO_3$ that could be dissolved by the acid, per kg solution. Since the mole weight of $CaCO_3$ is 100.09 g and each mole of carbonate can neutralize two equivalents of acid, the conversion is

$$\frac{\text{meq acid}}{\text{kg solution}} \times \frac{50.05 \text{ mg } CaCO_3}{\text{meq acid}} = \frac{\text{mg } CaCO_3}{\text{kg solution}} \qquad (13.6)$$

Note that there are two equivalents of acid per mole of H_2SO_4.

At the titration endpoint, most carbonate in solution is present as $CO_2(aq)$. We can expect that each $m$mol of HCO_3^- originally present in solution will neutralize one meq of acid, according to the reaction

$$HCO_3^- + H^+ \ \rangle \ CO_2(aq) + H_2O \tag{13.7}$$

Similarly, each $m$mol of CO_3^{--} originally present should neutralize two meq

$$CO_3^{--} + 2\,H^+ \ \rightarrow \ CO_2(aq) + H_2O \tag{13.8}$$

of acid. Therefore, knowing from the initial pH the proportions of HCO_3^- and CO_3^{--} in the fluid, we can estimate the total carbonate content.

Unfortunately, such simple estimations can be in error. Hydroxyl, borate, silicate, ammonia, phosphate, and organic species can contribute to the solution's ability to buffer acid. For example, each of the reactions

$$HPO_4^{--} + H^+ \ \rightarrow \ H_2PO_4^- \tag{13.9}$$

$$B(OH)_4^- + H^+ \ \rightarrow \ B(OH)_3 + H_2O \tag{13.10}$$

$$NH_3 + H^+ \ \rightarrow \ NH_4^+ \tag{13.11}$$

can consume hydrogen ions during the titration. Other possible complications include the effects of activity coefficients and complex species.

A more rigorous method for interpreting an alkalinity measurement is to use a reaction model to reproduce the titration. The technique is to calculate the effects of adding acid to the original solution, assuming various carbonate contents. When we produce a model that reaches the endpoint pH after adding the acid, we have found the correct carbonate concentration.

We now consider as an example an analysis (Table 13.1) of water from Mono Lake, California. The reported alkalinity of 34,818 mg/kg as $CaCO_3$ is equivalent to 700 meq of acid or 350 $m$mol of H_2SO_4. Since at this pH carbonate and bicarbonate species are present in roughly equal concentrations, we can quickly estimate the total carbonate concentration to be about 30,000 mg/kg. We take this value as a first guess and model the titration with REACT

```
TDS = 92540
pH = 9.68
Ca++    =    4.6 mg/kg
Mg++    =     42 mg/kg
Na+     = 37200 mg/kg
K+      =  1580 mg/kg
SO4--   = 12074 mg/kg
Cl-     = 20100 mg/kg
B(OH)3  =  2760 mg/kg
F-      =    54 mg/kg
HPO4--  =   120 mg/kg
HCO3-   = 30000 mg/kg
```

TABLE 13.1 Analysis of water from Mono
Lake, California, USA (James Bischoff,
personal communication)

pH	9.68
Ca^{++} (mg/kg)	4.6
Mg^{++}	42
Na^+	37 200
K^+	1 580
SO_4^-	12 074
Cl^-	20 100
$B(OH)_3$	2 760
F^-	54
HPO_4^-	120
Alkalinity, as $CaCO_3$	34 818
Dissolved solids	92 540

```
react 350 mmol of H2SO4

precip = off
go
```

Testing varying values for the total HCO_3^- concentration,

(cont'd)
```
HCO3-   = 25000 mg/kg
go
```

we find that 25,100 mg/kg gives a titration endpoint of 4.5, as shown in Fig. 13.4. The result differs from our initial guess primarily because of the protonation of the $B(OH)_4^-$ and $NaB(OH)_4$ species to form $B(OH)_3$ plus water, as shown in Fig. 13.5.

13.2 Minerals as Buffers

In a first example of how minerals can buffer a fluid's chemistry, we consider how a hypothetical groundwater that is initially in equilibrium with calcite ($CaCO_3$) at 25°C might respond to the addition of an acid. In REACT, we enter the commands

```
swap Calcite for HCO3-
10 free cm3 Calcite

pH = 8
Na+ = 100 mmolal
```

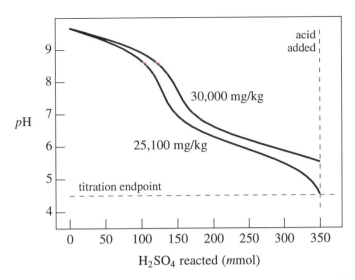

FIG. 13.4 Use of reaction modeling to derive a fluid's carbonate concentration from its titration alkalinity, as applied to an analysis of Mono Lake water. When the correct HCO_3^- total concentration (in this case, 25,100 mg/kg) is set, the final pH matches the titration endpoint.

```
Ca++ = 10 mmolal
balance on Cl-
```

to set the initial system containing a fluid of pH 8 in equilibrium with calcite. We then type

```
(cont'd)
dump
react 100 mmol of HCl
go
```

to define the reaction path and trigger the calculation. With the dump command, we specify that the calcite be separated from the fluid before the reaction path begins.

The only pH buffer in the calculation is the small concentration of carbonate species in solution. The buffer is quickly overwhelmed and the fluid shifts rapidly to acidic pH, as shown in Fig. 13.6. The dominant reaction,

$$HCl \rightarrow H^+ + Cl^- \tag{13.12}$$

is the dissociation of the HCl.

In a second calculation, we trace the same path while maintaining the fluid in equilibrium with the calcite. To do so, we enter the command

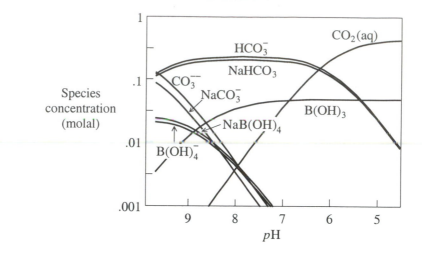

FIG. 13.5 Concentrations of species in buffer reactions that contribute to the titration alkalinity of Mono Lake water, plotted against *pH*.

(cont'd)
```
dump = off
```

to disable the dump option and type go to start the path. In the calculation results (Fig. 13.6), *pH* decreases by only a small amount and the fluid becomes just mildly acidic. The CO_2 fugacity rises steadily during the reaction, finally exceeding a value of one, at which point we would expect the gas to begin to effervesce against atmospheric pressure.

The overall reaction for the earliest portion of the reaction path, where HCO_3^- is the predominant carbonate species, is

$$2\ HCl + \underset{calcite}{CaCO_3}\ \rightarrow\ Ca^{++} + 2\ Cl^- + HCO_3^- + H^+ \qquad (13.13)$$

By this reaction, the fluid becomes more acidic with the addition of HCl. With decreasing *pH*, the CO_2(aq) species quickly comes to dominate HCO_3^-. At this point, the principal reaction becomes

$$2\ HCl + \underset{calcite}{CaCO_3}\ \rightarrow\ Ca^{++} + 2\ Cl^- + CO_2(aq) + H_2O \qquad (13.14)$$

According to this reaction, adding HCl to the fluid no longer affects *pH*. Instead, calcite dissolves to neutralize the acid, leaving Ca^{++} and CO_2(aq) in solution (Fig. 13.7).

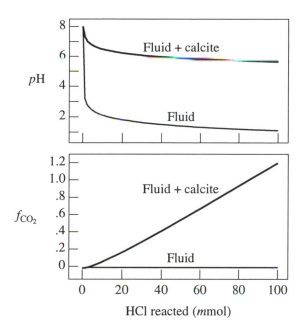

FIG. 13.6 Effects on pH (top) and CO_2 fugacity (bottom) of reacting HCl into a fluid not in contact with calcite (fine lines) and with the same fluid when it maintains equilibrium with calcite over the reaction path (bold lines).

In nature, the fluid would begin to exsolve $CO_2(g)$

$$CO_2(aq) \rightarrow CO_2(g) \tag{13.15}$$

once the $CO_2(aq)$ activity has built up sufficiently to drive the reaction forward. In this case, we can see that the acid introduced to the fluid is converted by reaction with calcite into CO_2 (or carbonic acid, equivalently) and then lost from the fluid by effervescence.

As a second example, we consider how the presence of pyrite (FeS_2) can serve to buffer a fluid's oxidation state. We set an initial system at 100°C containing a 1 molal NaCl solution in equilibrium with 0.2 moles (about 5 cm^3) of pyrite and a small amount of hematite (Fe_2O_3). In REACT, the procedure is

```
T = 100
swap Pyrite for Fe++
swap Hematite for O2(aq)

200 free mmol Pyrite
  1 free mmol Hematite

pH = 8
Cl- = 1 molal
```

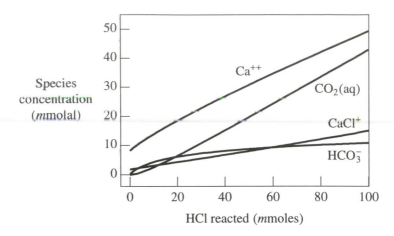

FIG. 13.7 Concentrations (*mmolal*) of the predominant carbonate and calcium species over the course of a reaction path in which HCl is added to a fluid in contact with calcite.

```
Na+ = 1 molal
SO4-- = 0.001 molal

go
```

The reaction between pyrite and hematite fixes the initial oxidation state to a reducing value ($\log f_{O_2} = -54$). We then set a reaction path in which we add oxygen to the fluid

```
(cont'd)
react 800 mmol of O2(aq)
go
```

simulating what might happen, for example, if O_2 were to diffuse into a reducing geologic formation.

Figure 13.8 shows the calculation results, and in Fig. 13.9 the reaction path is projected onto an f_{O_2}-pH diagram drawn for the Fe-S-H$_2$O system. (To project the path onto the diagram, we complete the reaction path, start ACT2, enter the commands

```
T = 100
swap O2(g) for O2(aq)
diagram Fe++ on O2(g) vs pH
log a Fe++ = -6
log a SO4-- = -3
speciate SO4-- over X-Y

x from 0 to 9
y from -65 to 2
trace
```

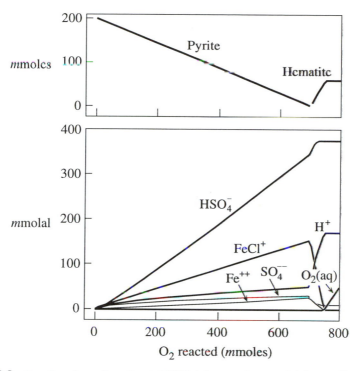

FIG. 13.8 Results of reacting O_2 at 100°C into a system containing pyrite. Pyrite dissolves (top) with addition of O_2. The reaction (bottom) produces bisulfate ions and ferric species ($FeCl^+$ and Fe^{++}), which in turn are consumed at the end of the path to form hematite

and type go.) In the earliest portion of the path, the system responds to the addition of oxygen by shifting quickly toward low pH. The system's rapid acidification results from the production of H^+ according to the reaction

$$\tfrac{7}{2} O_2(aq) + \underset{pyrite}{FeS_2} + H_2O \rightarrow Fe^{++} + 2\,SO_4^- + 2\,H^+ \qquad (13.16)$$

(lumping $FeCl^+$ together with Fe^{++}). As the system moves to lower pH, the bisulfate species HSO_4^- becomes more abundant than SO_4^{--} because of protonation of the sulfate species

$$SO_4^{--} + H^+ \rightarrow HSO_4^- \qquad (13.17)$$

At pH less than about 3, where bisulfate predominates, the dominant reaction is

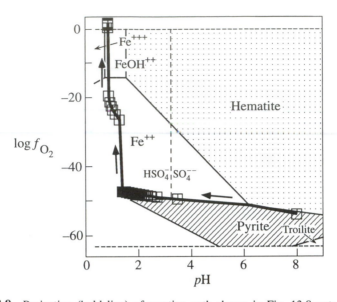

FIG. 13.9 Projection (bold line) of reaction path shown in Fig. 13.8 onto an f_{O_2}-pH diagram drawn for the Fe-S-H$_2$O system at 100°C. Horizontal dashed lines are stability limits for water at 1 atm pressure. Vertical dashed line is equal-activity line for SO_4^{--} and HSO_4^-. Boxes show reaction steps taken in tracing the path; each step represents the addition of an 8 $m$mol aliquot of O$_2$.

$$\frac{7}{2} O_2(aq) + FeS_2 + H_2O \rightarrow Fe^{++} + 2\,HSO_4^- \qquad (13.18)$$
$$pyrite$$

Since the reaction written in terms of HSO_4^- produces no H^+, the shift to lower pH slows and then ceases as the SO_4^{--} in solution is depleted.

At this point, the solution remains almost fixed in oxidation state and pH, accumulating Fe^{++} and HSO_4^- as the addition of oxygen causes pyrite to dissolve. When the pyrite is exhausted, the oxygen fugacity begins to rise rapidly. As oxygen is added, it is consumed in converting the dissolved ferrous iron to hematite

$$\frac{1}{2} O_2(aq) + 2\,Fe^{++} + 2\,H_2O \rightarrow Fe_2O_3 + 4\,H^+ \qquad (13.19)$$
$$hematite$$

which further acidifies the solution. Only when the ferrous iron is exhausted does the fluid become fully oxidized.

If we eliminate the pyrite and retrace the calculation

(cont'd)
dump

```
react 1 mmol of O2(aq)
go
```

we find that just a small amount of O_2 is sufficient to oxidize the system. In this case, pH changes little over the reaction path.

13.3 Gas Buffers

With a final example, we consider how the presence of a gas phase can serve as a chemical buffer. A fluid, for example, might maintain equilibrium with the atmosphere, soil gas in the root zone, or natural gas reservoirs in deep strata. Gases such as O_2 and H_2 can fix oxidation state, H_2S can set the activity of dissolved sulfide, and CO_2 (as we demonstrate in this section) can buffer pH.

In this experiment, we take an acidic water in equilibrium with atmospheric CO_2 and titrate NaOH into it. In REACT, the commands

```
pH = 2
swap CO2(g) for HCO3-
log f CO2(g) = -3.5
Na+ = .2 molal
balance on Cl-

react 1 mole NaOH
go
```

set up the calculation, assuming an atmospheric CO_2 fugacity of $10^{-3.5}$.

As shown in Fig. 13.10, the fluid quickly becomes alkaline, approaching a pH of 14. Since the fluid's carbonate content is small, about 10 μmolal, little beyond the fluid's initial H^+ content

$$NaOH + H^+ \rightarrow Na^+ + H_2O \qquad (13.20)$$

is available to buffer pH. Once the H^+ is exhausted, adding NaOH to the solution

$$NaOH \rightarrow Na^+ + OH^- \qquad (13.21)$$

simply produces OH^-, driving the pH to high values.

Now we consider the same reaction occurring in a water that maintains equilibrium with atmospheric CO_2. With the REACT commands

```
(cont'd)
fix fugacity CO2(g)
go
```

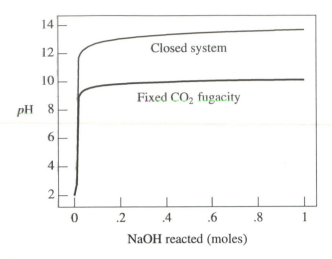

FIG. 13.10 Calculated effects on pH of reacting sodium hydroxide into an initially acidic solution that is either closed to mass transfer (fine line) or in equilibrium with atmospheric CO_2 (bold line).

we fix the CO_2 fugacity to its atmospheric value and retrace the path. In this case (Fig. 13.10), the pH rises initially but then levels off, approaching a value of 10.

The latter path differs from the closed system calculation because of the effect of $CO_2(g)$ dissolving into the fluid. In the initial part of the calculation, the $CO_2(aq)$ in solution reacts to form HCO_3^- in response to the changing pH. Since the fluid is in equilibrium with $CO_2(g)$ at a constant fugacity, however, the activity of $CO_2(aq)$ is fixed. To maintain this activity, the model transfers CO_2

$$CO_2(g) \rightarrow CO_2(aq) \tag{13.22}$$

from gas to fluid, replacing whatever $CO_2(aq)$ has reacted to form HCO_3^-.

The overall reaction for the earliest portion of the path, obtained by the slopes-of-the-lines method, is

$$2\,NaOH + CO_2(g) \rightarrow 2\,Na^+ + HCO_3^- + OH^- \tag{13.23}$$

In contrast to the unbuffered case (Reaction 13.21), two NaOHs are required to produce each OH^- ion. As pH continues to increase, the HCO_3^- reacts to produce CO_3^{--}, as shown in Fig. 13.11. At this point, a second overall reaction

$$2\,NaOH + CO_2(g) \rightarrow 2\,Na^+ + CO_3^{--} + H_2O \tag{13.24}$$

becomes increasingly important. According to this reaction, adding NaOH

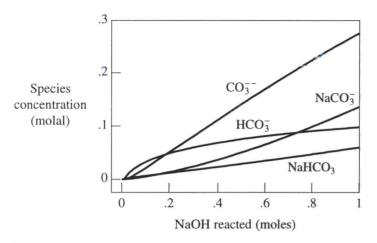

FIG. 13.11 Concentrations (molal) of the predominant carbonate species over the course of a reaction path in which NaOH is added to a fluid that maintains equilibrium with CO_2 in the atmosphere.

increases the fluid's sodium and carbonate contents but does not affect pH. The CO_2, an acid gas, neutralizes the alkaline NaOH as fast as it is added. Once Reaction 13.24 comes to dominate Reaction 13.23, pH ceases to change. As long as CO_2 is available in the gas reservoir, the fluid maintains the buffered pH.

14

Geochemical Kinetics

To this point we have measured reaction progress parametrically in terms of the reaction progress variable ξ, which is dimensionless. When in Chapter 11 we reacted feldspar with water, for example, we tied reaction progress to the amount of feldspar that had reacted and expressed our results along that coordinate. Studying reactions in this way is in many cases perfectly acceptable. But what if we want to know how much time it took to reach a certain point along the reaction path? Or, when modeling the reaction of granite with rainwater, how can we set the relative rates at which the various minerals in the granite dissolve? In such cases, we need to incorporate reaction rate laws from the field of geochemical kinetics.

The differences between the study of thermodynamics and kinetics might be illustrated (e.g., Lasaga, 1981a) by the analogy of rainfall on a mountain. On the mountaintop, the rainwater contains a considerable amount of potential energy. With time, it flows downhill, losing energy (to be precise, losing hydraulic potential, the mechanical energy content of a unit mass of water; Hubbert, 1940), until it eventually reaches the ocean, its lowest possible energy level. The thermodynamic interpretation of the process is obvious: the water seeks to minimize its energy content.

But how long will it take for the rainfall to reach the ocean? The rain might enter a swift mountain stream, flow into a river, and soon reach the sea. It might infiltrate the subsurface and migrate slowly through deep aquifers until it discharges in a distant valley, thousands of years later. Or, perhaps it will find a faster route through a fracture network or flow through an open drill hole. There are many pathways, just as there are many mechanisms by which a chemical reaction can proceed. Clearly, the questions addressed by geochemical kinetics are more difficult to answer than are those posed in thermodynamics.

In geochemical kinetics, the rates at which reactions proceed are given (in units such as moles/sec or moles/yr) by rate laws, as discussed in the next section. Kinetic theory can be applied to study reactions among the species in solution. We might, for example, study the rate at which the ferrous ion Fe^{++} oxidizes to produce the ferric species Fe^{+++}. Since the reaction occurs within a single phase, it is termed *homogeneous*. Reactions involving more than one phase (including the reactions by which minerals precipitate and dissolve) are called *heterogeneous*.

The kinetics of heterogeneous reactions have received the most attention in reaction modeling, primarily because of the slow rates at which many minerals react and the resulting tendency of fluids, especially at low temperature, to be out of equilibrium with the minerals they contact. In this chapter, we first discuss how rate laws for heterogeneous reactions can be integrated into reaction models and then calculate some simple kinetic reaction paths. The kinetics of homogeneous reactions, especially redox reactions, are also important in geochemistry; this subject deserves further attention, but is beyond the scope of our discussion. In Chapter 20, we explore a number of examples in which we apply kinetic reaction paths to problems of geochemical interest.

14.1 Kinetic Rate Laws

Despite the authority apparent in its name, no single "rate law" describes how quickly a mineral precipitates or dissolves. The mass action equation, which describes the equilibrium point of a mineral's dissolution reaction, is independent of reaction mechanism. A rate law, on the other hand, reflects our idea of how a reaction proceeds on a molecular scale. Rate laws, in fact, quantify the slowest or "rate-limiting" step in a hypothesized reaction mechanism.

Different reaction mechanisms can predominate in fluids of differing composition, since species in solution can serve to catalyze or inhibit the reaction mechanism. For this reason, there may be a number of valid rate laws that describe the reaction of a single mineral (e.g., Brady and Walther, 1989). It is not uncommon to find that one rate law applies under acidic conditions, another at neutral pH, and a third under alkaline conditions. We may discover, furthermore, that a rate law measured for reaction with deionized water fails to describe how a mineral reacts with electrolyte solutions.

In studying dissolution and precipitation, geochemists commonly consider that a reaction proceeds in five generalized steps:

    (1) diffusion of reactants from the bulk fluid to the mineral surface,

    (2) adsorption of the reactants onto reactive sites,

    (3) a chemical reaction involving the breaking and creation of bonds,

    (4) desorption of the reaction products, and

    (5) diffusion of the products from the mineral surface to the bulk fluid

(e.g., Brezonik, 1994). The adsorption and desorption processes (steps 2 and 4) are almost certainly rapid, so two classes of rate-limiting steps are possible (e.g., Lasaga, 1984). If the reaction rate depends on how quickly reactants can reach the surface by aqueous diffusion and the products can move away from it (steps 1 and 5), the reaction is said to be "transport controlled." If, on the other hand, the speed of the surface reaction (step 3) controls the rate, the reaction is termed "surface controlled." A reaction may be revealed to be transport controlled if its rate in the laboratory varies with stirring speed, or if a low value is found for its activation energy (defined later in this section).

Reactions for common minerals fall in both categories, but many important cases tend, except under acidic conditions, to be surface controlled (e.g., Aagaard and Helgeson, 1982; Stumm and Wollast, 1990). For this reason and because of their relative simplicity, we will consider in this chapter rate laws for surface-controlled reactions. The problem of integrating rate laws for transport-controlled reactions into reaction path calculations, nonetheless, is complex and interesting (Steefel and Lasaga, 1994), and warrants further attention.

Almost all published rate laws for surface controlled reactions are presented in a form derived from transition state theory (Lasaga, 1981a, 1981b, 1984; Aagaard and Helgeson, 1982). According to the theory, a mineral dissolves by a mechanism involving the creation and subsequent decay of an "activated complex," which is less stable (of higher free energy per mole) than either the bulk mineral or product species. The rate at which the activated complex decays controls how quickly the mineral dissolves.

The dissolution rate, according to the theory, does not depend on the mineral's saturation state. The precipitation rate, on the other hand, varies strongly with saturation, exceeding the dissolution rate only when the mineral is supersaturated. At the point of equilibrium, the dissolution rate matches the rate of precipitation so that the net rate of reaction is zero. There is, therefore, a strong conceptual link between the kinetic and thermodynamic interpretations: equilibrium is the state in which the forward and reverse rates of a reaction balance.

To formulate a kinetic reaction path, we consider one or more minerals $A_{\vec{k}}$ whose rates of dissolution and precipitation are to be controlled by kinetic rate laws. We wish to avoid assuming that the minerals $A_{\vec{k}}$ are in equilibrium with the system, so they do not appear in the basis (i.e., $A_{\vec{k}} \notin A_k$). We can write a reaction

$$A_{\vec{k}} \rightleftarrows \nu_{w\vec{k}} A_w + \sum_i \nu_{i\vec{k}} A_i + \sum_k \nu_{k\vec{k}} A_k + \sum_m \nu_{m\vec{k}} A_m \qquad (14.1)$$

for $A_{\vec{k}}$ in terms of the current basis (A_w, A_i, A_k and A_m) and calculate the reaction's equilibrium constant $K_{\vec{k}}$.

Following transition state theory, we can write a rate law giving the time rate of change $r_{\vec{k}}$ in the mole number $n_{\vec{k}}$ of mineral $A_{\vec{k}}$. The law takes the form

$$r_{\vec{k}} = \frac{dn_{\vec{k}}}{dt} = (A_S \, k_+)_{\vec{k}} \prod_{}^{i} (a_i)^{P_{i\vec{k}}} \prod_{}^{j} (a_j)^{P_{j\vec{k}}} \left[\frac{Q_{\vec{k}}}{K_{\vec{k}}} - 1 \right] \qquad (14.2)$$

Here, A_S is the mineral's surface area (cm^2) and k_+ is its intrinsic rate constant ($mol/cm^2 sec$), a_i and a_j are the activities of the basis species A_i and secondary species A_j, and $P_{i\vec{k}}$ and $P_{j\vec{k}}$ are exponents for each species, which are derived empirically. In practice, only a few species (if any) appear in the Π terms; the values of $P_{i\vec{k}}$ and $P_{j\vec{k}}$ for most species in the system are zero. $Q_{\vec{k}}$ is the activity product for Reaction 14.1 (Eqn. 3.35).

There are three functional parts of Eqn. 14.2. The first grouping ($A_S \, k_+$) requires that the reaction proceed at a rate proportional to the surface area and the rate constant. The surface area of a sample can be measured by a nitrogen adsorption technique (the BET method) or estimated from geometric considerations, and the rate constant is determined experimentally. It is interesting to note that whenever A_S is zero, the reaction rate vanishes. A mineral that does not exist, therefore, cannot begin to precipitate until crystal nuclei form. Various theories have been suggested for describing the rate at which nuclei might develop spontaneously or on the surfaces of other minerals (e.g., Berner, 1980) and it is possible to integrate the theories into reaction models (Steefel and Van Cappellen, 1990). Considerable uncertainties exist in applying nucleation theory to practical cases, however, and we will not include the theory in the scope of our discussion.

The Π grouping in Eqn. 14.2 represents the role that species in solution play in catalyzing the reaction or inhibiting its progress. A species can catalyze the reaction by promoting formation of the activated complex (in which case the corresponding $P_{i\vec{k}}$ or $P_{j\vec{k}}$ is positive) or it can inhibit the reaction by impeding its formation ($P_{i\vec{k}}$ or $P_{j\vec{k}}$ is negative). The final grouping represents the thermodynamic drive for reaction. When mineral $A_{\vec{k}}$ is supersaturated, $Q_{\vec{k}} > K_{\vec{k}}$ and the mineral precipitates. When the mineral is undersaturated, it dissolves because $Q_{\vec{k}} < K_{\vec{k}}$.

The rate constant in Eqn. 14.2 can be related to temperature by the phenomenological Arrhenius equation

$$k_+ = A \, e^{-E_A/RT_K} \qquad (14.3)$$

(e.g., Lasaga, 1981a). Here, A is the preexponential factor ($mol/cm^2 sec$), E_A is the activation energy (J/mol), R is the gas constant (8.3143 J/K·mol), and T_K is absolute temperature (K). The values of A and E_A are determined for a given reaction by measuring k_+ at several temperatures and fitting the data in semilog coordinates.

In an example of the forms rate laws can take, we consider the reaction of albite ($NaAlSi_3O_8$), the dissolution of which was studied by Knauss and Wolery (1986). They found that the reaction proceeds according to different rate laws, depending on pH. From their results, we can write a rate law valid at pH values more acidic than about 1.5 as

$$r_{\text{alb}} = A_S \, k_+ \, a_{H^+} \left[\frac{Q}{K} - 1 \right] \quad (14.4)$$

where at 70°C $k_+ = 10^{-12.2}$ mol/cm²sec. In this case, just one species (H^+) appears in the Π terms of Eqn. 14.2; the corresponding value of the exponent P_{H^+} is one. In the *pH* range of about 1.5 to 8, a second reaction mechanism is predominant. In the corresponding rate law

$$r_{\text{alb}} = A_S \, k_+ \left[\frac{Q}{K} - 1 \right] \quad (14.5)$$

in which $k_+ = 10^{-15.1}$ mol/cm²sec at 70°C, there are no species in the Π terms. The reaction rate, therefore, is not affected by solution composition. A third law

$$r_{\text{alb}} = A_S \, k_+ \, a_{H^+}^{-\frac{1}{2}} \left[\frac{Q}{K} - 1 \right] \quad (14.6)$$

with $k_+ = 10^{-19.5}$ at 70°C, describes the dominant mechanism at higher *pH*. The exponent P_{H^+} in this case is −0.5.

Some caveats about the form presented for the rate law (Eqn. 14.2) are worth noting. First, although Eqn. 14.2 is linear in $Q_{\vec{k}}$, transition state theory does not demand that rate laws take such a form. There are nonlinear forms of the rate law that are equally valid (e.g., Merino et al., 1993; Lasaga et al., 1994) and that in some cases may be required to explain observations. Specifically, the term $Q_{\vec{k}}/K_{\vec{k}}$ can appear raised to an arbitrary (not necessarily integer) exponent, as can the entire $(Q_{\vec{k}}/K_{\vec{k}} - 1)$ term (provided that its original sign is preserved). Such rate expressions have seldom been invoked in geochemistry, if only because most experiments have been designed to study the dissolution reaction under conditions far from equilibrium, where there is no basis for observing nonlinear effects. Nonlinear rate laws can, nonetheless, be readily incorporated into reaction models, as described in Appendix 5.

Second, in deriving Eqn. 14.2 from transition state theory, it is necessary to assume that the overall reaction proceeds on a molecular scale as a single elementary reaction or a series of elementary reactions (e.g., Lasaga, 1984; Nagy et al., 1991). In general, the elementary reactions that occur as a mineral dissolves and precipitates are not known. Thus, even though the form of Eqn. 14.2 is convenient and broadly applicable for explaining experimental results, it is not necessarily correct in the strictest sense.

14.2 From Laboratory to Application

The great value of kinetic theory is that it frees us from many of the constraints of the equilibrium model and its variants (partial equilibrium, local equilibrium, and so on; see Chapter 2). In early studies (e.g., Lasaga, 1984), geochemists were openly optimistic that the results of laboratory experiments could be applied directly to the study of natural systems. Transferring the laboratory results to field situations, however, has proved to be much more challenging than many first imagined.

Many minerals have been found to dissolve and precipitate in nature at dramatically different rates than they do in laboratory experiments. As first pointed out by Paces (1983) and confirmed by subsequent studies, for example, albite weathers in the field much more slowly than predicted on the basis of reaction rates measured in the laboratory. The discrepancy can be as large as four orders of magnitude (Brantley, 1992, and references therein). As we calculate in Chapter 20, furthermore, the measured reaction kinetics of quartz (SiO_2) suggest that water should quickly reach equilibrium with this mineral, even at low temperatures. Equilibrium between groundwater and quartz, however, is seldom observed, even in aquifers composed largely of quartz sand.

Geochemists (e.g., Aagaard and Helgeson, 1982) commonly attribute such discrepancies to difficulties in representing the surface area A_S of minerals in natural samples. In the laboratory, the mineral is fresh and any surface coatings have been removed. The same mineral in the field, however, may be shielded with oxide, hydroxide, or organic coatings. It may be occluded by contact with other materials, including reaction products, organic matter, and other grains. In addition, the aged surface of the natural sample is probably smoother than the laboratory material and hence contains fewer kinks and sharp edges, which are highly reactive.

Even where it is not occluded, the mineral surface may not be reactive. In the vadose zone, the surface may not be fully in contact with water or may contact water only intermittently. In the saturated zone, a mineral may touch virtually immobile water within isolated portions of the sediment's pore structure. Fluid chemistry in such microenvironments may bear little relationship to the bulk chemistry of the pore water. Since groundwater flow tends to be channeled through the most permeable portions of the subsurface, furthermore, fluids may bypass many or most of the mineral grains in a sediment or rock. The latter phenomenon is especially pronounced in fractured rocks, where only the mineral surfaces lining the fracture may be reactive.

There are other important factors beyond the state of the surface that may lead to discrepancies between laboratory and field studies. Measurement error in the laboratory, first of all, is considerable. Brantley (1992) notes that rate constants determined by different laboratories generally agree to within only a factor of about 30. Agreement to better than a factor of 5, she reasons, might not be an attainable goal.

There is no certainty, furthermore, that the reaction or reaction mechanism studied in the laboratory will predominate in nature. Data for reaction in deionized water, for example, might not apply if aqueous species present in nature catalyze a different reaction mechanism, or if they inhibit the mechanism that operated in the laboratory. Dove and Crerar (1990), for example, showed that quartz dissolves into dilute electrolyte solutions up to 30 times more quickly than it does in pure water. Laboratory experiments, furthermore, are nearly always conducted under conditions in which the fluid is far from equilibrium with the mineral, although reactions in nature proceed over a broad range of saturation states across which the laboratory results may not apply.

Further error is introduced if reactions distinct from those for which data is available affect the chemistry of a natural fluid. Consider as an example the problem of predicting the silica content of a fluid flowing through a quartz sand aquifer. There is little benefit in modeling the reaction rate for quartz if the more reactive minerals (such as clays and zeolites) in the aquifer control the silica concentration.

Finally, whereas most laboratory experiments have been conducted in largely abiotic environments, the action of bacteria may control reaction rates in nature (e.g., Chapelle, 1993). In the production of acid drainage (see Chapter 23), for example, bacteria such as *Thiobacillus ferrooxidans* control the rate at which pyrite (FeS_2) oxidizes (Taylor et al., 1984; Okereke and Stevens 1991). Laboratory observations of how quickly pyrite oxidizes in abiotic systems (e.g., Williamson and Rimstidt, 1994, and references therein), therefore, might poorly reflect the oxidation rate in the field.

14.3 Numerical Solution

The procedure for tracing a kinetic reaction path differs from the procedure for paths with simple reactants (Chapter 11) in two principal ways. First, progress in the simulation is measured in units of time t rather than by the reaction progress variable ξ. Second, the rates of mass transfer, instead of being set explicitly by the modeler (Eqns. 11.3a–c), are computed over the course of the reaction path by a kinetic rate law (Eqn. 14.2).

From Eqns. 14.1 and 14.2, we can write the instantaneous rate of change in the system's bulk composition as

$$\frac{dM_w}{dt} = -\sum_{\vec{k}} \nu_{w\vec{k}}\, r_{\vec{k}} \tag{14.7a}$$

$$\frac{dM_i}{dt} = -\sum_{\vec{k}} \nu_{i\vec{k}}\, r_{\vec{k}} \tag{14.7b}$$

$$\frac{dM_k}{dt} = -\sum_{\vec{k}} \nu_{k\vec{k}}\, r_{\vec{k}} \tag{14.7c}$$

In stepping forward from t' to a new point in time t, the instantaneous rate will change as the fluid's chemistry evolves. Rather than carrying the rate at t' over the step, it is more accurate (e.g., Richtmyer, 1957; Peaceman, 1977) to take the average of the rates at t' and t. In this case, the new bulk composition (at t) is given from its previous value (at t') and Eqns. 14.7a–c by

$$M_w(t) = M_w(t') - \frac{(t-t')}{2} \sum_{\vec{k}} v_{w\vec{k}} \left[r_{\vec{k}}(t) + r_{\vec{k}}(t') \right] \tag{14.8a}$$

$$M_i(t) = M_i(t') - \frac{(t-t')}{2} \sum_{\vec{k}} v_{i\vec{k}} \left[r_{\vec{k}}(t) + r_{\vec{k}}(t') \right] \tag{14.8b}$$

$$M_k(t) = M_k(t') - \frac{(t-t')}{2} \sum_{\vec{k}} v_{k\vec{k}} \left[r_{\vec{k}}(t) + r_{\vec{k}}(t') \right] \tag{14.8c}$$

We use these relations instead of Eqns. 11.3a–c when tracing a kinetic path.

To solve for the chemical system at t, we use Newton-Raphson iteration to minimize a set of residual functions, as discussed in Chapter 5. For a kinetic path, the residual functions are derived by combining Eqns. 5.19a–b with Eqns. 14.8a–b, giving

$$R_w = n_w \left[55.5 + \sum_j v_{wj} m_j \right] - M_w(t') + \frac{(t-t')}{2} \sum_{\vec{k}} v_{w\vec{k}} \left[r_{\vec{k}}(t) + r_{\vec{k}}(t') \right] \tag{14.9a}$$

$$R_i = n_w \left[m_i + \sum_j v_{ij} m_j \right] - M_i(t') + \frac{(t-t')}{2} \sum_{\vec{k}} v_{i\vec{k}} \left[r_{\vec{k}}(t) + r_{\vec{k}}(t') \right] \tag{14.9b}$$

In order to derive the corresponding Jacobian matrix, we need to differentiate $r_{\vec{k}}$ (Eqn. 14.2) with respect to n_w and m_i. Taking advantage of the relations

$$\frac{\partial m_j}{\partial n_w} = 0 \quad \text{and} \quad \frac{\partial m_j}{\partial m_i} = \frac{v_{ij} m_j}{m_i} \tag{14.10a}$$

(Eqn. 5.21), and

$$\frac{\partial Q_{\vec{k}}}{\partial n_w} = 0 \quad \text{and} \quad \frac{\partial Q_{\vec{k}}}{\partial m_i} = \frac{v_{i\vec{k}} Q_{\vec{k}}}{m_i} \tag{14.10b}$$

(following from the definition of the activity product; Eqn. 3.35), we can show that $dr_{\vec{k}}/dn_w = 0$ and

$$\frac{dr_{\vec{k}}}{dm_i} = \frac{(A_S k_+)_{\vec{k}}}{m_i} \prod^i (a_{i'})^{P_{i'\vec{k}}} \prod^j (a_j)^{P_{j\vec{k}}} \times$$

$$\left[\frac{v_{i\vec{k}} Q_{\vec{k}}}{K_{\vec{k}}} + \left(P_{i\vec{k}} + \sum_j v_{ij} P_{j\vec{k}} \right) \left[\frac{Q_{\vec{k}}}{K_{\vec{k}}} - 1 \right] \right] \tag{14.11}$$

As discussed in Section 5.3, the entries in the Jacobian matrix are given by differentiating the residual functions (Eqns. 14.9a–b) with respect to the independent variables n_w and m_i. The resulting entries are

$$J_{ww} = \frac{\partial R_w}{\partial n_w} = 55.5 + \sum_j \nu_{wj} m_j \qquad (14.12a)$$

$$J_{wi} = \frac{\partial R_w}{\partial m_i} = \frac{n_w}{m_i} \sum_j \nu_{wj} \nu_{ij} m_j + \frac{(t - t')}{2} \sum_{\vec{k}} \nu_{w\vec{k}} \frac{dr_{\vec{k}}}{dm_i} \qquad (14.12b)$$

$$J_{iw} = \frac{\partial R_i}{\partial n_w} = m_i + \sum_j \nu_{ij} m_j \qquad (14.12c)$$

$$J_{ii'} = \frac{\partial R_i}{\partial m_{i'}} = n_w \left[\delta_{ii'} + \sum_j \nu_{ij} \nu_{i'j} m_j / m_{i'} \right] + \frac{(t - t')}{2} \sum_{\vec{k}} \nu_{i\vec{k}} \frac{dr_{\vec{k}}}{dm_{i'}} \qquad (14.12d)$$

where the values $dr_{\vec{k}}/dm_i$ are given by Eqn. 14.11. Equations 14.12a–d take the place of Eqns. 5.22a–d in a kinetic model. As before (Eqn. 5.23), $\delta_{ii'}$ is the Kronecker delta function.

The time-stepping proceeds as previously described (Chapter 11), with the slight complication that the surface areas A_S of the kinetic minerals must be evaluated after each iteration to account for changing mineral masses. For polythermal paths, each rate constant k_+ must be set before beginning a time step according to the Arrhenius equation (Eqn. 14.3) to a value corresponding to the temperature at the new time level.

14.4 Example Calculations

In an example of a kinetic reaction path, we calculate how quartz sand reacts at 100°C with deionized water. According to Rimstidt and Barnes (1980), quartz reacts according to the rate law

$$r_{\text{qtz}} = A_S \, k_+ \left[\frac{Q}{K} - 1 \right] \qquad (14.13)$$

with a rate constant k_+ at this temperature of about 2×10^{-15} mol/cm^{2}sec. From their data, we assume that the sand has a specific surface area of 1000 cm^2/g.

In REACT, we set a system containing 1 kg of water and 5 kg of quartz

```
time begin = 0 days, end = 5 days
T = 100

SiO2(aq) = 1 umolal

react 5000 g Quartz
kinetic Quartz rate_con = 2.e-15   surface = 1000
go
```

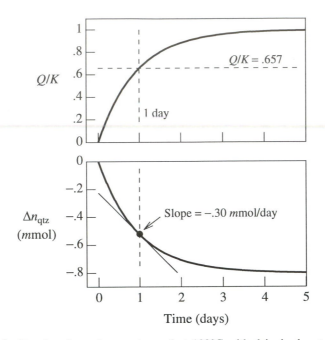

FIG. 14.1 Results of reacting quartz sand at 100°C with deionized water, calculated according to a kinetic rate law. Top diagram shows how the saturation state Q/K of quartz varies with time; bottom plot shows change in amount (*m*mol) of quartz in system (bold line). The slope of the tangent to the curve (fine line) is the instantaneous reaction rate, shown at one day of reaction.

and allow it to react for five days. Figure 14.1 shows the calculation results. Quartz dissolves with time, adding a small amount of $SiO_2(aq)$ to solution. The increasing silica content of the solution causes the saturation state Q/K of quartz to increase, slowing the reaction rate. After five days, the reaction has approached equilibrium ($Q/K \approx 1$) and the reaction nearly ceases.

We can calculate the reaction rate according to Eqn. 14.13 and compare it to the calculation results. The required quantities at one day of reaction are

$$A_S = 5000 \text{ g} \times 1000 \text{ cm}^2/\text{g} = 5 \times 10^6 \text{ cm}^2$$

$$k_+ = 2 \times 10^{-15} \text{ mol/cm}^2\text{sec}$$

$$Q/K = 10^{-0.182} = .657 \tag{14.14}$$

Entering these values into the rate equation (Eqn. 14.13) gives the reaction rate

$$r_{qtz} = (5 \times 10^6 \text{ cm}^2) \times (2 \times 10^{-15} \text{mol/cm}^2\text{sec}) \times (.657 - 1)$$

$$= -3.43 \times 10^{-9} \text{ mol/sec} = -.30 \text{ } m\text{mol/day} \tag{14.15}$$

which is negative since the quartz is dissolving. We can confirm that on a plot of the mole number n_{qtz} for quartz versus time (Fig. 14.1), this value is the slope of the tangent line and hence the reaction rate dn_{qtz}/dt we expect.

In a slightly more complicated example, we calculate the rate at which albite dissolves at 70°C into an acidic NaCl solution. We use the rate law shown in Eqn. 14.4, which differs from Eqn. 14.13 by the inclusion of the a_{H^+} term, and a rate constant of 6.3×10^{-13} mol/cm²sec determined for this temperature by Knauss and Wolery (1986). We set *pH* to a constant value of 1.5, as if this value were maintained by an internal buffer or an external control such as a *pH*-stat, and allow 250 grams of the mineral to react with 1 kg of water for thirty days.

The procedure in REACT is

```
time begin = 0 days, end = 30 days
T = 70

pH = 1.5
0.1 molal Cl-
0.1 molal Na+
1 umolal SiO2(aq)
1 umolal Al+++

react 250 grams of "Albite low"
kinetic "Albite low"  rate_con = 6.3e-13, power(H+) = 1
kinetic "Albite low"  surface = 1000
fix pH
go
```

We can quickly verify the calculation results (Fig. 14.2). Choosing day 15 of the reaction,

$$A_S = 250 \text{ g} \times 1000 \text{ cm}^2/\text{g} = 2.5 \times 10^5 \text{ cm}^2$$

$$k_+ = 6.3 \times 10^{-13} \text{ mol/cm}^2 \text{sec}$$

$$a_{H^+} = 10^{-1.5} = 3.16 \times 10^{-2}$$

$$(Q/K - 1) = (10^{-10.1} - 1) \approx -1 \qquad (14.16)$$

Substituting into Eqn. 14.4,

$$r_{alb} = (2.5 \times 10^5 \text{ cm}^2) \times (6.3 \times 10^{-13} \text{mol/cm}^2 \text{sec}) \times (3.16 \times 10^{-2}) \times (-1)$$

$$= -5.0 \times 10^{-9} \text{ mol/sec} = -.43 \text{ mmol/day} \qquad (14.17)$$

we find the reaction rate dn_{alb}/dt shown in Fig. 14.2. Note that the fluid in this example remains so undersaturated with respect to albite that the value of Q/K is nearly zero and hence has virtually no influence on the reaction rate. For this reason, the reaction rate remains nearly constant over the course of the calculation.

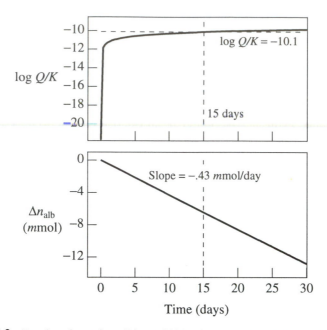

FIG. 14.2 Results of reacting albite at 70°C with an NaCl solution maintained at pH 1.5, calculated as a kinetic reaction path. Top diagram shows how the saturation index of albite varies with time; bottom plot shows change in amount ($m$mol) of albite.

14.5 Modeling Strategy

A practical consideration in reaction modeling is choosing the extent to which reaction kinetics should be integrated into the calculations. On the one hand, kinetic theory is an important generalization of the equilibrium model that lets us account for the fact that fluids and minerals do not necessarily coexist at equilibrium. On the other hand, the theory can add considerable complexity to developing and evaluating a reaction model.

We might take a purist's approach and attempt to use kinetic theory to describe the dissolution and precipitation of each mineral that might appear in the calculation. Such an approach, although appealing and conceptually correct, is seldom practical. The database required to support the calculation would have to include rate laws for every possible reaction mechanism for each of perhaps hundreds of minerals. Even unstable minerals that can be neglected in equilibrium models would have to be included in the database, since they might well form in a kinetic model (see Section 20.4, Ostwald's Step Rule). If we are to allow new minerals to form, furthermore, it will be necessary to describe how quickly each mineral can nucleate on each possible substrate.

The modeling software would have to trace a number of reactions occurring at broadly different rates. Although certainly feasible, such a calculation can present practical difficulties, especially at the onset of a reaction path, if the software must take very small steps to accurately trace the progress of the faster reactions. For each calculation, furthermore, we would need to be able to set initial conditions, requiring knowledge of hydrologic factors (flow rates, residence times, and so on) that in real life are not always available.

A worthwhile strategy for conceptualizing kinetic reaction paths is to divide mineral reactions into three groups. The first group contains the reactions that proceed quickly over the time span of the calculation. We can safely assume that these minerals remain in equilibrium with the fluid. A second group consists of minerals that react negligibly over the calculation and hence may be ignored or "suppressed." The reactions for the remaining minerals fall in the third group, for which we need to account for reaction kinetics. Our time is best spent attempting to define the rates at which minerals in the latter group react.

15

Stable Isotopes

Stable isotopes serve as naturally occurring tracers that can provide much information about how chemical reactions proceed in nature, such as which reactants are consumed and at what temperatures reactions occur. The stable isotopes of several of the lighter elements are sufficiently abundant and fractionate strongly enough to be of special usefulness. Foremost in importance are hydrogen, carbon, oxygen, and sulfur.

The strong conceptual link between stable isotopes and chemical reaction makes it possible to integrate isotope fractionation into reaction modeling, allowing us to predict not only the mineralogical and chemical consequences of a reaction process, but also the isotopic compositions of the reaction products. By tracing the distribution of isotopes in our calculations, we can better test our reaction models against observation and perhaps better understand how isotopes fractionate in nature.

Bowers and Taylor (1985) were the first to incorporate isotope fractionation into a reaction model. They used a modified version of EQ3/EQ6 (Wolery, 1979) to study the convection of hydrothermal fluids through the oceanic crust, along midocean ridges. Their calculation method is based on evaluating mass balance equations, as described in this chapter.

As originally derived, however, the mass balance model has an important (and well acknowledged) limitation: implicit in its formulation is the assumption that fluid and minerals in the modeled system remain in isotopic equilibrium over the reaction path. This assumption is equivalent to assuming that isotope exchange between fluid and minerals occurs rapidly enough to maintain equilibrium compositions.

We know, however, that isotope exchange in nature tends to be a slow process, especially at low temperature (e.g., O'Neil, 1987). This knowledge

217

comes from experimental study (e.g., Cole and Ohmoto, 1986) as well as from the simple observation that, unless they have reacted together, groundwaters and minerals are seldom observed to be in isotopic equilibrium with each other. In fact, if exchange were a rapid process, it would be very difficult to interpret the origin of geologic materials from their isotopic compositions: the information would literally diffuse away.

Lee and Bethke (1996) presented an alternative technique, also based on mass balance equations, in which the reaction modeler can segregate minerals from isotopic exchange. By segregating the minerals, the model traces the effects of the isotope fractionation that would result from dissolution and precipitation reactions alone. Not unexpectedly, segregated models differ broadly in their results from reaction models that assume isotopic equilibrium.

The segregated model works by defining a subset of the chemical equilibrium system called the isotope equilibrium system. The isotope system contains (1) the fluid, (2) any minerals not held segregated from isotope exchange, (3) any segregated minerals that dissolve over the current reaction step, and (4) the increments in mass of any segregated minerals that precipitate over the step. By holding the components of just the isotope system in equilibrium, the calculation procedure assures that the compositions of segregated minerals change only when new mass precipitates. Conversely, the segregated minerals affect the composition of the fluid only when they dissolve. In this way, chemical reaction is the only force driving isotope fractionation.

These ideas might be further developed to model situations in which exchange among aqueous species occurs slowly. The oxygen in SO_4^{--} species might not be allowed to exchange with water, or H_2S sulfur might be held segregated from sulfur in SO_4^{--}. To date, however, no description of isotope disequilibrium within the fluid phase has been implemented within a reaction model.

In this chapter, we develop a mass balance model of the fractionation in reacting systems of the stable isotopes of hydrogen, carbon, oxygen, and sulfur. We then demonstrate application of the model by simulating the isotopic effects of the dolomitization reaction of calcite.

15.1 Isotope Fractionation

Each species and phase of an element with two or more stable isotopes consists of light and heavy isotopes in proportions that can be measured by using a mass spectrometer (e.g., Faure, 1986). The isotope ratio R is the quotient of the number of moles of a heavy isotope (such as ^{18}O) to the number of moles of a light isotope (such as ^{16}O).

It is not especially practical, however, to express isotopic composition in terms of R. One isotope (e.g., ^{16}O) greatly dominates the others (Table 15.1), so values of R are small, rather inconvenient numbers. In addition, mass

TABLE 15.1 The stable isotopes of hydrogen, carbon, oxygen, and sulfur, and their approximate terrestrial abundances (Wedepohl et al., 1978)

Isotope	Abundance (%)	Isotope	Abundance (%)
Hydrogen		*Carbon*	
^1H	99.984	^{12}C	98.89
^2H	.016	^{13}C	1.11
Oxygen		*Sulfur*	
^{16}O	99.763	^{32}S	94.94
^{17}O	.0375	^{33}S	.77
^{18}O	.1995	^{34}S	4.27
		^{36}S	.02

spectrometers can measure the difference in isotopic composition between two samples much more accurately than they can determine an absolute ratio R. For these reasons, isotope geochemists express isotopic composition in a δ notation

$$\delta = \left[\frac{R_{Sample} - R_{Standard}}{R_{Standard}} \right] \times 1000 \qquad (15.1)$$

as the permil (‰) deviation from a standard. For example, the δ value for ^{18}O in a sample is

$$\delta^{18}O = \left[\frac{(^{18}O/^{16}O)_{Sample} - (^{18}O/^{16}O)_{SMOW}}{(^{18}O/^{16}O)_{SMOW}} \right] \times 1000 \qquad (15.2)$$

expressed relative to the SMOW standard. SMOW is the composition of "standard mean ocean water," the usual standard for this element. In this notation, a sample with a positive $\delta^{18}O$ is enriched in the heavy isotope ^{18}O relative to the standard, whereas a negative value shows that the sample is depleted in the isotope.

When species or phases are in isotopic equilibrium, their isotopic ratios differ from one another by predictable amounts. The segregation of heavier isotopes into one species and light isotopes into the other is called isotope fractionation. The fractionation among species is represented by a fractionation factor α, which is determined empirically. The fractionation factor between species A and B is the ratio

$$\alpha_{A-B} = \frac{R_A}{R_B} = \frac{\delta_A + 1000}{\delta_B + 1000} \qquad (15.3)$$

By rearranging this equation, we can express the isotopic composition of species A,

$$\delta_A = \alpha_{A-B}\,(1000 + \delta_B) - 1000 \qquad (15.4)$$

from the fractionation factor and the composition of B. The useful expression

$$1000 \ln \alpha_{A-B} \approx \delta_A - \delta_B \qquad (15.5)$$

follows from the approximation $\ln x \approx x - 1$ for $x \rangle 1$. Since this form gives the difference in δ values between A and B directly, fractionation factors are commonly compiled expressed as $1000 \ln \alpha$.

As a rule, isotopes fractionate more strongly at low temperatures than they do at high temperature. Variation of the fractionation factor with absolute temperature T_K can be fit to a polynomial, such as

$$1000 \ln \ \alpha_{A-B} = a + b\,T_K + c\left[\frac{10^3}{T_K}\right] + d\left[\frac{10^6}{T_K{}^2}\right] + e\left[\frac{10^9}{T_K{}^3}\right] + f\left[\frac{10^{12}}{T_K{}^4}\right] \qquad (15.6)$$

Here, coefficients a through f are determined by fitting measurements to the polynomial. Equations of this form fit the data well in most cases, but may fail to describe hydrogen fractionation at high temperature in hydrated minerals or in minerals containing hydroxyl groups.

In REACT, dataset "isotope.data" contains polynomial coefficients that define temperature functions for the fractionation factors of species, minerals, and gases. The factors describe fractionation relative to a reference species chosen for each element. The reference species for oxygen and hydrogen is solvent water, H_2O. CO_2 and H_2S, in either aqueous or gaseous form, serve as reference species for carbon and sulfur.

15.2 Mass Balance Equations

The key to tracing isotope fractionation over a reaction path, as shown by Bowers and Taylor (1985), is writing a mass balance equation for each fractionating element. We begin by specifying a part of the chemical system to be held in isotopic equilibrium, as already described. The isotope system, of course, excludes segregated minerals. At each step over the course of the reaction path, we apply the mass balance equations to determine, from the bulk composition of the isotope system, the isotopic composition of each species and phase. In this section, we derive the mass balance equation for each element; in the next we show how the equations can be integrated into reaction modeling.

We begin with oxygen. The total number of moles $n^T_{^{18}O}$ of ^{18}O in the isotope system is the sum of the mole numbers for this isotope in (1) the solvent, $n^w_{^{18}O}$, (2) the aqueous species, $n^j_{^{18}O}$, and (3) whatever mineral mass appears in the isotope system, $n^{k'}_{^{18}O}$. Expressed mathematically,

$$n_{^{18}O}^{T} = n_{^{18}O}^{w} + \sum_{j} n_{^{18}O}^{j} + \sum_{k'} n_{^{18}O}^{k'} \tag{15.7a}$$

A parallel expression

$$n_{^{16}O}^{I} = n_{^{16}O}^{w} + \sum_{j} n_{^{16}O}^{J} + \sum_{k'} n_{^{16}O}^{k'} \tag{15.7b}$$

applies to ^{16}O. Note that we use the notation k' to identify the mineral mass carried in the isotope system; as before, subscript k denotes minerals in the chemical system, whether in isotopic equilibrium or not.

Using these expressions and Eqn. 15.2, and recognizing that the mole ratio $n_{^{16}O}/n_O$ of ^{16}O to elemental oxygen is nearly constant, we can show that the total (or bulk) isotopic composition of the isotope system

$$\delta^{18}O_T \approx \frac{1}{n_O^T} \left[n_O^w \, \delta^{18}O_w + \sum_j n_O^j \, \delta^{18}O_j + \sum_{k'} n_O^{k'} \, \delta^{18}O_{k'} \right] \tag{15.8}$$

can be calculated from the δ values of solvent ($\delta^{18}O_w$), species ($\delta^{18}O_j$), and minerals ($\delta^{18}O_{k'}$).

This equation can be expanded by expressing the compositions of species and minerals in terms of their fractionation factors and the composition $\delta^{18}O_w$ of solvent water, the reference species. From Eqn. 15.4,

$$\delta^{18}O_j = \alpha_{j-w} (1000 + \delta^{18}O_w) - 1000 \tag{15.9a}$$

and

$$\delta^{18}O_{k'} = \alpha_{k'-w} (1000 + \delta^{18}O_w) - 1000 \tag{15.10a}$$

By substituting these relations,

$$n_O^T \, \delta^{18}O_T = n_O^w \, \delta^{18}O_w + \sum_j n_O^j \, [\alpha_{j-w} (1000 + \delta^{18}O_w) - 1000]$$
$$+ \sum_{k'} n_O^{k'} \, [\alpha_{k'-w} (1000 + \delta^{18}O_w) - 1000] \tag{15.11}$$

and solving for the composition of solvent water, we arrive at

$$\delta^{18}O_w = \frac{n_O^T \delta^{18}O_T - 1000 \left[\sum_j n_O^j \, (\alpha_{j-w}-1) + \sum_{k'} n_O^{k'} \, (\alpha_{k'-w}-1) \right]}{n_O^w + \sum_j n_O^j \, \alpha_{j-w} + \sum_{k'} n_O^{k'} \, \alpha_{k'-w}} \tag{15.12a}$$

which is the mass balance equation for oxygen.

Equation 15.12a is useful because it allows us to use the system's total isotopic composition, $\delta^{18}O_T$, to determine the compositions of the solvent and each species and mineral. The calculation proceeds in two steps. First, given $\delta^{18}O_T$ and the fractionation factors α_{j-w} and $\alpha_{k'-w}$ for the various species and minerals, we compute the composition of the solvent, applying Eqn. 15.12a.

Second, we use this result to calculate the composition of each species and mineral directly, according to Eqns. 15.9a and 15.10a.

The equations for the isotope pairs ^{2}H/^{1}H, ^{13}C/^{12}C, and ^{34}S/^{32}S parallel the relations for ^{18}O/^{16}O, except that the reference species for carbon and sulfur are CO_2 and H_2S, rather than solvent water. Carbon and sulfur compositions are many times reported with respect to the PDB (Pee Dee belemnite) and CDT (Canyon Diablo troilite) standards, instead of SMOW. It makes little difference which standard we choose in applying these equations, however, as long as we carry a single standard for each element through the calculation.

Fractionation of hydrogen, carbon, and sulfur isotopes among the aqueous species is set by the relations

$$\delta^2 H_j = \alpha_{j-w} (1000 + \delta^2 H_w) - 1000 \tag{15.9b}$$

$$\delta^{13} C_j = \alpha_{j-CO_2} (1000 + \delta^{13} C_{CO_2}) - 1000 \tag{15.9c}$$

$$\delta^{34} S_j = \alpha_{j-H_2S} (1000 + \delta^{34} S_{H_2S}) - 1000 \tag{15.9d}$$

Similarly, the equations

$$\delta^2 H_{k'} = \alpha_{k'-w} (1000 + \delta^2 H_w) - 1000 \tag{15.10b}$$

$$\delta^{13} C_{k'} = \alpha_{k'-CO_2} (1000 + \delta^{13} C_{CO_2}) - 1000 \tag{15.10c}$$

$$\delta^{34} S_{k'} = \alpha_{k'-H_2S} (1000 + \delta^{34} S_{H_2S}) - 1000 \tag{15.10d}$$

define fractionation among minerals.

Compositions of the reference species are given by

$$\delta^2 H_w = \frac{n_H^T \delta^2 H_T - 1000 \left[\sum_j n_H^j (\alpha_{j-w} - 1) + \sum_{k'} n_H^{k'} (\alpha_{k'-w} - 1) \right]}{n_H^w + \sum_j n_H^j \alpha_{j-w} + \sum_{k'} n_H^{k'} \alpha_{k'-w}} \tag{15.12b}$$

$$\delta^{13} C_{CO_2} = \frac{n_C^T \delta^{13} C_T - 1000 \left[\sum_j n_C^j (\alpha_{j-CO_2} - 1) + \sum_{k'} n_C^{k'} (\alpha_{k'-CO_2} - 1) \right]}{\sum_j n_C^j \alpha_{j-CO_2} + \sum_{k'} n_C^{k'} \alpha_{k'-CO_2}} \tag{15.12c}$$

$$\delta^{34} S_{H_2S} = \frac{n_S^T \delta^{34} S_T - 1000 \left[\sum_j n_S^j (\alpha_{j-H_2S} - 1) + \sum_{k'} n_S^{k'} (\alpha_{k'-H_2S} - 1) \right]}{\sum_j n_S^j \alpha_{j-H_2S} + \sum_{k'} n_S^{k'} \alpha_{k'-H_2S}} \tag{15.12d}$$

which constitute the mass balance equations for these elements.

15.3 Fractionation in Reacting Systems

Integrating isotope fractionation into the reaction path calculation is a matter of applying the mass balance equations while tracing over the course of the reaction path the system's total isotopic composition. Much of the effort in programming an isotope model consists of devising a careful accounting of the mass of each isotope.

The calculation begins with an initial fluid of specified isotopic composition. The model, by mass balance, assigns the initial compositions of the solvent and each aqueous species. The model then sets the composition of each unsegregated mineral to be in equilibrium with the fluid. The modeler specifies the composition of each segregated mineral as well as that of each reactant to be added to the system.

The model traces the reaction path by taking a series of steps along reaction progress, moving forward each step from ξ_1 to ξ_2. Over a step, the system's isotopic composition can change in two ways: reactants can be added or removed, and segregated minerals can dissolve.

Using ^{18}O as an example, we can calculate the composition $\delta^{18}O_T(\xi_2)$ at the end of the step according to the equation

$$\delta^{18}O_T(\xi_2) = \frac{1}{n_O^T(\xi_2)} \left[n_O^T \, \delta^{18}O_T(\xi_1) + \sum_r \Delta n_r \, n_O^r \, \delta^{18}O_r - \sum_{k \neq k'} \Delta n_k^{(-)} \, n_O^k \, \delta^{18}O_k \right]$$

$$(15.13a)$$

Here, $\delta^{18}O_T(\xi_1)$ is the composition at the beginning of the reaction step. Δn_r is the number of moles of reactant r added over the step (its value is negative for a reactant being removed from the system) and n_O^r is the number of oxygen atoms in the reactant's stoichiometry. For each segregated mineral $k \neq k'$ that dissolves over the step, $\Delta n_k^{(-)}$ is the mineral's change in mass (in moles) and n_O^k is its stoichiometric oxygen content. $\delta^{18}O_r$ and $\delta^{18}O_k$ are the isotopic compositions of reactant and segregated mineral. The mole number $n_O^T(\xi_2)$ of oxygen at the end of the step,

$$n_O^T(\xi_2) = n_O^T(\xi_1) + \sum_r \Delta n_r \, n_O^r - \sum_{k \neq k'} \Delta n_k^{(-)} \, n_O^k \qquad (15.14a)$$

is given by summing over the contributions of reactants and dissolving segregated minerals.

The parallel equations for hydrogen, carbon, and sulfur isotopes are

$$\delta^2 H_T(\xi_2) = \frac{1}{n_H^T(\xi_2)} \left[n_H^T \, \delta^2 H_T(\xi_1) + \sum_r \Delta n_r \, n_H^r \, \delta^2 H_r - \sum_{k \neq k'} \Delta n_k^{(-)} \, n_H^k \, \delta^2 H_k \right]$$

$$(15.13b)$$

$$n_H^T(\xi_2) = n_H^T(\xi_1) + \sum_r \Delta n_r \, n_H^r - \sum_{k \neq k'} \Delta n_k^{(-)} \, n_H^k \qquad (15.14b)$$

$$\delta^{13}C_T(\xi_2) = \frac{1}{n_C^T(\xi_2)} \left[n_C^T \, \delta^{13}C_T(\xi_1) + \sum_r \Delta n_r \, n_C^r \, \delta^{13}C_r - \sum_{k \neq k'} \Delta n_k^{(-)} \, n_C^k \, \delta^{13}C_k \right]$$

$$(15.13c)$$

$$n_C^T(\xi_2) = n_C^T(\xi_1) + \sum_r \Delta n_r \, n_C^r - \sum_{k \neq k'} \Delta n_k^{(-)} \, n_C^k \qquad (15.14c)$$

$$\delta^{34}S_T(\xi_2) = \frac{1}{n_S^T(\xi_2)} \left[n_S^T \, \delta^{34}S_T(\xi_1) + \sum_r \Delta n_r \, n_S^r \, \delta^{34}S_r - \sum_{k \neq k'} \Delta n_k^{(-)} \, n_S^k \, \delta^{34}S_k \right]$$

$$(15.13d)$$

$$n_S^T(\xi_2) = n_S^T(\xi_1) + \sum_r \Delta n_r \, n_S^r - \sum_{k \neq k'} \Delta n_k^{(-)} \, n_S^k \qquad (15.14d)$$

Once we have computed the total isotopic compositions, we calculate the compositions of the reference species using the mass balance equations (Eqns. 15.12a–d). We can then use the isotopic compositions of the reference species to calculate the compositions of the other species (Eqns. 15.9a–d) and the unsegregated minerals (Eqns. 15.10a–d).

To update the composition of each segregated mineral, we average the composition (e.g., $\delta^{18}O_{k'}$) of the mass that precipitated over the reaction step and the composition (e.g., $\delta^{18}O_k(\xi_1)$) of whatever mass was present at the onset. The averaging equation for oxygen, for example, is

$$\delta^{18}O_k(\xi_2) = \frac{n_k \, \delta^{18}O_k(\xi_1) + \Delta n_k^{(+)} \, \delta^{18}O_{k'}}{n_k(\xi_1) + \Delta n_k^{(+)}} \qquad (15.15)$$

where n_k is the mole number of mineral k and $\Delta n_k^{(+)}$ is the increase in this value over the reaction step. The compositions of segregated minerals that dissolve over the step are unchanged.

The precise manner in which we apply the mass transfer equations (Eqns. 15.13a–d and 15.14a–d) depends on how we have configured the reaction path. Table 15.2 provides an overview of the process of assigning isotopic compositions to the various constituents of the model. When a simple reactant is added to the system (see Chapter 11), the increment Δn_r added is the reaction rate n_r multiplied by the step length, $\xi_2 - \xi_1$. The modeler explicitly prescribes the reactant's isotopic composition ($\delta^{18}O_r$, and so on). When a simple reactant is removed from the system, the value of Δn_r is negative and the reactant's isotopic composition is the value in equilibrium with the system at the beginning of the

TABLE 15.2 Assigning isotopic compositions in a
reaction model (Lee and Bethke, 1996)

Initial system

Water & aqueous species	In equilibrium with each other, reflecting fluid's bulk composition, as set by modeler.
Unsegregated minerals & gases	In equilibrium with fluid.
Segregated minerals	Set by modeler.

Reactants being added to system

| Species, minerals, & gases | Set by modeler. |
| Buffered gases | In equilibrium with initial system. |

Reactants being removed from system

| Species, minerals, gases, & buffered gases | In equilibrium with system at beginning of current step. |

Chemical system, over course of reaction path

Water, species, & unsegregated minerals	In equilibrium with each other, reflecting system's bulk composition.
Segregated minerals	Increment precipitated over reaction step forms in equilibrium with fluid; composition of preexisting mass is unaffected.
Gases	In equilibrium with current system.

---

step. A mineral reactant that precipitates and dissolves according to a kinetic rate law (Chapter 14) is treated in the same fashion, except that the model must calculate the increment Δn_r (which is positive when the mineral dissolves and negative when it precipitates) from the rate law.

In fixed and sliding fugacity paths, the model transfers gas into and out of an external buffer to obtain the fugacity desired at each step along the path (see Chapter 12). The increment Δn_r is the change in the total mole number M_m of the gas component as it passes to and from the buffer (see Chapter 3). When gas passes from buffer to system (Δn_r is positive), it is probably most logical to take its isotopic composition as the value in equilibrium with the initial system, at the start of the reaction path. Gas passing from system to buffer (i.e., Δn_r is negative) should be in isotopic equilibrium with the system at the start of the current reaction step. For polythermal paths (Chapter 12), it is necessary to update the fractionation factors α at each step, according to Eqn. 15.6, before evaluating the fractionation and mass balance equations (Eqns. 15.9a–d, 15.10a–d, and 15.12a–d).

15.4 Dolomitization of a Limestone

As an example of how we might integrate isotope fractionation into reaction modeling (borrowing from Lee and Bethke, 1996), we consider the dolomitization of a limestone as it reacts after burial with a migrating pore fluid. When dolomite $[CaMg(CO_3)_2]$ forms by alteration of a carbonate mineral, geochemists commonly assume that the dolomite reflects the isotopic composition of carbon in the precursor mineral (e.g., Mattes and Mountjoy, 1980; Meyers and Lohmann, 1985). This assumption seems logical since the reservoir of carbon in the precursor mineral, in this case calcite $(CaCO_3)$, is likely to exceed that available from dissolved carbonate.

Some groundwaters in sedimentary basins, on the other hand, are charged with CO_2. In this case, if dolomite forms by reaction of a limestone with large volumes of migrating groundwater, the reservoir of dissolved carbon might be considerable. As we have noted, furthermore, the dolomitization reaction is best considered to occur in the presence of the carbon provided by dissolving calcite, not the entire reservoir of carbon present in the rock. Hence, it is interesting to use reaction modeling to investigate the factors controlling the isotopic composition of authigenic dolomite.

In our example, we test the consequences of reacting an isotopically light (i.e., nonmarine) limestone at 60°C with an isotopically heavier groundwater that is relatively rich in magnesium. We start by defining the composition of a hypothetical groundwater that is of known CO_2 fugacity (we initially set f_{CO_2} to 1) and in equilibrium with dolomite:

```
T = 60
swap CO2(g) for H+
swap Dolomite-ord for HCO3-
f CO2(g) = 1
1 free mol Dolomite-ord

Na+ = .1 molal
Cl- = .1 molal
Ca++ = .01 molal
Mg++ = .01 molal

go
```

The resulting fluid has an activity ratio $a_{Mg^{++}}/a_{Ca^{++}}$ of 1.26. By the reaction

$$CaMg(CO_3)_2 + Ca^{++} \rightleftarrows 2\,CaCO_3 + Mg^{++} \qquad (15.16)$$

$$\qquad \text{\textit{dolomite}} \qquad\qquad\qquad \text{\textit{calcite}}$$

a fluid in equilibrium with calcite and dolomite has a ratio of 0.08 (as we can quickly show with program RXN). Hence, the groundwater is undersaturated with respect to calcite.

We then "pick up" the fluid from the previous step as a reactant and define a system representing the limestone and its pore fluid. We specify that the rock contain 3000 cm^3 of calcite, implying a porosity of about 25% since the extent of the system is 1 kg (about 1 liter) of fluid. The pore fluid is similar to the reactant fluid, except that it contains less magnesium. The procedure is

```
(cont'd)
pickup reactants = fluid

swap Calcite for HCO3-
3000 free cm3 Calcite

pH = 6
Ca++ = .02   molal
Na+ =   .1   molal
Cl- =   .1   molal
Mg++ = .001 molal

reactants times 10
flush
```

The final commands define a reaction path in which 10 kg of reactant fluid gradually migrate through the system, displacing the existing (reacted) pore fluid. Typing go triggers the calculation.

As expected, the fluid, as it migrates through the limestone, converts calcite into dolomite. For each kg of fluid reacted, about .65 cm^3 (17.5 *m*mol) of calcite is consumed and .56 cm^3 (8.7 *m*mol) of dolomite forms.

To trace isotope fractionation over the reaction, we need to set the composition of the carbon in the initial system and the reactant fluid. We set $\delta^{13}C_{PDB}$ of the initial fluid to $-10\%o$. By equilibrium with this value, the calcite has a composition of $-6.1\%o$, as might be observed in a nonmarine limestone. We then set the composition of dissolved carbonate (HCO_3^- is the carbon-bearing component, as we can verify by typing show) in the reactant fluid to $0\%o$, a value typical for marine carbonate rocks. Finally, we specify that calcite and dolomite be held segregated from isotopic exchange. The procedure for tracing fractionation at low water-rock ratios is

```
(cont'd)
carbon initial = -10, HCO3- = 0
segregate Calcite, Dolomite-ord
go
```

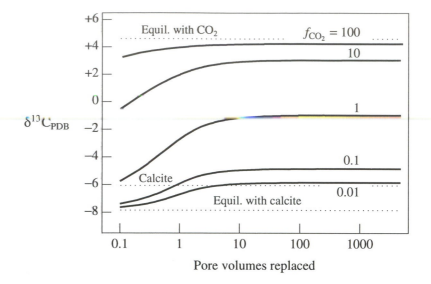

FIG. 15.1 Carbon isotopic composition of dolomite formed by reaction between a limestone and migrating groundwater, calculated by holding minerals segregated from isotope exchange. In the calculation, calcite in the limestone reacts at 60°C with the migrating groundwater, producing dolomite. Solid lines show results calculated assuming differing CO_2 fugacities (from 0.01 to 100) for the migrating fluid, plotted against the number of times the pore fluid was displaced. Dotted lines show compositions of calcite in the limestone, and of dolomite in isotopic equilibrium with the calcite and with CO_2 in the migrating groundwater.

To carry the calculation to high water-rock ratios, we enter the commands

```
(cont'd)
reactants times 4700
go
```

In this case, enough fluid passes through the system to completely transform the limestone to dolomite. We can then repeat the entire procedure (taking care each time to first type reset) for differing CO_2 fugacities.

The calculation results (Fig. 15.1) show that the isotopic composition of the product dolomite depends strongly on the value assumed for CO_2 fugacity. At low fugacity ($f_{CO_2} < 0.1$), the dolomite forms at compositions similar to the $\delta^{13}C$ of the limestone, consistent with common interpretations in sedimentary geochemistry. At moderate to high fugacity ($f_{CO_2} > 1$), however, the dolomite $\delta^{13}C$ more closely reflects the heavier carbon in the migrating pore fluid. Because of the fractionation of carbon between dolomite and CO_2 ($1000 \ln \alpha \approx 4.6$), the isotopic composition of dolomite formed from groundwater rich in CO_2 can be as much as about 4‰ heavier than the groundwater $\delta^{13}C$.

For each CO_2 fugacity chosen, the dolomite composition during the reaction approaches a steady state, as can be seen in Fig. 15.1. The steady state reflects the isotopic composition that balances sources (the unreacted fluid that enters the system and the calcite that dissolves over a reaction step) and sinks (the fluid displaced and dolomite precipitated over the step) of carbon to the isotope system. When the CO_2 fugacity is low, the carbon source is isotopically light (being dominated by the dissolving calcite), and so are the sinks. At high fugacity, conversely, the dominant carbon source is the heavy carbon in the migrating fluid. The pore fluid in the reacting system and the mineral mass that precipitates from it are therefore isotopically heavy.

How would the results differ if we had assumed isotopic equilibrium among minerals instead of holding them segregated from isotopic exchange? To find out, we enter the commands

```
(cont'd)
unsegregate ALL
reactants times 10
go
```

Then, to carry the run to high water-rock ratios, we type

```
(cont'd)
reactants times 4700
go
```

In the unsegregated case (Fig. 15.2), a family of curves, one for each CO_2 fugacity chosen, represents dolomite composition. At low water-rock ratios, regardless of CO_2 fugacity, the product dolomite forms at the composition in isotopic equilibrium with the original calcite. Because of the fractionation between the minerals, the dolomite is about 1.8‰ lighter than the calcite. As the reaction proceeds, the compositions of each component of the system (fluid, calcite, and dolomite) become isotopically heavier, reflecting the introduction of heavy carbon by the migrating fluid. Transition from light to heavy compositions occurs after a small number of pore volumes have been displaced by a fluid of high CO_2 fugacity, or a large quantity of fluid of low CO_2 fugacity.

These predictions differ qualitatively from the results of the segregated model (Fig. 15.1), as can be seen by comparing the bold and fine lines in Fig. 15.2. In contrast to the segregated model, the initial and final $\delta^{13}C$ values of the dolomite in the unsegregated model are independent of f_{CO_2}, and the final value does not depend on the original composition of the calcite. Dolomite compositions predicted by the two models coincide only when CO_2 fugacity is quite high ($f_{CO_2} \approx 100$), and then only at high water-rock ratios Under other conditions, the predictions diverge. Since the unsegregated model depends on the unrealistic

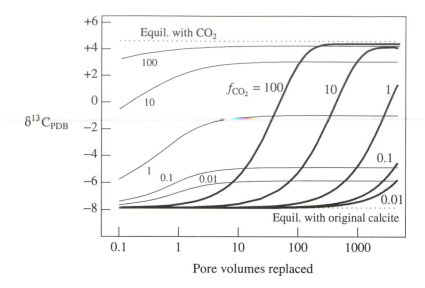

FIG. 15.2 Isotopic composition (bold lines) of dolomite formed by reaction between a limestone and migrating groundwater, assuming that minerals maintain isotopic equilibrium over the simulation. Fine lines show results of simulation holding minerals segregated from isotopic exchange, as already presented (Fig. 15.1).

premise of maintaining isotopic equilibrium at low temperature, these results argue strongly for the importance of holding minerals segregated from isotope exchange when modeling many types of reaction processes.

16

Hydrothermal Fluids

Hydrothermal fluids, hot groundwaters that circulate within the Earth's crust, play central roles in many geological processes, including the genesis of a broad variety of ore deposits, the chemical alteration of rocks and sediments, and the origin of hot springs and geothermal fields. Many studies have been devoted to modeling how hydrothermal fluids react chemically as they encounter wall rocks, cool, boil, and mix with other fluids. Such modeling proliferated in part because hydrothermal fluids are highly reactive and because the reaction products are commonly well preserved, readily studied, and likely to be of economic interest. Further impetus was provided by the development of reliable modeling software in the 1970s, a period of concern over the availability of strategic and critical minerals and of heightened interest in economic geology and the exploitation of geothermal energy.

As a result, many of the earliest and most imaginative applications of geochemical modeling, beginning with Helgeson's (1970) simulation of ore deposition in hydrothermal veins and the alteration of nearby country rock, have addressed the reaction of hydrothermal fluids. For example, Reed (1977) considered the origin of a precious metal district; Garven and Freeze (1984), Sverjensky (1984, 1987), and Anderson and Garven (1987) studied the role of sedimentary brines in forming Mississippi Valley-type and other ore deposits; Wolery (1978), Janecky and Seyfried (1984), Bowers et al. (1985), and Janecky and Shanks (1988) simulated hydrothermal interactions along the midocean ridges; and Drummond and Ohmoto (1985) and Spycher and Reed (1988) modeled how fluid boiling is related to ore deposition.

In this chapter, we develop geochemical models of two hydrothermal processes: the formation of fluorite veins in the Albigeois ore district and the

TABLE 16.1 Composition of ore-forming fluids
at the Albigeois district, as determined by
analysis of fluid inclusions (Deloule, 1982)

Component	Concentration (molar)
Na^+	3.19
K^+	.45
Ca^{++}	.45
Mg^{++}	$(0.2-1) \times 10^{-2}$
Fe^{++}	$(0.75-3) \times 10^{-2}$
Cu^+	$(0.2-3) \times 10^{-2}$
Cl^-	4.5
HCO_3^-	.2
SO_4^{--}	0.004–0.2

origin of "black smokers," a name given to hydrothermal vents found along the ocean floor at midocean ridges.

16.1 Origin of a Fluorite Deposit

As a first case study, we borrow from the modeling work of Rowan (1991), who considered the origin of fluorite (CaF_2) veins in the Albigeois district of the southwest Massif Central, France. Production and reserves for the district as a whole total about 7 million metric tons, making it comparable to the more famous deposits of southern Illinois and western Kentucky, USA.

Like other fluorite deposits, the Albigeois ores are notable for their high grade. In veins of the Le Burc deposit, for example, fluorite comprises 90% of the ore volume (Deloule, 1982). Accessory minerals include quartz (SiO_2), siderite ($FeCO_3$), chalcopyrite ($CuFeS_2$), and small amounts of arsenopyrite ($AsFeS$). The deposits occur in a tectonically complex terrain dominated by metamorphic, plutonic, and volcanic rocks and sediments.

Deloule (1982) studied fluid inclusions from the Montroc and Le Burc deposits. He found that the ore-forming fluid was highly saline at both deposits; Table 16.1 summarizes its chemical composition. Homogenization temperatures in fluorite from the Le Burc veins range from about 110°C to 150°C, and most of the measurements fall between 120°C to 145°C. Assuming burial at the time of ore deposition to depths as great as 1 km, these values should be corrected upward for the effect of pressure by perhaps 10°C and no more than 30°C. We will assume here that the ore was deposited at temperatures between 125°C and 175°C.

To model the process of ore formation in the district, we first consider the effects of simply cooling the ore fluid. We begin by developing a model of the fluid at 175°C, the point at which we assume that it begins to precipitate ore. To constrain the model, we use the data in Table 16.1 and assume that the fluid was in equilibrium with minerals in and near the vein. Specifically, we assume that the fluid's silica and aluminum contents were controlled by equilibrium with the quartz (SiO_2) and muscovite [$KAl_3Si_3O_{10}(OH)_2$] in the wall rocks, and that equilibrium with fluorite set the fluid's fluorine concentration.

We further specify equilibrium with kaolinite [$Al_2Si_2O_5(OH)_4$], which occurs in at least some of the veins as well as in the altered wall rock. Since we know the fluid's potassium content (Table 16.1), assuming equilibrium with kaolinite fixes pH according to the reaction

$$\frac{3}{2} Al_2Si_2O_5(OH)_4 + K^+ \rightleftarrows KAl_3Si_3O_{10}(OH)_2 + \frac{3}{2} H_2O + H^+ \quad (16.1)$$

$$\underset{kaolinite}{} \qquad\qquad\qquad \underset{muscovite}{}$$

By this reaction, we can expect the modeled fluid to be rather acidic, since it is rich in potassium. We could have chosen to fix pH by equilibrium with the siderite, which also occurs in the veins. It is not clear, however, that the siderite was deposited during the same paragenetic stages as the fluorite. It is difficult on chemical grounds, furthermore, to reconcile coexistence of the calcium-rich ore fluid and siderite with the absence of calcite ($CaCO_3$) in the district. In any event, assuming equilibrium with kaolinite leads to a fluid rich in fluorine and, hence, to an attractive mechanism for forming fluorite ore.

The model calculated in this manner predicts that two minerals, alunite [$KAl_3(OH)_6(SO_4)_2$] and anhydrite ($CaSO_4$), are supersaturated in the fluid at 175°C, although neither mineral is observed in the district. This result is not surprising, given that the fluid's salinity exceeds the correlation limit for the activity coefficient model (Chapter 7). The observed composition in this case (Table 16.1), furthermore, actually represents the average of fluids from many inclusions and hence a mixture of hydrothermal fluids present over a range of time. As noted in Chapter 6, mixtures of fluids tend to be supersaturated, even if the individual fluids are not.

To avoid starting with a supersaturated initial fluid, we use alunite to constrain the SO_4 content to the limiting case, and anhydrite to similarly set the calcium concentration. The resulting SO_4 concentration will fall near the lower end of the range shown in Table 16.1, and the calcium content will lie slightly above the reported value.

In the program REACT, the procedure to model the initial fluid is

```
T = 175

swap Quartz    for SiO2(aq)
swap Muscovite for Al+++
```

```
swap Fluorite   for F-
swap Alunite    for SO4--
swap Kaolinite for H+
swap Anhydrite for Ca++

density = 1.14
TDS = 247500
Na+   =  3.19  molar
K+    =   .45  molar
Mg++  =   .006 molar
Fe++  =  2.e-2 molar
Cu+   =  1.e-2 molar
HCO3- =  0.2   molar
Cl-   =  4.5   molar

1 free mole Quartz
1 free mole Muscovite
1 free mole Fluorite
1 free mole Alunite
1 free mole Kaolinite
1 free mole Anhydrite
go
```

Since the input constraints are in molar (instead of molal) units, we have specified the dissolved solid content and the fluid density under laboratory conditions, the latter estimated from the correlation of Phillips et al. (1981) for NaCl solutions. The resulting fluid is, as expected, acidic, with a predicted pH of 2.9. Neutral pH at 175°C, for reference, is 5.7.

The fluid is extraordinarily rich in fluorine, as shown in Table 16.2, primarily because of the formation of the complex species AlF_3 and AlF_2^+. The importance of these species results directly from the assumed mineral assemblage

$$AlF_3 + \frac{3}{4} \underset{muscovite}{KAl_3Si_3O_{10}(OH)_2} + \frac{3}{2} \underset{anhydrite}{CaSO_4} + \frac{5}{2} H_2O \rightleftarrows$$

$$\frac{3}{2} \underset{fluorite}{CaF_2} + \frac{5}{4} \underset{quartz}{SiO_2} + \frac{3}{4} \underset{alunite}{KAl_3(OH)_6(SO_4)_2} + \frac{1}{2} \underset{kaolinite}{Al_2Si_2O_5(OH)_4} \quad (16.2)$$

and, in the case of AlF_2^+, the activity of K^+

$$AlF_2^+ + \frac{3}{2} \underset{muscovite}{KAl_3Si_3O_{10}(OH)_2} + \underset{anhydrite}{CaSO_4} + 4\,H_2O \rightleftarrows$$

$$\underset{fluorite}{CaF_2} + \frac{1}{2} \underset{quartz}{SiO_2} + \frac{1}{2} \underset{alunite}{KAl_3(OH)_6(SO_4)_2} + 2\,\underset{kaolinite}{Al_2Si_2O_5(OH)_4} + K^+ \quad (16.3)$$

TABLE 16.2 Calculated concentrations
of fluorine-bearing species in the
Albigeois ore fluid at 175°C

Species	Concentration (*m*molal)
AlF_3	227.5
AlF_2^+	38.8
AlF^{++}	5.54
HF	1.16
CaF^+	.44
AlF_4^-	.0443
F^-	.0275
total F	767.

(as can be verified quickly with the program RXN). These complex species make the solution a potent ore-forming fluid. The fluid's acidity and fluorine content might reflect the addition of HF(g) derived from a magmatic source, as Plumlee et al. (1995) suggested for the Illinois and Kentucky deposits.

To model the consequences of cooling the fluid, we enter the commands

```
(con'd)
pickup fluid
T final = 125
```

and type go to trigger the calculation. REACT carries the modeled fluid over a polythermal path, incrementally cooling it from 175°C to 125°C.

Figure 16.1 shows the mineralogic results of tracing the reaction path. As the fluid cools by 50°C, it produces about 1.8 cm^3 of fluorite and .02 cm^3 of quartz. No other minerals form. From a plot of the concentration of fluorine-bearing species (Fig. 16.2), it is clear that the fluorite forms in response to progressive breakdown of the AlF_3 complex with decreasing temperature. The complex sheds fluorine to produce AlF_2^+ according to the reaction

$$AlF_3 + \frac{1}{2} CaCl^+ \rightarrow AlF_2^+ + \underset{fluorite}{\frac{1}{2} CaF_2} + \frac{1}{2} Cl^- \qquad (16.4)$$

yielding fluorite.

As an alternative to simple cooling, we can test the consequences of quenching the ore fluid by incrementally mixing cooler water into it. The commands

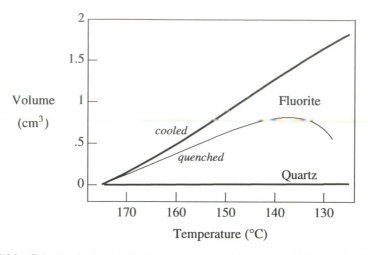

FIG. 16.1 Calculated mineralogical consequences of cooling (bold lines) the Albigeois ore fluid from 175°C to 125°C, and of quenching it (fine line) with 125°C water.

```
(cont'd)
react 20 kg H2O
T 175, reactants = 125
```

set up a polythermal path in which temperature is set by the mixing proportions of the fluids, assuming each has a constant heat capacity. Typing go triggers the calculation.

As shown in Fig. 16.1, quenching is effective at producing fluorite. The process, however, is somewhat less efficient than simple cooling because of the counter-effect of dilution, which limits precipitation and eventually begins to cause the fluorite that formed early in the reaction path to redissolve. Each mechanism of fluorite deposition is quite efficient, nonetheless, producing — over a temperature drop of just 50°C — about a cm^3 or more mineral per kg of water in the ore fluid.

There are considerable uncertainties in the calculation. The fluid has an ionic strength of about 5 molal, far in excess of the range considered in the correlations for estimating species' activity coefficients, as discussed in Chapter 7. Stability constants for aluminum-bearing species are, in general, rather difficult to determine accurately (e.g., May, 1992). And the model relies on the equilibria assumed between the fluid and minerals in the vein and country rock. Nonetheless, the modeling suggests an attractive explanation for how veins of almost pure fluorite might form and provides an excellent example of the importance of complex species in controlling mineral solubility in hydrothermal fluids.

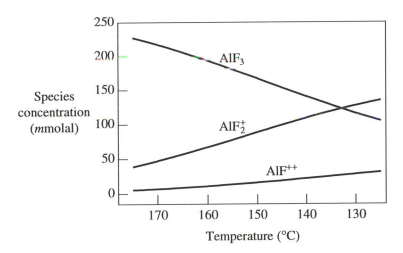

FIG. 16.2 Calculated concentrations of predominant fluorine-bearing species in the Albigeois ore fluid as it cools from 175°C to 125°C.

16.2 Black Smokers

In the spring of 1977, researchers on the submarine ALVIN discovered hot springs on the seafloor of the Pacific Ocean, along the Galapagos spreading center. Later expeditions to the East Pacific Rise and Juan de Fuca spreading center found more springs, some discharging fluids as hot as 350°C. The hot springs are parts of large scale hydrothermal systems in which seawater descends into the oceanic crust, circulates near magma bodies where it warms and reacts extensively with deep rocks, and then, under the force of its buoyancy, discharges back into the ocean.

Discovery of the hot springs has had an important impact in the geosciences. Geologists today recognize the importance of hydrothermal systems in controlling the thermal structure of the ocean crust and the composition of ocean waters, as well as their role in producing ore deposits. The expeditions, in fact, discovered a massive sulfide deposit along the Galapagos spreading center. The springs also created excitement in the biologic sciences because of the large number of previously unknown species, such as tube worms, discovered near the vents.

Where fluids discharge from hot springs and mix with seawater, they cool quickly and precipitate clouds of fine-grained minerals. The clouds are commonly black with metal sulfides, giving rise to the term "black smokers." Some vents give off clouds of white anhydrite; these are known as "white smokers." Structures composed of chemical precipitates tend to form at the vents, where the hot fluids discharge into the ocean. The structures can extend

TABLE 16.3 Endmember compositions of a hydrothermal fluid and seawater, East Pacific Rise near 21°N (Von Damm et al., 1985; Drever, 1988; Mottl and McConachy, 1990)

	NGS field	Seawater
	Concentration ($m$molal)	
Na^+	529	480
K^+	26.7	10.1
Mg^{++}	~ 0	54.5
Ca^{++}	21.6	10.5
Cl^-	600	559
HCO_3^-	2*	2.4
SO_4^{--}	~ 0	29.5
H_2S	6.81	~ 0
SiO_2	20.2	.17
	Concentration (μmolal)	
Sr^{++}	100.5	90.
Ba^{++}	>15	.20
Al^{+++}	4.1	.005
Mn^{++}	1039	<0.001
Fe^{++}	903	<0.001
Cu^+	<0.02	.007
Zn^{++}	41	.01
O_2(aq)	~ 0	123
T (°C)	273	4
pH	3.8 (25°C)	7.8 (25°C)
	4.2 (273°C)	8.1 (4°C)

*Estimated from titration alkalinity.

upward into the ocean for several meters or more, and are composed largely of anhydrite and, in some cases, sulfide minerals.

The chemical processes occurring within a black smoker are certain to be complex because the hot, reducing hydrothermal fluid mixes quickly with cool, oxidizing seawater, allowing the mixture little chance to approach equilibrium. Despite this obstacle, or perhaps because of it, we bravely attempt to construct a chemical model of the mixing process. Table 16.3 shows chemical analyses of fluid from the NGS hot spring, a black smoker along the East Pacific Rise near 21°N, as well as ambient seawater from the area.

To model the mixing of the hydrothermal fluid with seawater, we begin by equilibrating seawater at 4°C, "picking up" this fluid as a reactant, and then reacting it into the hot hydrothermal fluid. In REACT, we start by suppressing several minerals:

```
suppress Quartz, Tridymite, Cristobalite, Chalcedony
suppress Hematite
```

According to Mottl and McConachy (1990), amorphous silica (SiO_2) is the only silica polymorph present in the "smoke" at the site. To allow it to form in the calculation, we suppress each of the more stable silica polymorphs. We also suppress hematite (Fe_2O_3) in order to give the iron oxy-hydroxide goethite (FeOOH) a chance to form.

We then equilibrate seawater, using data from Table 16.3,

```
(cont'd)
T = 4
pH = 8.1

Cl-        559.       mmolal
Na+        480.       mmolal
Mg++        54.5      mmolal
SO4--       29.5      mmolal
Ca++        10.5      mmolal
K+          10.1      mmolal
HCO3-        2.4      mmolal
SiO2(aq)      .17     mmolal
Sr++          .09     mmolal

Ba++          .20     umolal
Zn++          .01     umolal
Al+++         .005    umolal
Cu+           .007    umolal
Fe++          .001    umolal
Mn++          .001    umolal
O2(aq)     123.       free umolal

go

pickup reactant = fluid
reactants times 10
```

and pick it up as a reactant. By multiplying the extent of the reactant system tenfold, we prescribe mixing of seawater into the hydrothermal fluid in ratios as great as 10:1.

Finally, we define a polythermal path by equilibrating the hot hydrothermal fluid (Table 16.3)

```
(cont'd)
pH = 4.2
swap H2S(aq) for O2(aq)

Cl-        600.      mmolal
Na+        529.      mmolal
K+          26.7     mmolal
Ca++        21.6     mmolal
SiO2(aq)    20.2     mmolal
H2S(aq)      6.81    mmolal
HCO3-        2.      mmolal

Mn++      1039.      umolal
Fe++       903.      umolal
Sr++       100.5     umolal
Zn++        41.      umolal
Ba++        15.      umolal
Al+++        4.1     umolal
Cu+           .02    umolal
Mg++          .01    umolal
SO4--         .01    umolal

dump
T initial 273, reactants = 4
```

and reacting into it the cold seawater. The dump command causes the program to remove any minerals present in the initial system, before beginning to trace the reaction path. The final command sets a polythermal path in which temperature depends on the proportion of hot and cold fluids in the mixture, assuming that each has a constant heat capacity. Typing go triggers the calculation.

The calculation results (Fig. 16.3) show that anhydrite is the most abundant mineral to form. The mineral forms rapidly during the initial mixing

$$Ca^{++} + SO_4^{--} \rightarrow \underset{anhydrite}{CaSO_4} \tag{16.5}$$

from reaction of the calcium in the hydrothermal fluid with sulfate in the seawater, eventually forming about a half cm^3 per kg of hydrothermal fluid. At a mixing ratio somewhat less than 1:1, however, anhydrite begins to redissolve, reflecting dilution of the hydrothermal calcium as mixing proceeds. The mineral has completely dissolved by a mixing ratio of about 1.5:1. These results seem in accord with the observation that anhydrite forms in abundance near the vent sites, where the hydrothermal fluid is least diluted.

After anhydrite, the most voluminous minerals to form are amorphous silica, talc [$Mg_3Si_4O_{10}(OH)_2$], and pyrite (FeS_2). Amorphous silica forms because its solubility decreases with cooling faster than it is diluted. Other minerals that

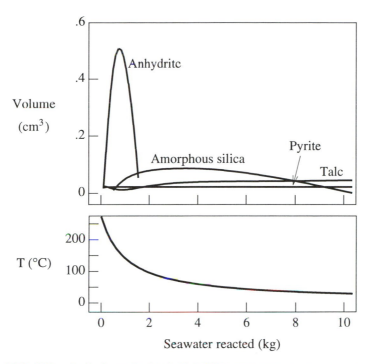

FIG. 16.3 Mineralogical results (top) of mixing cold seawater into one kg of hot hydrothermal fluid from the NGS field. Minerals present in volumes less than 0.01 cm³ are not shown. These minerals, in order of decreasing abundance, are: sphalerite, barite, potassium mordenite, calcium clinoptilolite, and covellite. Also shown (bottom) is temperature of the fluid during mixing.

form in small quantities over the simulation include sphalerite (ZnS), barite $(BaSO_4)$, the zeolites mordenite $(KAlSi_5O_{12} \cdot 3H_2O)$ and clinoptilolite $(CaAl_2Si_{10}O_{24} \cdot 8H_2O)$, and covellite (CuS). According to the data of Mottl and McConachy (1990), each of these minerals (lumping the talc and zeolites into the category of unidentified aluminosilicates) is observed suspended above the NGS vent. Even though the smoke at this site is described as black, sulfide minerals make up just a small fraction of the mineral volume precipitated over the course of the simulation.

A number of the observed minerals (formulae given in Table 16.4) do not form in the simulation. Wurtzite is metastable with respect to sphalerite, so it cannot be expected to appear in the calculation results. Similarly, the formation of pyrite in the simulation probably precludes the possibility of pyrrhotite precipitating. In the laboratory, and presumably in nature, pyrite forms slowly, allowing less stable iron sulfides to precipitate. Elemental sulfur at the site probably results from incomplete oxidation of $H_2S(aq)$, a process not accounted

TABLE 16.4 Minerals in samples taken above
black smokers of the East Pacific Rise near
21°N (Mottl and McConachy, 1990)

"Smoke" in plume

Pyrrhotite	$Fe_{1-x}S$
Sphalerite	ZnS
Pyrite	FeS_2
Unidentified Fe-S-Si phases	
Chalcopyrite	$CuFeS_2$
Amorphous silica	SiO_2
Elemental sulfur	S
Goethite, etc.	$FeOOH$
Anhydrite	$CaSO_4$
Barite (trace)	$BaSO_4$
Cubanite (trace)	$CuFe_2S_3$
Wurtzite (trace)	ZnS
Covellite (trace)	CuS
Marcasite (trace)	FeS_2
Unidentified silicates,	
aluminosilicates (traces)	

Particles dispersed in local seawater

Anhydrite	$CaSO_4$
Pyrite	FeS_2
Gypsum	$CaSO_4 \cdot 2H_2O$
Chalcopyrite	$CuFeS_2$
Sphalerite	ZnS
Sulfur	S
Pyrrhotite	$Fe_{1-x}S$

---

for in the simulation. There is no data in the LLNL database for marcasite or cubanite. Finally, goethite forms after we run the simulation to higher ratios of seawater to hydrothermal fluid than shown in Fig. 16.3.

An interesting aspect of the calculation is that when oxidizing seawater mixes into the reduced hydrothermal fluid, the oxygen fugacity decreases (Fig. 16.4). The capacity of seawater to oxidize the large amount of hydrothermal $H_2S(aq)$ is limited by the supply of $O_2(aq)$ in seawater, which is small. Given the reaction

$$H_2S(aq) + 2\,O_2(aq) \rightarrow SO_4^{--} + 2\,H^+ \qquad (16.6)$$

and the data in Table 16.3, more than 100 kg of seawater are needed to oxidize each kg of hydrothermal fluid. The decrease in f_{O_2} shown in Fig. 16.4 results

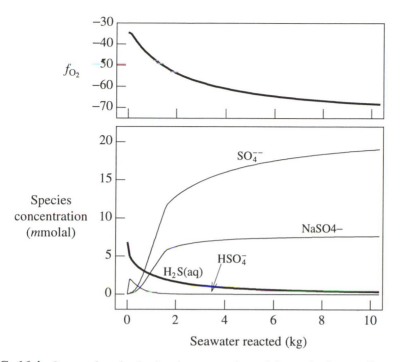

FIG. 16.4 Oxygen fugacity (top) and concentrations of the predominant sulfur species (bottom) during the mixing simulation shown in Fig. 16.3. Decrease in the $H_2S(aq)$ concentration is mostly in response to dilution with seawater, rather than oxidation.

from the shift of the sulfide-sulfate buffer with temperature. As long as any significant amount of $H_2S(aq)$ remains, the oxygen fugacity of the mixed fluid tracks the redox buffer as it moves to lower oxygen fugacity during cooling. The ability of the hydrothermal fluid to remain reducing as it mixes with seawater helps explain Mottl and McConachy's (1990) observation that sulfide minerals sampled from the "smoke" show no evidence under the electron microscope of beginning to redissolve.

17

Geothermometry

Geothermometry is the use of a fluid's (or, although not discussed here, a rock's) chemical composition to estimate the temperature at which it equilibrated in the subsurface. The specialty is important, for example, in exploring for and exploiting geothermal fields, characterizing deep groundwater flow systems, and understanding the genesis of ore deposits.

Several chemical geothermometers are in widespread use. The silica geothermometer (Fournier and Rowe, 1966) works because the solubilities of the various silica minerals (e.g., quartz and chalcedony, SiO_2) increase monotonically with temperature. The concentration of dissolved silica, therefore, defines a unique equilibrium temperature for each silica mineral. The Na-K (White, 1970) and Na-K-Ca (Fournier and Truesdell, 1973) geothermometers take advantage of the fact that the equilibrium points of cation exchange reactions among various minerals (principally, the feldspars) vary with temperature.

In applying these methods, it is necessary to make a number of assumptions or corrections (e.g., Fournier, 1977). First, the minerals with which the fluid reacted must be known. Applying the silica geothermometer assuming equilibrium with quartz, for example, would not give the correct result if the fluid's silica content is controlled by reaction with chalcedony. Second, the fluid must have attained equilibrium with these minerals. Many studies have suggested that equilibrium is commonly approached in geothermal systems, especially for ancient waters at high temperature, but this may not be the case in young sedimentary basins like the Gulf of Mexico basin (Land and Macpherson, 1992). Third, the fluid's composition must not have been altered by separation of a gas phase, mineral precipitation, or mixing with other fluids. Finally, corrections may be needed to account for the influence of certain dissolved components, including CO_2 and Mg^{++}, which affect the equilibrium composition (Paces, 1975; Fournier and Potter, 1979; Giggenbach, 1988).

Using geochemical modeling, we can apply chemical geothermometry in a more generalized manner. By utilizing the entire chemical analysis rather than just a portion of it, we avoid some of the restricting assumptions mentioned in the preceding paragraph (see Michard et al., 1981; Michard and Roekens, 1983; and especially Reed and Spycher, 1984). Having constructed a theoretical model of the fluid in question, we can calculate the saturation state of each mineral in the database, noting the temperature at which each is in equilibrium with the fluid. Hence, we need make no *a priori* assumption about which minerals control the fluid's composition in the subsurface.

Given sufficient field data, we can use modeling techniques to restore flashed gases or precipitated minerals to the fluid, minimizing another potential source of error. In the final section of this chapter, for example, we use production data from wet-steam wells to reconstitute geothermal fluids as they existed before gas separation. Finally, since the geochemical model is a relatively complete description of the fluid's chemistry, we avoid the necessity of the various corrections that might have to be applied had we used a simpler calculation method. A disadvantage of applying geochemical modeling, however, is that the technique requires a reasonably complete and accurate chemical analysis of the fluid in question, which is not always available.

In this chapter, we explore how we can use chemical analyses and *p*H determinations made at room temperature to deduce details about the origins of natural fluids. These same techniques are useful in interpreting laboratory experiments performed at high temperature, since analyses made at room temperature need to be projected to give *p*H, oxidation state, gas fugacity, saturation indices, and so on under experimental conditions.

17.1 Principles of Geothermometry

The most direct way to demonstrate the principles of geothermometry is to construct a synthetic example on the computer. We start by "sampling" a hypothetical geothermal water at 250°C and letting it cool to room temperature as a closed system. We assume a water that is initially in equilibrium with albite ($NaAlSi_3O_8$), muscovite [$KAl_3Si_3O_{10}(OH)_2$], quartz (SiO_2), potassium feldspar ($KAlSi_3O_8$; "maximum microcline" in the LLNL database), and calcite ($CaCO_3$).

In REACT, the commands

```
T = 250
swap Albite for Na+
swap "Maximum Microcline" for K+
swap Muscovite for Al+++
swap Quartz for SiO2(aq)
swap Calcite for HCO3-
```

```
pH = 5
Ca++ =  .05 molal
Cl-  = 3.   molal
1 free cm3 Albite
1 free cm3 "Maximum Microcline"
1 free cm3 Muscovite
1 free cm3 Quartz
1 free cm3 Calcite
```

define the initial system. To describe sampling and cooling of the fluid, we enter the commands

```
(cont'd)
dump
precip = off
T final = 25
go
```

which set up a polythermal reaction path.

With the dump command, we cause the program to discard the minerals present in the initial system before beginning the reaction path. In this way, we simulate the separation of the fluid from reservoir minerals as it flows into the wellbore. The precip = off command prevents the program from allowing minerals to precipitate as the fluid cools. In practice, samples are acidified immediately after they have been sampled and their pH determined. Preservation by this procedure helps to prevent solutes from precipitating, which would alter the fluid's composition before it is analyzed.

Since we have provided initial and final temperatures but have not specified any reactants, the program traces a polythermal path for a closed system (see Chapter 12). The fluid's pH (Fig. 17.1) changes with temperature from its initial value of 5 at 250°C to less than 4 at 25°C. The change is entirely due to variation in the stabilities of the aqueous species in solution. As shown in Fig. 17.2, the H^+ concentration increases in response to the dissociation of the HCl ion pair

$$HCl \rightarrow H^+ + Cl^- \tag{17.1}$$

and the breakdown of $CO_2(aq)$

$$CO_2(aq) + H_2O \rightarrow H^+ + HCO_3^- \tag{17.2}$$

to form HCO_3^-.

The CO_2 fugacity decreases sharply during cooling (Fig. 17.3), as would be expected, since gas solubility increases as temperature decreases. In the calculation, the fugacity decrease results almost entirely from variation in the equilibrium constant for the reaction

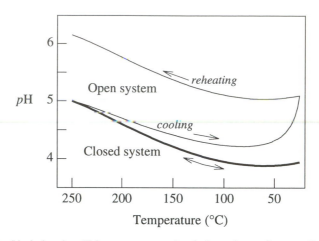

FIG. 17.1 Variation in pH in a computer simulation of sampling, cooling, and then reheating a hypothetical geothermal fluid. Bold line shows path followed when system is held closed; fine lines show variations in pH when fluid is allowed to degas CO_2 as it cools.

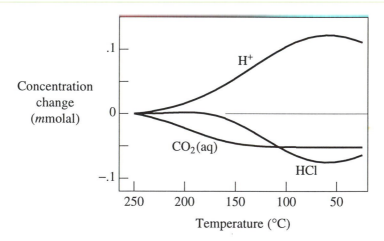

FIG. 17.2 Changes in concentration of aqueous species H^+, $CO_2(aq)$, and HCl with temperature during cooling of a geothermal fluid as a closed system. A positive value indicates an increase in concentration relative to 250°C; a negative value represents a decrease.

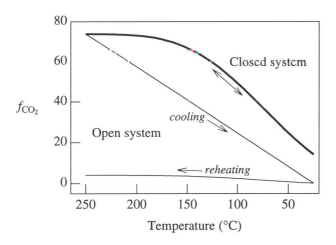

FIG. 17.3 Variation in CO_2 fugacity in a computer simulation of sampling, cooling, and then reheating a hypothetical geothermal fluid. Bold line shows path followed when system is held closed. Fine lines show effects of an open system in which fluid is allowed to degas CO_2 as it cools.

$$CO_2(g) \rightleftarrows CO_2(aq) \qquad (17.3)$$

The log K for this reaction increases from -2.12 at 250°C to -1.45 at 25°C. The final CO_2 fugacity is about 15, corresponding to a partial pressure considerably in excess of atmospheric pressure. We would certainly need to take extraordinary measures to prevent the fluid from effervescing, if we were actually performing this experiment instead of simulating it.

Now that we have simulated sampling the fluid and letting it cool, let us predict the fluid's original temperature (which we already know to be 250°C). The REACT commands

```
(cont'd)
pickup
T final = 300
go
```

cause the program to "pick up" the results at the end of the previous path as the starting point for the current calculation. In other words, the cooled fluid will constitute the new initial system. We specify that the program heat the fluid to 300°C and trigger the calculation.

The values predicted for pH and CO_2 fugacity retrace the paths followed during the cooling calculation (Figs. 17.1 and 17.3). Since the system is closed to mass transfer, its equilibrium state depends only on temperature. Figure 17.4 shows the saturation indices calculated for various minerals. There are two

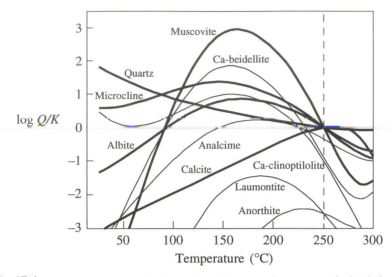

FIG. 17.4 Mineral saturation indices ($\log Q/K$) over the course of simulating the reheating of a hypothetical geothermal fluid. Bold lines show indices for minerals assumed to be present in the initial formation; fine lines show values for other minerals. Dashed line marks sampling temperature (250°C).

salient points to consider in this plot. First, each of the minerals (albite, muscovite, quartz, potassium feldspar, and calcite) present in the formation when the fluid was sampled is in equilibrium (i.e., $\log Q/K = 0$) at 250°C. The minerals are in equilibrium together at no other temperature. Second, minerals in the database that were not present in the formation appear undersaturated at the equilibrium temperature. These two criteria allow us to uniquely identify the fluid's original temperature and hence form the basis of our generalized chemical geothermometer.

As a second experiment, let us simulate the sampling of the same fluid as an open system. This time, we allow it to effervesce CO_2 as we bring it to the surface and let it cool. We start as before, but include a `slide` command to vary CO_2 fugacity from its initial value to one, corresponding to the fugacity of this gas in the atmosphere. The procedure (starting anew in REACT) is

(Enter initial system as before, at the beginning of this section)

```
dump
precip = off
T final = 25
slide f CO2(g) to 1
go
```

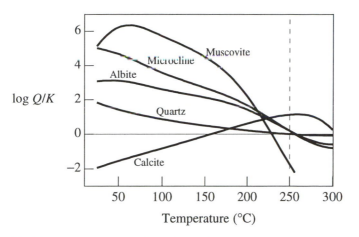

FIG. 17.5 Mineral saturation indices (log Q/K) over the course of simulating the reheating of a hypothetical geothermal fluid that degassed CO_2 during sampling. Dashed line marks original formation temperature (250°C). Although each of the minerals shown was present in the formation, the saturation profiles do not clearly identify the formation temperature because of the CO_2 loss.

In the open system, the CO_2 fugacity (Fig. 17.3) varies linearly over the reaction path from about 75 to a value of one, as we prescribed, tracing a path of lower fugacity than predicted for the closed system. To maintain the lower fugacity, the program allows CO_2 to escape from the fluid into the external gas buffer (see Chapter 12). The pH (Fig. 17.1) follows a path of higher values than in closed system, reflecting the loss of CO_2, an acid gas.

Now we apply our geothermometer by simulating the reheating of the fluid:

```
(cont'd)
pickup
T final = 300
go
```

The system was not closed during cooling, so neither the pH nor CO_2 fugacity (Figs. 17.2 and 17.3) returns to its original value at 250°C.

The predicted saturation states of the formation minerals (Fig. 17.5), furthermore, no longer identify a unique formation temperature. Whereas the temperatures suggested by albite, quartz, and potassium feldspar are quite close to the 250°C formation temperature, those predicted by assuming that the fluid was in equilibrium with muscovite and calcite are too low, respectively, by margins of about 25°C and 100°C. To avoid error of this sort, we would need to determine the amount of gas lost from the sample and reintroduce it to the equilibrium system before calculating saturation indices.

TABLE 17.1 Chemical composition of water
emanating from a hot sping at Gjögur,
Hveravik, Iceland (Arnorsson et al., 1983)

SiO_2 (mg/kg)	49
Na^+	715.5
K^+	17.3
Ca^{++}	759.4
Mg^{++}	3.68
Fe^{++}	0.018
Al^{+++}	0.01
$CO_2(aq)$	13.3
SO_4^-	297.6
$H_2S(aq)$	<0.01
Cl^-	2460
F^-	0.82
Dissolved solids	4366
Sampling temperature (°C)	72
pH at 11°C	7.10

17.2 Hot Spring at Hveravik, Iceland

To see how we might apply geochemical modeling to the geothermometry of natural waters, we consider the effluent of a hot spring at Gjögur, Hveravik, Iceland. The hot spring is part of the surface expression of Iceland's well-known geothermal resources, which are developed within basaltic rocks in or near active volcanic belts. Arnorsson et al. (1983), part of a group noted for its high quality analyses of geothermal waters, provide the water's chemical composition (Table 17.1).

The spring yields a Na-Ca-Cl water at about 72°C with a pH (measured at 11°C) of 7.1. The water appears to be oxidized, since there is abundant sulfate but no detected dissolved sulfide. Details of the origin of the spring water are unknown. Does it circulate deeply, reaching high temperatures only to cool near the discharge point? Is it a hot saline water from depth that has mixed with local groundwater near the spring (and, hence, is likely to be out of equilibrium with most minerals)? Did it degas significantly as it discharged? Or is it a relatively shallow groundwater that has reached equilibrium with its host rock near its discharge temperature?

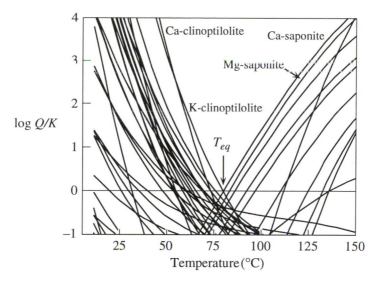

FIG. 17.6 Calculated saturation indices (log Q/K) of aluminum-bearing minerals plotted versus temperature for a hot spring water from Gjögur, Hveravik, Iceland. Lines for most of the minerals are not labeled, due to space limitations. Sampling temperature is 72°C and predicted equilibrium temperature (arrow) is about 80°C. Clinoptilolite (zeolite) minerals are the most supersaturated minerals below this temperature and saponite (smectite clay) minerals are the most supersaturated above it.

In REACT, the commands

```
swap CO2(aq) for  HCO3-
pH =   7.10
SiO2(aq) =    49.0    mg/kg
Na+      =   715.7    mg/kg
K+       =    17.3    mg/kg
Ca++     =   759.4    mg/kg
Mg++     =     3.68   mg/kg
Fe++     =      .018  mg/kg
Al+++    =     0.01   mg/kg
CO2(aq)  =    13.3    mg/kg
SO4--    =   297.6    mg/kg
Cl-      =  2460.     mg/kg
F-       =      .82   mg/kg

precip = off
T initial = 11, final = 150
go
```

set up the geothermometry calculation, as discussed in the previous section.

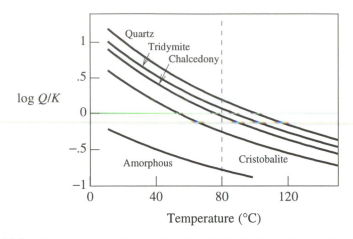

FIG. 17.7 Calculated saturation indices (log Q/K) of silica minerals for Gjögur hot spring water. Chalcedony is approximately in equilibrium at 80°C, but quartz is supersaturated at this temperature.

Figure 17.6 shows the saturation states of the aluminosilicate minerals in the LLNL database, plotted against temperature. Many minerals are supersaturated at low and high temperature, but a clear equilibrium point appears at about 80°C, slightly warmer than the discharge temperature of 72°C. At this point, the fluid's composition seems to be controlled by calcium and potassium clinoptilolite (zeolites; e.g., $CaAl_2Si_{10}O_{24} \cdot 8H_2O$) and calcium and magnesium saponite [smectite clays; e.g., $Ca_{.165}Mg_3Al_{.33}Si_{3.67}O_{10}(OH)_2$] minerals.

From a plot of the saturation states of the silica polymorphs (Fig. 17.7), the fluid's equilibrium temperature with quartz is about 100°C. Quartz, however, is commonly supersaturated in geothermal waters below about 150°C and so can give erroneously high equilibrium temperatures when applied in geothermometry (Fournier, 1977). Chalcedony is in equilibrium with the fluid at about 76°C, a temperature consistent with that suggested by the aluminosilicate minerals.

According to the calculation, the CO_2 fugacity varies in the range 10^{-2} to 10^{-3}, somewhat higher than the atmospheric partial pressure of this gas ($10^{-3.5}$ atm). The fugacity, however, is much lower than the total atmospheric pressure, suggesting that the fluid has not effervesced CO_2. In light of these results, we might reasonably argue that the fluid is a fresh water that has circulated to relatively shallow depths into the geothermal area, obtaining a moderate solute content but remaining oxidized. The fluid probably circulated through fractures in the basalt, reacting with zeolites, smectite clays, and chalcedony lining the fracture surfaces. The fluid last attained equilibrium with these minerals at about 75°C to 80°C, a temperature just slightly higher than the 72°C discharge temperature at the hot spring.

17.3 Geothermal Fields in Iceland

In a final application, following Reed and Spycher (1984), we consider fluids produced from wet-steam wells (i.e., wells that produce both vapor and liquid phases) at three geothermal fields in Iceland. Arnorsson et al. (1983) again supply the analytical data, given in Table 17.2. The calculations in this case are more complicated than those for the spring water considered in the previous section because, before applying our geothermometer, we must recombine the vapor and liquid phases sampled at the wellhead to find the composition of the original fluid.

To find the mass ratio of vapor to liquid produced, we note the discharge enthalpy H_{tot} and sampling pressure P_s from Table 17.2. From the steam tables (Keenan et al., 1969), we find the sampling temperature T_s corresponding to the boiling point at P_s, and the enthalpies H_{liq} and H_{vap} of liquid water and steam at this temperature. The mass fraction X_{vap} of vapor produced by the well is given (e.g., Henley, 1984) by energy balance

$$X_{vap} = \frac{H_{tot} - H_{liq}}{H_{vap} - H_{liq}} \qquad (17.4)$$

The resulting values for the three wells are

	H_{tot}	P_s	T_s	H_{liq}	H_{vap}	X_{vap}
Reykjanes #8	275	20	213.1	217.8	668.7	12.7%
Hveragerdi #4	183	6.8	164.4	165.9	659.9	3.5%
Namafjall #8	261	9.8	179.6	181.9	663.5	16.4%

Here, enthalpy is given in kcal/kg, temperature in °C, and pressure in atm. Note that the sampling temperatures are considerably lower than subsurface temperatures because of the energy used to produce the vapor phase.

Next, we need to calculate the amount of each component in the vapor phase. At room temperature, the vapor separates into a condensate that is mostly water and a gas phase that is mostly CO_2. Table 17.2 provides the composition of each. The mole number of each component (H_2O, CO_2, and H_2S) in the condensate, expressed per kg H_2O in the liquid, is derived by multiplying the concentration (g/kg) by the vapor fraction X_{vap} and dividing by the component's mole weight.

The mole numbers per kg liquid for the gases (H_2O, CO_2, H_2S, H_2, and CH_4) that separated from the condensate are obtained by multiplying the volume fraction of each by (1) the volume of gas per kg condensate, (2) the mass fraction X_{vap} of the vapor phase produced, and (3) the number of moles per liter of gas. The latter value can be calculated from the ideal gas law $PV = n\,RT_K$; at 20°C and 1 atm pressure, there are 0.0416 moles per liter of gas. The final values for each well

TABLE 17.2 Chemical compositions of water and steam discharged from wet-steam geothermal wells in Iceland (Arnorsson et al., 1983)

	Reykjanes Well #8	Hveragerdi Well #4	Namafjall Well #8
Fluid			
pH; °C	6.38; 20	8.82; 20	8.20; 22
SiO_2 (mg/kg)	631.1	281.0	446.3
Na^+	11150	153.3	154.8
K^+	1720	13.4	24.0
Ca^{++}	1705	1.73	4.52
Mg^{++}	1.44	0.002	0.085
Fe^{++}	.329	0.008	0.019
Al^{+++}	0.07	0.14	0.10
CO_2(aq)	63.1	74.2	88.2
SO_4^-	28.4	43.7	48.7
H_2S(aq)	2.21	19.2	132.6
Cl^-	22835	109.5	16.6
F^-	.21	1.82	.43
Dissolved solids	39124	765	902
Condensate			
CO_2 (mg/kg)	584	627	172
H_2S	65.6	84.5	277
Gas with condensate			
CO_2 (vol. %)	96.2	84.5	36.8
H_2S	2.9	3.0	17.0
H_2	0.2	2.8	37.4
CH_4	0.1	0.3	2.9
N_2	0.6	9.4	5.9
l gas/kg condensate; °C	2.63; 20	1.06; 20	6.25; 20
Sampling pressure (bars abs.)	20	6.8	9.8
Discharge enthalpy (kcal/kg)	275	183	261

	Reykjanes #8	Hveragerdi #4	Namafjall #8
H_2O	7.045	1.941	9.099
CO_2	.01505	1.80×10^{-3}	.01633
H_2S	6.47×10^{-4}	1.33×10^{-4}	8.58×10^{-3}
H_2	2.78×10^{-5}	4.32×10^{-5}	.01595
CH_4	1.39×10^{-5}	4.63×10^{-6}	3.59×10^{-3}

are sums of the mole numbers for the condensate and gas, expressed in moles per kg of H_2O in the liquid phase.

To run REACT for the Reykjanes #8 well, we start by defining the initial system

```
TDS = 39124
swap H2S(aq) for O2(aq)
swap CO2(aq) for HCO3-

pH = 6.38
SiO2(aq) =    631.1   mg/kg
Na+      = 11150      mg/kg
K+       = 1720       mg/kg
Ca++     = 1705       mg/kg
Mg++     =    1.44    mg/kg
Fe++     =     .329   mg/kg
Al+++    =     .07    mg/kg
CO2(aq)  =   63.1     mg/kg
SO4--    =   28.4     mg/kg
H2S(aq)  =    2.21    mg/kg
Cl-      = 22835      mg/kg
F-       =    0.21    mg/kg
```

Here, we use the ratio of sulfate to sulfide to constrain oxidation state in the fluid.

To invoke our geothermometer, we need to recombine the vapor and fluid phases and then heat the mixture to determine saturation indices as functions of temperature. We could do this in two steps, first titrating the vapor phase into the liquid and then picking up the results as the starting point for a polythermal path. We will employ a small trick, however, to accomplish these steps in a single reaction path. The trick is to add the vapor phase quickly during the first part of the reaction path but use the `cutoff` option to prevent mass transfer over the remainder of the path. The commands to set the mass transfer are

```
(cont'd)
react 70.045   moles H2O      cutoff = 7.045
react    .1505 moles CO2(g)   cutoff =  .01505
react  6.47e-3 moles H2S(g)   cutoff = 6.47e-4
react  2.78e-4 moles H2(g)    cutoff = 2.78e-5
react  1.39e-4 moles CH4(g)   cutoff = 1.39e-5
```

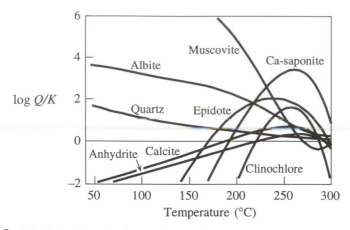

FIG. 17.8 Calculated saturation indices (log Q/K) of various minerals for the Reykjanes #8 wet-steam geothermal well. The well produces from seven intervals at temperatures varying from 274°C to 292°C. The calculated saturation indices suggest an equilibrium temperature between about 285°C and slightly above 300°C.

The cutoff values are the mole numbers calculated above, and the corresponding reaction rates are the mole numbers augmented by a factor of ten. We type the commands

```
(cont'd)
T initial = 20, final = 300
precip = off
go
```

to set up the polythermal path and trigger the calculation.

In the calculation results (Fig. 17.8), a number of minerals converge to equilibrium with the fluid in the range of about 285°C to somewhat above 300°C (slightly beyond the calculation's high-temperature limit). The well, for comparison, produces fluid from seven zones at temperatures ranging from 274°C to 292°C (Arnorsson et al., 1983). Several minerals [i.e, epidote, $Ca_2FeAl_2Si_3O_{12}OH$, and calcium saponite, $Ca_{.165}Mg_3Al_{.33}Si_{3.67}O_{10}(OH)_2$] are supersaturated at temperatures less than about 300°C. This result might reasonably be interpreted to reflect a fluid that equilibrated at a temperature somewhat higher than observed in the well. Alternatively, the supersaturation may be due to the mixing in the wellbore of fluids from the well's various producing zones. As discussed in Chapter 6, fluid mixing tends to leave minerals supersaturated.

For the Hveragerdi #4 well, we follow the same procedure, using the data in Table 17.2 and the calculations already shown. In this case, the model predicts that a number of minerals in the LLNL database are supersaturated near the inflow

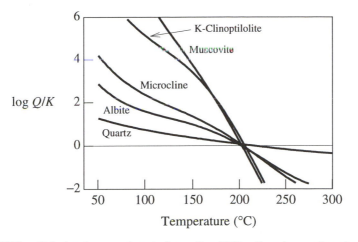

FIG. 17.9 Calculated saturation indices (log Q/K) of various minerals for the Hveragerdi #4 wet-steam geothermal well. The inflow temperature is 181°C. The calculated saturation indices suggest an equilibrium temperature near 200°C.

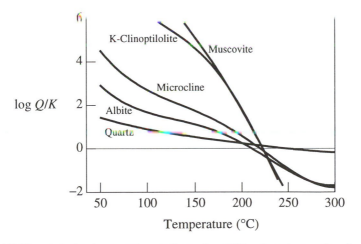

FIG. 17.10 Calculated saturation indices (log Q/K) of various minerals for the Namafjall #8 wet-steam geothermal well. The inflow temperature is 246°C. The calculated saturation indices suggest an equilibrium temperature in the range 205°C to 250°C.

temperature of 181°C. Close examination reveals that each of the supersaturated minerals contains either Mg^{++}, Ca^{++}, or Fe^{++}, components that are characteristically depleted in geothermal fluids. The Mg^{++} concentration in this fluid, for example, is just 2 μg/kg.

The minor amounts of Mg^{++}, Ca^{++}, and Fe^{++} in the fluid might easily be accounted for by contamination by a small volume of a shallow groundwater or even by dissolution of concrete and steel in the wellbore. If we discount the results for minerals containing these components, we arrive at a well-defined equilibrium temperature slightly above 200°C (Fig. 17.9). Again, the equilibrium temperature is slightly higher than the temperature measured in the wellbore.

The analysis for Namafjall well #8 is similar to that of the Hveragerdi well in that a number of Ca, Mg, and Fe-bearing minerals appear supersaturated over the temperature range of interest. Again, this result probably results from mixing or contamination. The equilibrium temperatures for quartz, albite, potassium feldspar, potassium clinoptilolite, and muscovite (Fig. 17.10) bracket a relatively broad temperature range of 205°C to 250°C, which can be compared to the well's inflow temperature of 246°C. In this case, the equilibrium temperature is notably less well defined than in the previous example, perhaps reflecting the mixing of a significant amount of shallow groundwater into the geothermal fluid.

18

Evaporation

The process of evaporation, including transpiration (evaporation from plants), returns to the atmosphere more than half of the water reaching the Earth's land surface; thus, it plays an important role in controlling the chemistry of surface water and groundwater, especially in relatively arid climates. Geochemists study the evaporation process to understand the evolution of water in desert playas and lakes as well as the origins of evaporite deposits. They also investigate environmental aspects of evaporation (e.g., Appelo and Postma, 1993), such as its effects on the chemistry of rainfall and, in areas where crops are irrigated, the quality of groundwater and runoff.

To model the chemical effects of evaporation, we construct a reaction path in which H_2O is removed from a solution, thereby progressively concentrating the solutes. We also must account in the model for the exchange of gases such as CO_2 and O_2 between fluid and atmosphere. In this chapter we construct simulations of this sort, modeling the chemical evolution of water from saline alkaline lakes and the reactions that occur as seawater evaporates to desiccation.

18.1 Springs and Saline Lakes of the Sierra Nevada

We choose as a first example the evaporation of spring water from the Sierra Nevada mountains of California and Nevada, USA, as modeled by Garrels and Mackenzie (1967). Their hand calculation, the first reaction path traced in geochemistry (see Chapter 1), provided the inspiration for Helgeson's (1968 and later) development of computerized methods for reaction modeling.

Garrels and Mackenzie wanted to test whether simple evaporation of groundwater discharging from the mountains, which is the product of the

261

TABLE 18.1 Mean composition of spring water
from the Sierra Nevada, California and Nevada,
USA (Garrels and Mackenzie, 1967)

	mg/kg	mmolal
$SiO_2(aq)$	24.6	.410
Ca^{++}	10.4	.260
Mg^{++}	1.70	.070
Na^+	5.95	.259
K^+	1.57	.040
HCO_3^-	54.6	.895
SO_4^{--}	2.38	.025
Cl^-	1.06	.030
pH		6.8*

*Median value.

reaction of rainwater and CO_2 with igneous rocks, could produce the water compositions found in the saline alkaline lakes of the adjacent California desert. They began with the mean of analyses of perennial springs from the Sierra Nevada (Table 18.1). The springs are Na-Ca-HCO_3 waters, rich in dissolved silica. Using REACT to distribute species

```
26.4  mg/kg SiO2(aq)
10.4  mg/kg Ca++
 1.7  mg/kg Mg++
 5.95 mg/kg Na+
 1.57 mg/kg K+
54.6  mg/kg HCO3-
 2.38 mg/kg SO4--
 1.06 mg/kg Cl-
pH = 6.8

balance on HCO3-
go
```

we calculate a CO_2 fugacity for the water of $10^{-2.1}$, typical of soil waters but somewhat higher than the atmospheric value of $10^{-3.5}$. In their model, Garrels and Mackenzie assumed that the evaporating water remained in equilibrium with the CO_2 in the atmosphere. To prepare the reaction path, therefore, they computed the effects of letting the spring water exsolve CO_2 until it reached atmospheric fugacity.

They made several assumptions about which minerals could precipitate from the fluid. The alkaline lakes tend to be supersaturated with respect to each of the silica polymorphs (quartz, tridymite, and so on) except amorphous silica, so they suppressed each of the other silica minerals. They assumed that dolomite [$CaMg(CO_3)_2$], a highly ordered mineral known not to precipitate at 25°C except from saline brines, would not form. Finally, they took the clay mineral sepiolite [$Mg_4Si_6O_{15}(OH)_2 \cdot 6H_2O$] in preference to minerals such as talc [$Mg_3Si_4O_{10}(OH)_2$] as the magnesium silicate likely to precipitate during evaporation.

The procedure to suppress these minerals and adjust the fluid's CO_2 fugacity is

```
(cont'd)
suppress Quartz, Tridymite, Cristobalite, Chalcedony
suppress Dolomite, Dolomite-ord, Dolomite-dis
suppress Talc, Tremolite, Antigorite

slide f CO2(g) to 10^-3.5
go
```

The resulting fluid, which provides the starting point for modeling the effects of evaporation, has a *p*H of 8.2.

To model the process of evaporation, we take the fluid resulting from the previous calculation and, while holding the CO_2 fugacity constant, remove almost all (999.9 grams of the original kg) of its water. The procedure is

```
(cont'd)
pickup fluid
react -999.9 g H2O
fix f CO2(g)
delxi  = .001
dxplot = .001
go
```

The `delxi` and `dxplot` commands set a small reaction step to provide increased detail near the end of the reaction path.

In the calculation results (Fig. 18.1), amorphous silica, calcite ($CaCO_3$), and sepiolite precipitate as water is removed from the system. The fluid's *p*H and ionic strength increase with evaporation as the water evolves toward a $Na-CO_3$ brine (Fig. 18.2). The concentrations of the components Na^+, K^+, Cl^-, and SO_4^{--} rise monotonically (Fig. 18.2), since they are not consumed by mineral precipitation. The HCO_3^- and $SiO_2(aq)$ concentrations increase sharply but less regularly, since they are taken up in forming the minerals. The components Ca^{++} and Mg^{++} are largely consumed by the precipitation of calcite and sepiolite. Their concentrations, after a small initial rise, decrease with evaporation.

Two principal factors drive reaction in the evaporating fluid. First, the loss of solvent concentrates the species in solution, causing the saturation states of many minerals to increase. The precipitation of amorphous silica, for example,

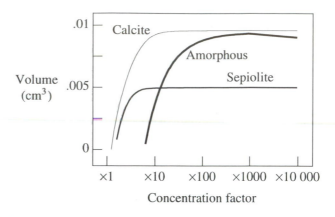

FIG. 18.1 Volumes of minerals (amorphous silica, calcite, and sepiolite) precipitated during a reaction model simulating at 25°C the evaporation of Sierra Nevada spring water in equilibrium with atmospheric CO_2, plotted against the concentration factor. For example, a concentration factor of ×100 means that of the original 1 kg of water, 10 grams remain.

$$SiO_2(aq) \rightarrow \underset{amorphous}{SiO_2} \qquad (18.1)$$

results almost entirely from the increase in $SiO_2(aq)$ concentration as water evaporates.

A second and critical factor is the loss of CO_2 from the fluid to the atmosphere. Evaporation concentrates $CO_2(aq)$, driving $CO_2(g)$ to exsolve

$$CO_2(aq) \rightarrow CO_2(g) \qquad (18.2)$$

About 0.5 *m*mol of CO_2, or about 60% of the fluid's carbonate content at the onset of the calculation, is lost in this way.

The escape of CO_2, an acid gas, affects the fluid's *p*H. By mass action, the loss of $CO_2(aq)$ from the system causes readjustment among the carbonate species in solution

$$HCO_3^- + H^+ \rightarrow H_2O + CO_2(aq) \qquad (18.3)$$

$$CO_3^{--} + 2\,H^+ \rightarrow H_2O + CO_2(aq) \qquad (18.4)$$

The reactions consume H^+, driving the fluid toward alkaline *p*H, as shown in Fig. 18.2. This effect explains the alkalinity of saline alkaline lakes.

Calcite and sepiolite precipitate in large part because of the effects of the escaping CO_2. The corresponding reactions are

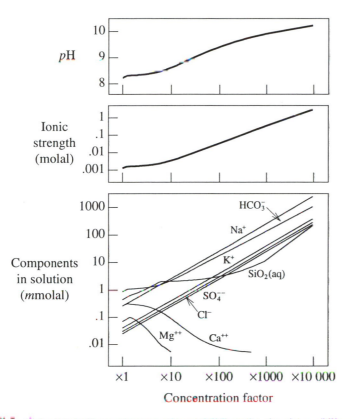

FIG. 18.2 Calculated effects of evaporation at 25°C on the chemistry of Sierra Nevada spring water. Top figures show how *p*H and ionic strength vary over the reaction path in Fig. 18.1; bottom figure shows variation in the fluid's bulk composition.

$$Ca^{++} + 2\,HCO_3^- \rightarrow \underset{calcite}{CaCO_3} + H_2O + CO_2(g) \qquad (18.5)$$

and

$$4\,Mg^{++} + 6\,SiO_2(aq) + 8\,HCO_3^- + 3\,H_2O \rightarrow$$
$$\underset{sepiolite}{Mg_4Si_6O_{15}(OH)_2 \cdot 6H_2O} + 8\,CO_2(g) \qquad (18.6)$$

Figure 18.3 compares the calculated composition of the evaporated water, concentrated 100-fold and 1000-fold, with analyses of waters from six saline alkaline lakes (compiled by Garrels and Mackenzie, 1967). The field for the modeled water overlaps that for the analyzed waters, except that Ca^{++} and Mg^{++} are more depleted in the model than in the lake waters. This discrepancy might

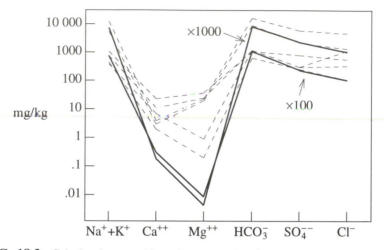

FIG. 18.3 Calculated composition of evaporated spring water (from the reaction path shown in Figs. 18.1 and 18.2) concentrated 100-fold and 1000-fold (solid lines) compared with the compositions of six saline alkaline lakes (dashed lines), as compiled by Garrels and Mackenzie (1967).

be explained if in nature the calcite and sepiolite begin to precipitate but remain supersaturated in the fluid.

In the reaction path we calculated (Fig. 18.2), the precipitation of calcite consumes nearly all of the calcium originally in solution so that no further calcium-bearing minerals form. Calcite precipitation, on the other hand, does not deplete the dissolved carbonate, because the original fluid was considerably richer in carbonate than in calcium ($M_{HCO_3^-} > M_{Ca^{++}}$). The carbonate concentration, in fact, increases during evaporation.

If the fluid had been initially richer in calcium than carbonate ($M_{Ca^{++}} > M_{HCO_3^-}$), as noted by Hardie and Eugster (1970), it would have followed a distinct reaction path. In such a case, calcite precipitation would deplete the fluid in carbonate, allowing the calcium concentration to increase until gypsum ($CaSO_4 \cdot 2H_2O$) saturates and forms. The point at which the calcium and carbonate are present at equal initial concentration ($M_{Ca^{++}} = M_{HCO_3^-}$) is known as a *chemical divide*.

According to Hardie and Eugster's (1970) model and its later variants (see discussions in Eugster and Jones, 1979; Drever, 1988, pp. 232–250; and Jankowski and Jacobson, 1989), a natural water, as it evaporates, encounters a series of chemical divides that controls the sequence of minerals that precipitate. The reaction pathway specific to the evaporation of a water of any initial composition can be traced in detail using a reaction model like the one applied in this section to Sierra spring water.

18.2 Chemical Evolution of Mono Lake

In a second example, we consider the changing chemistry of Mono Lake, a saline alkaline lake that occupies a closed desert basin in California, USA, and why gaylussite [$CaNa_2(CO_3)_2 \cdot 5H_2O$], a rare hydrated carbonate mineral, has begun to form there. The lake has shrunk dramatically since 1941, when the Los Angeles Department of Water and Power began to divert tributary streams to supply water for southern California.

As it shrank, the lake became more saline. Salinity has almost doubled from 50,000 mg/kg in 1940 to about 90,000 mg/kg in recent years. The change in the lake's chemistry threatens to damage a unique ecosystem that supports large flocks of migratory waterfowl.

In 1988, Bischoff et al. (1991) discovered gaylussite crystals actively forming in the lake. The crystals were found growing on hard surfaces, especially in the lake's deeper sections. Gaylussite had also formed earlier, because psuedomorphs after gaylussite were observed. The earlier gaylussite has been replaced by aragonite, leaving the porous skeletal psuedomorphs.

Bischoff et al. (1991) attributed the occurrence of gaylussite to the lake's increase in Na^+ content and pH since diversion began. Table 18.2 shows chemical analyses of lake water sampled at ten points in time from 1956 to 1988. We can use REACT to calculate for each sample the saturation state of gaylussite. The procedure for the 1956 sample, for example, is

```
Ca++    =     4.3 mg/kg
Mg++    =      38 mg/kg
Na+     =   22540 mg/kg
K+      =    1124 mg/kg
SO4--   =   12000 mg/kg
Cl-     =   13850 mg/kg
B(OH)3  =    1720 mg/kg
HCO3-   =   17600 mg/kg

TDS = 67686
pH  =  9.49

precip = off
go
```

Figure 18.4 shows the trend in gaylussite saturation plotted against time and against the salinity of the lake water.

In the calculation results, gaylussite appears increasingly saturated in the lake water as salinity increases irregularly over time. The calculations suggest that between about 1975 and 1980, as salinities reached about 75,000 to 80,000 mg/kg, the mineral became supersaturated at summer temperatures.

The solubility of gaylussite, however, varies strongly with temperature. Unlike calcite and aragonite, gaylussite grows less soluble (or more stable) as

TABLE 18.2 Chemical analyses (mg/kg) of surface water from Mono Lake, 1956 to 1988 (James Bischoff, personal communication*)

	1956	1957	1974	1978	1979
Ca^{++}	4.3	4.2	3.5	4.5	4.6
Mg^{++}	38	37	30	34	42
Na^+	22 540	22 990	31 100	21 500	37 200
K^+	1 124	1 140	1 500	1 170	1 580
SO_4^-	12 000	7 810	11 000	7 380	12 074
Cl^-	13 850	14 390	18 000	13 500	20 100
$B(OH)_3$	1 720	2 000	1 850	—	2 760
Alkalinity†	21 854	21 600	30 424	21 617	34 818
HCO_3^-‡	17 600	15 900	22 700	17 000	25 000
TDS	67 686	59 500	80 370	59 312	92 540
pH	9.49	9.7	9.66	9.6	9.68

	1982	1983	1984	1985	1988
Ca^{++}	4.4	4.3	3.1	3.6	3.3
Mg^{++}	37	36	30	34	37
Na^+	34 000	29 300	26 600	28 900	32 000
K^+	1 980	1 600	1 240	1 300	1 500
SO_4^-	11 100	10 700	9 590	10 190	11 200
Cl^-	20 400	18 860	17 673	20 370	19 700
$B(OH)_3$	2 460	2 230	1 720	1 720	2 120
Alkalinity†	34 700	31 240	27 240	29 170	31 900
HCO_3^-‡	25 300	22 800	18 600	19 600	20 500
TDS	92 200	83 800	74 700	80 600	85 600
pH	9.66	9.68	9.90	9.93	10.03

*From published (see Bischoff et al., 1991) and unpublished sources.
†As $CaCO_3$.
‡Calculated from alkalinity, using method described in Section 13.1.

temperature decreases. We can recalculate mineral solubility under winter conditions by setting temperature to 0°C

(cont'd)
T = 0
go

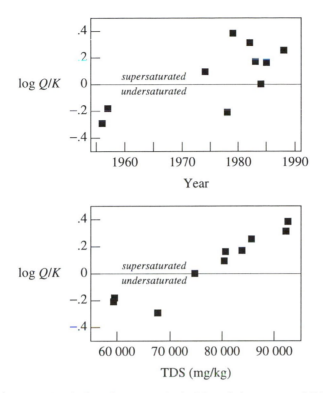

FIG. 18.4 Saturation indices for gaylussite in Mono Lake water at 25°C for various years between 1956 and 1988, calculated from analyses in Table 18.2 and plotted against time (top diagram) and salinity (bottom).

These calculations (Fig. 18.5) suggest that, to the extent that the analyses in Table 18.2 are representative of the wintertime lake chemistry, gaylussite has been supersaturated in Mono Lake during the winter months since sampling began.

The saturation state of aragonite (Fig. 18.5), on the other hand, is affected little by temperature. Aragonite remains supersaturated by a factor of about ten (one log unit) over the gamut of analyses. The supersaturation probably arises from the effect of orthophosphate, present at concentrations of about 100 mg/kg in Mono Lake water; orthophosphate is observed in the laboratory (Bischoff et al., 1993) to inhibit the precipitation of calcite and aragonite.

We can use our results to predict the conditions favorable for the transformation of gaylussite to aragonite. The porous nature of the psuedomorphs and the small amounts of calcium available in the lake water (Table 18.2) suggest that the replacement occurs by the incongruent dissolution of gaylussite, according to the reaction

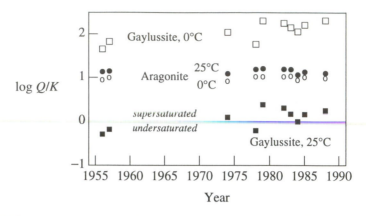

FIG. 18.5 Saturation indices of gaylussite (□, ■) and aragonite (○, •) calculated for Mono Lake water in various years from analyses in Table 18.2. Open symbols (□, ○) represent values calculated for 0°C and solid symbols (■, •) show those for 25°C.

$$CaNa_2(CO_3)_2 \cdot 5H_2O \rightarrow CaCO_3 + 2\,Na^+ + CO_3^- + 5\,H_2O \qquad (18.7)$$
$$\textit{gaylussite} \qquad\qquad \textit{aragonite}$$

(Bischoff et al., 1991). Figure 18.6 compares the activity product for this reaction to its equilibrium constant, showing which mineral is favored to form at the expense of the other. According to the calculations, gaylussite is prone to transform into aragonite during the summer (even though gaylussite is supersaturated in the lake water then), but is not likely to be replaced during the winter months.

18.3 Evaporation of Seawater

Since the experimental studies of van't Hoff at the turn of the century, geochemists have sought a quantitative basis for describing the chemical evolution of seawater and other complex natural waters, including the minerals that precipitate from them, as they evaporate. The interest has stemmed in large part from a desire to understand the origins of ancient deposits of evaporite minerals, a goal that remains mostly unfulfilled (Hardie, 1991).

The results of the early experimental studies, although of great significance, were limited by the complexity of a chemical system that can be portrayed on paper within a phase diagram. There was little possibility, furthermore, of calculating a useful reaction model of the evaporation process using conventional correlations to compute activity coefficients for the aqueous species (see Chapter 7), given the inherent inaccuracy of the correlations at high ionic strength.

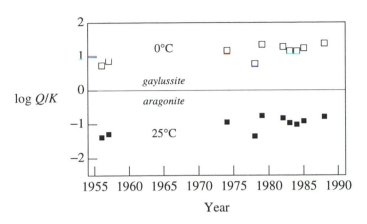

FIG. 18.6 The state (log Q/K) of React. 18.7 between gaylussite and aragonite at 0°C (□) and 25°C (■), showing which mineral is favored to form at the expense of the other.

In a series of papers, Harvie and Weare (1980), Harvie et al. (1980), and Eugster et al. (1980) attacked this problem by presenting a virial method for computing activity coefficients in complex solutions (see Chapter 7) and applying it to construct a reaction model of seawater evaporation. Their calculations provided the first quantitative description of this process that accounted for all of the abundant components in seawater.

To reproduce their results, we trace the reaction path taken by seawater at 25°C as it evaporates to desiccation. Our calculations follow those of Harvie et al. (1980) and Eugster et al. (1980), except that we employ the more recent Harvie-Møller-Weare activity model (Harvie et al., 1984), which accounts for bicarbonate. We include a HCO_3^- component in our calculations, assuming that the fluid as it evaporates remains in equilibrium with the CO_2 in the atmosphere.

In a first calculation, we specify that the fluid maintains equilibrium with whatever minerals precipitate. Minerals that form, therefore, can redissolve into the brine as evaporation proceeds. In REACT, we set the Harvie-Møller-Weare model and specify that our initial system contains seawater

```
hmw

swap CO2(g) for H+
log f CO2(g) = -3.5

TDS = 35080
Cl-   =  19350 mg/kg
Ca++  =    411 mg/kg
Mg++  =   1290 mg/kg
Na+   =  10760 mg/kg
K+    =    399 mg/kg
```

```
SO4-- =   2710 mg/kg
HCO3- =    142 mg/kg
```

just as we did in Chapter 6. We then set a reaction path in which we fix the CO_2 fugacity and remove solvent from the system

```
(cont'd)
fix fugacity of CO2(g)
react -996 grams of H2O

delxi = .001
dxplot = 0
dump
go
```

The dump command serves to eliminate the small mineral masses that precipitate when, at the onset of the calculation, the program brings seawater to its theoretical equilibrium state (see Chapter 6). The delxi and dxplot commands serve to set a small reaction step, assuring that the results are rendered in sufficient detail.

By removing 996 grams of H_2O, we eliminate all of the 1 kg of solvent initially present in the system; the remaining 4 grams are consumed by the precipitation of hydrated minerals. In fact, just slightly less than 996 grams of water can be removed before the system is completely desiccated. The program continues until less than 1 µg of solvent remains and then abandons its efforts to trace the path. At this point, it gives a warning message, which can be ignored.

Figure 18.7 shows the minerals that precipitate over the reaction path (Table 18.3 lists their compositions), and Fig. 18.8 shows how fluid chemistry in the calculation varies. Initially, dolomite and gypsum precipitate. When the fluid is concentrated about ten-fold, the decreasing water activity causes the gypsum to dehydrate

$$CaSO_4 \cdot 2H_2O \rightarrow CaSO_4 + 2H_2O \qquad (18.8)$$
$$\text{gypsum} \qquad \text{anhydrite}$$

forming anhydrite.

Shortly afterwards, halite becomes saturated and begins to precipitate

$$Na^+ + Cl^- \rightarrow NaCl \qquad (18.9)$$
$$\text{halite}$$

Halite forms in the calculation in far greater volume than any other mineral, reflecting the fact that seawater is dominantly a NaCl solution. The precipitation reaction represents a chemical divide, as discussed in Section 18.1. Since Na^+ is less concentrated (on a molal basis) in seawater than Cl^- ($M_{Na^+} < M_{Cl^-}$), it becomes depleted in solution. As a result (Fig. 18.8), seawater evolves with evaporation from a dominantly NaCl solution into a $MgCl_2$ bittern.

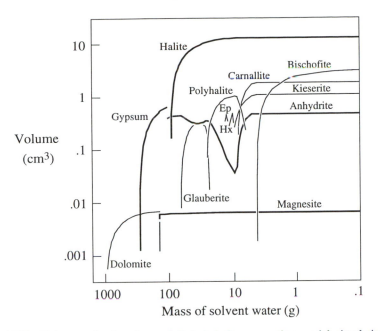

FIG. 18.7 Volumes of minerals precipitated during a reaction model simulating the evaporation of seawater as an equilibrium system at 25°C, calculated using the Harvie-Møller-Weare activity model. Abbreviations: Ep = Epsomite, Hx = Hexahydrite.

With further evaporation, a small amount of glauberite precipitates

$$\underset{anhydrite}{CaSO_4} + 2\,Na^+ + SO_4^{--} \rightarrow \underset{glauberite}{Na_2Ca(SO_4)_2} \qquad (18.10)$$

at the expense of anhydrite. The increasing activities of K^+, Mg^{++}, and SO_4^{--} in solution, however, soon drive the glauberite and some of the anhydrite to form polyhalite according to the reaction

$$\underset{glauberite}{Na_2Ca(SO_4)_2} + \underset{anhydrite}{CaSO_4} + 2\,K^+ + Mg^{++} + SO_4^{--} + 2\,H_2O \rightarrow$$
$$\underset{polyhalite}{K_2MgCa_2(SO_4)_4\cdot 2H_2O} + 2\,Na^+ \qquad (18.11)$$

In accord with these predictions (Harvie et al., 1980), psuedomorphs of glauberite after gypsum and anhydrite are observed in marine evaporites, and polyhalite, in turn, is known to replace glauberite.

As it evolves toward a dominantly $MgCl_2$ solution, the fluid becomes supersaturated with respect to kieserite and carnallite. The reaction to form these minerals

TABLE 18.3 Minerals formed during the simulated evaporation of seawater

Anhydrite	$CaSO_4$
Bischofite	$MgCl_2 \cdot 6H_2O$
Bloedite	$Na_2Mg(SO_4)_2 \cdot 4H_2O$
Carnallite	$KMgCl_3 \cdot 6H_2O$
Dolomite	$CaMg(CO_3)_2$
Epsomite	$MgSO_4 \cdot 7H_2O$
Glauberite	$Na_2Ca(SO_4)_2$
Gypsum	$CaSO_4 \cdot 2H_2O$
Halite	$NaCl$
Hexahydrite	$MgSO_4 \cdot 6H_2O$
Kainite	$KMgClSO_4 \cdot 3H_2O$
Kieserite	$MgSO_4 \cdot H_2O$
Magnesite	$MgCO_3$
Polyhalite	$K_2MgCa_2(SO_4)_4 \cdot 2H_2O$

$$\underset{polyhalite}{K_2MgCa_2(SO_4)_4 \cdot 2H_2O} + Mg^{++} + 6\ Cl^- + 12\ H_2O \ \rightarrow$$

$$2\ \underset{kieserite}{MgSO_4 \cdot H_2O} + 2\ \underset{carnallite}{KMgCl_3 \cdot 6H_2O} + 2\ \underset{anhydrite}{CaSO_4} \qquad (18.12)$$

consumes the polyhalite. Finally, when about 4.5 g of solvent remain, bischofite forms

$$Mg^{++} + 2\ Cl^- + 6\ H_2O \ \rightarrow \ \underset{bischofite}{MgCl_2 \cdot 6H_2O} \qquad (18.13)$$

in response to the high Mg^{++} and Cl^- activities in the residual fluid.

With the precipitation of bischofite, the system reaches an invariant point at which the mineral assemblage (magnesite, anhydrite, kieserite, carnallite, bischofite, and halite) fully constrains the fluid composition. Further evaporation causes more of these phases (principally bischofite) to form, but the fluid chemistry no longer changes, as can be seen in Fig. 18.8.

In a second calculation, we model the reaction path taken when the minerals, once precipitated, cannot redissolve into the fluid. In this model, the solutes in seawater fractionate into the minerals as they precipitate, irreversibly altering the fluid composition. As discussed in Chapter 2, we set up such a model using the "flow-through" configuration. The procedure is

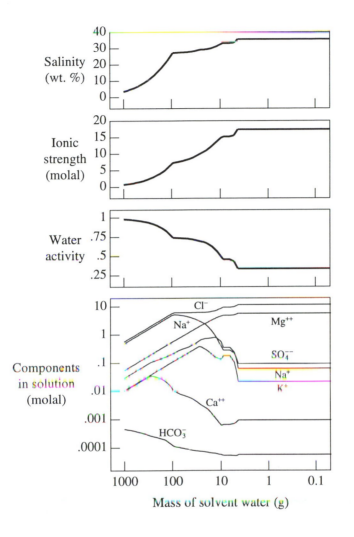

FIG. 18.8 Evolution of fluid chemistry during the simulated evaporation of seawater as an equilibrium system at 25°C, calculated using the Harvie-Møller-Weare activity model. Upper figures show variation in salinity, water activity (a_w), and ionic strength (I) over the reaction path in Fig. 18.7; bottom figure shows how the fluid's bulk composition varies.

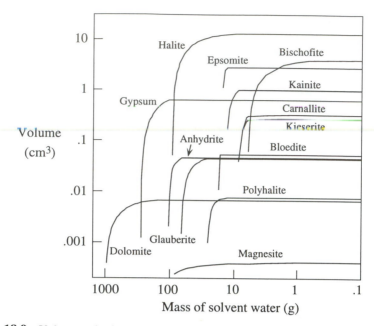

FIG. 18.9 Volumes of minerals precipitated during a reaction model simulating the evaporation of seawater as a fractionating system (the ''flow-through'' configuration) at 25°C, calculated using the Harvie-Møller-Weare activity model.

(cont'd)
flow-through
go

The results of the fractionation model (Fig. 18.9) differ from the equilibrium model in two principal ways. First, the mineral masses can only increase in the fractionation model, since they are protected from resorption into the fluid. Therefore, the lines in Fig. 18.9 do not assume negative slopes. Second, in the equilibrium calculation the phase rule limits the number of minerals present at any point along the reaction path. In the fractionation calculation, on the other hand, no limit to the number of minerals present exists, since the minerals do not necessarily maintain equilibrium with the fluid. Therefore, the fractionation calculation ends with twelve minerals in the system, whereas the equilibrium calculation reaches an invariant point at which only six minerals are present.

The fractionation calculation is notable in that it predicts the formation of two minerals (bloedite and kainite) that did not precipitate in the equilibrium model. As well, hexahydrite, which appeared briefly in the equilibrium model, does not form in the fractionation model. The two classes of models, therefore, represent qualitatively distinct pathways by which an evaporating water can evolve.

19

Sediment Diagenesis

Diagenesis is the set of processes by which sediments evolve after they are deposited and begin to be buried. Diagenesis includes physical effects such as compaction and the deformation of grains in the sediment (or sedimentary rock), as well as chemical reactions such as the dissolution of grains and the precipitation of minerals to form cements in the sediment's pore space. The chemical aspects of diagenesis are of special interest here.

Formerly, geologists considered chemical diagenesis to be a process by which the minerals and pore fluid in a sediment reacted with each other in response to changes in temperature, pressure, and stress. As early as the 1960s and especially since the 1970s, however, geologists have recognized that many diagenetic reactions occur in systems open to groundwater flow and mass transfer. The reactions proceed in response to a supply of reactants introduced into the sediments by flowing groundwater, which also serves to remove reaction products.

Hay (1963, 1966), in studies of the origin of diagenetic zeolite, was perhaps the first to emphasize the effects of mass transport on sediment diagenesis. He showed that sediments open to groundwater flow followed reaction pathways different from those observed in sediments through which flow was restricted. Sibley and Blatt (1976) used cathodoluminescence microscopy to observe the Tuscarora orthoquartzite of the Appalachian basin. The almost nonporous Tuscarora had previously been taken as a classic example of pressure welding, but the microscopy demonstrated that the rock is not especially well compacted but, instead, tightly cemented. The rock consists of as much as 40% quartz (SiO_2) cement that was apparently deposited by advecting groundwater.

By the end of the decade, Hayes (1979) and Surdam and Boles (1979) argued forcefully that the extent to which diagenesis has altered sediments in sedimentary basins can be explained only by recognition of the role of groundwater flow in transporting dissolved mass. This view has become largely accepted among geoscientists, although it is clear that the scale of groundwater flow might range from the regional (e.g., Bethke and Marshak, 1990) to circulation cells perhaps as small as tens of meters (e.g., Bjorlykke and Egeberg, 1993; Aplin and Warren, 1994). Since the possible reactions occurring in open geochemical systems are numerous and complex, the study of diagenesis has become a fertile field for applying reaction modeling (e.g., Bethke et al., 1988; Baccar and Fritz, 1993).

In this chapter we consider how reaction modeling applied to open systems might be used to study the nature of diagenetic alteration. We develop examples in which modeling of this type can aid in interpreting the diagenetic reactions observed to have occurred in sedimentary rocks.

19.1 Dolomite Cement in the Gippsland Basin

As a first example, we consider the diagenesis of clastic sandstones in the Gippsland basin, southeastern Australia, basing our model on the work of Harrison (1990). The Gippsland basin is the major offshore petroleum province in Australia. Oil production is from the Latrobe group, a fluvial to shallow marine sequence of Late Cretaceous to early Eocence age that partly fills a Mesozoic rift valley.

In the fluvial sandstones, the distribution of diagenetic cements in large part controls reservoir quality and the capacity for petroleum production. These sandstones are composed of quartz and potassium feldspar ($KAlSi_3O_8$) grains and detrital illite [which we will represent by muscovite, $KAl_3Si_3O_{10}(OH)_2$], kaolinite [$Al_2Si_2O_5(OH)_4$], and lithic fragments. Where cementation is minor, reservoir properties are excellent. Porosity can exceed 25%, and permeabilities greater than 2 darcys (2×10^{-8} cm^2) have been noted.

In more diagenetically altered areas, however, cements including dolomite [$CaMg(CO_3)_2$], clay minerals (principally kaolinite), and quartz nearly destroy porosity and permeability. In these facies, potassium feldspar grains are strongly leached and pyrite (FeS_2) is corroded. Dolomite cement occupies up to 40% of the rock's volume, and quartz cement takes up an average of several percent. Understanding the processes that control the distribution of cements, therefore, is of considerable practical importance in petroleum exploration.

Harrison (1990) proposed that the diagenetic alteration observed in the Latrobe group resulted from the mixing within the formation of two types of groundwaters. Table 19.1 shows analyses of waters sampled from two oil wells, which she took to be representative of the two water types as they exist in the producing areas of the basin.

TABLE 19.1 Analyses of formation water sampled at two
oil wells producing from the Latrobe group,
Gippsland basin (Harrison, 1990)

	"Fresh" water (Barracouta A-3)	Saline water (Kingfish A-19)
Ca^{++} (mg/*l*)	32	220
Mg^{++}	9	1 000
Na^+	2 943	11 000
SO_4^{--}	1 461	900
HCO_3^-	1 135	198
Cl^-	2 953	19 000
Dissolved solids	8 530	32 320
*p*H (measured)	7.0	5.6
*p*H (corrected to 60°C)	6.95	5.55

The first water is considerably less saline than seawater and hence is termed a
"fresh" water, although it is far too saline to be potable. This water is
apparently derived from meteoric water that recharges the Latrobe group where
it outcrops onshore. The water flows basinward through an aquifer that extends
60 km offshore and 2 km subsea. A second, more saline water exists in deeper
strata and farther offshore. This water is very similar in composition to seawater,
although slightly depleted in Ca^{++} and SO_4 , as can be seen by comparing the
analysis to Table 6.2. On the basis of isotopic evidence, the diagenetic alteration
probably occurred at temperatures of 60°C or less.

To test whether the mixing hypothesis might explain the diagenetic alteration
observed, we begin by equilibrating the fresh water, assuming equilibrium with
the potassium feldspar ("maximum microcline" in the database), quartz, and
muscovite (a proxy for illite) in the formation. In REACT, we enter the commands

```
swap "Maximum Microcline" for Al+++
swap Quartz for SiO2(aq)
swap Muscovite for K+

T = 60
TDS = 8530

pH = 6.95
Ca++  =    32 mg/l
Mg++  =     9 mg/l
Na+   = 2943 mg/l
HCO3- = 1135 mg/l
```

```
SO4-- = 1461 mg/l
Cl-   = 2953 mg/l

1 free cm3 Muscovite
1 free cm3 Quartz
1 free cm3 "Maximum Microcline"

go
```

causing the program to iterate to a description of the fluid's equilibrium state.

To model fluid mixing, we will use the fresh water as a reactant, titrating it into a system containing the saline water and formation minerals. To do so, we "pick up" the fluid from the previous step to use as a reactant:

```
(cont'd)
pickup reactants = fluid
reactants times 100
```

The latter command multiplies the amount of fluid (1 kg) to be used as a reactant by 100. Hence, we will model mixing in ratios from zero to as high as 100 parts fresh to one part saline water.

To prepare the initial system, we use the analysis in Table 19.1 for the saline water, which we assume to be in equilibrium with potassium feldspar, quartz, muscovite, and dolomite ("dolomite-ord" is the most stable variety in the database). The commands

```
(cont'd)
swap "Maximum Microcline" for Al+++
swap Quartz for SiO2(aq)
swap Muscovite for K+
swap Dolomite-ord for HCO3-

T = 60
TDS = 32320

pH = 5.55
Ca++  =    220 mg/l
Mg++  =   1000 mg/l
Na+   =  11000 mg/l
SO4-- =    900 mg/l
Cl-   =  19000 mg/l

10 free cm3 Muscovite
10 free cm3 Quartz
10 free cm3 "Maximum Microcline"
10 free cm3 Dolomite-ord
```

set the initial system, with the minerals present in excess amounts. Typing go triggers the reaction path. Figure 19.1 shows the calculation results.

The mixing calculation is interesting in that it demonstrates a common ion effect by which dolomite precipitation drives feldspar alteration. In the model,

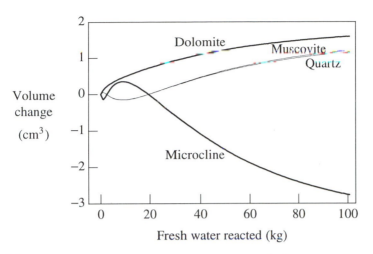

FIG. 19.1 Mineralogical consequences of mixing the two fluids shown in Table 19.1 at 60°C in the presence of microcline, muscovite, quartz, and dolomite. Results shown as the volume change for each mineral (precipitation is positive, dissolution negative), expressed per kg of pore water.

dolomite forms because the saline water is rich in Ca^{++} and Mg^{++}, whereas the fresh water contains abundant HCO_3^-. As the fluids mix, dolomite precipitates according to the reaction

$$Ca^{++} + Mg^{++} + 2\,HCO_3^- \rightarrow CaMg(CO_3)_2 + 2\,H^+ \qquad (19.1)$$
$$dolomite$$

producing H^+. Many of the hydrogen ions produced are consumed by driving the reaction of potassium feldspar

$$\underset{microcline}{3\,KAlSi_3O_8} + 2\,H^+ \rightarrow \underset{muscovite}{KAl_3Si_3O_{10}(OH)_2} + \underset{quartz}{6\ SiO_2} + 2\,K^+ \qquad (19.2)$$

to produce muscovite (illite) and quartz. Hence, the diagenetic reactions for carbonate and silicate minerals in the formation are closely linked.

Strangely, Reaction 19.2 proceeds backward in the early part of the calculation (Fig. 19.1), producing a small amount of potassium feldspar at the expense of muscovite and quartz. This result, quite difficult to explain from the perspective of mass transfer, is an activity coefficient effect. As seen in Fig. 19.2, the activity coefficient for K^+ increases rapidly as the fluid is diluted over the initial segment of the reaction path, whereas that for H^+ remains nearly constant. (The activity coefficients differ because the $\overset{\circ}{a}$ parameter in the Debye-

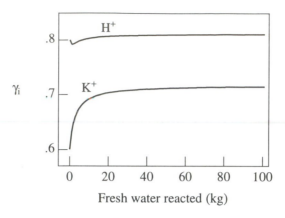

FIG. 19.2 Activity coefficients γ_i for the aqueous species H^+ and K^+ over the course of the mixing reaction shown in Fig. 19.1.

Hückel model is 3 Å for K^+ and 9 Å for H^+.) As a result, a_{K^+} increases more quickly than a_{H^+}, temporarily driving Reaction 19.2 from right to left.

The model shown is quite simple and, although certainly useful from a conceptual point of view, might be expanded to better describe petrographic observations. The reaction path shown does not predict that kaolinite forms, because we assumed rather arbitrarily that the fresh water begins in equilibrium with potassium feldspar and muscovite. If we chose a lower value for the initial activity ratio a_{K^+}/a_{H^+} (or select a less evolved meteoric water than shown in Table 19.1), the reaction eventually produces kaolinite, once the available microcline is consumed. We could also account for the oxidation of pyrite as it reacts with dissolved oxygen carried by the fresh water. Pyrite oxidation produces hydrogen ions, which might further drive the reactions to produce clay minerals (Harrison, 1990), but the fact that the formation fluid is depleted in sulfate relative to seawater (Table 19.1) suggests that sulfide oxidation plays a minor role in the overall diagenetic reaction.

19.2 Lyons Sandstone, Denver Basin

As a second example, we consider the origin of anhydrite ($CaSO_4$) and dolomite cements of the Permian Lyons sandstone in the Denver basin, which lies in Colorado and Wyoming, USA, to the east of the Front Range of the Rocky Mountains. The Lyons is locally familiar as a red building stone that outcrops in hogbacks (known as the Flatirons) along the Front Range. This red facies is a quartz sand in which the sand grains are coated with iron oxides and clays and cemented by quartz overgrowths and at least two generations of calcite ($CaCO_3$;

Hubert, 1960). The facies, where shallowly buried, provides a source of potable water.

Petroleum reservoirs, however, occur in a gray facies of the Lyons found in the deep basin (Levandowski et al., 1973). This facies contains no ferric oxides or calcite. Many grains in the facies are coated with bitumen, the remnants of oil that migrated through the rock, and the rock is cemented with anhydrite and dolomite. The anhydrite and dolomite cements occupy as much as 25% and 15%, respectively, of the rock's precement pore volume. The origin of these cements is of special interest because of their relationship to the distribution of petroleum reservoirs in the basin.

Anhydrite and dolomite cements are known to occur together in sediments that, shortly after burial in a sabkha environment, were invaded by evaporated seawater (e.g., Butler, 1969). The Denver basin contained evaporite subbasins in the late Paleozoic (Martin, 1965), but textural and isotopic evidence argues that the cements are unlikely to have formed in a sabkha. Cements in the gray facies overlie bitumen, so they must have formed after basin strata were buried deeply enough to generate oil. The oxygen isotopic composition of the dolomite (Levandowski et al., 1973), for which $\delta^{18}O_{SMOW}$ values are as low as +8.8 ‰, argues that the cement precipitated after the formation was buried from a fluid containing varying amounts of meteoric water. The dolomite would be composed of isotopically heavier oxygen if it had formed at surface temperatures from evaporated seawater.

The cements also might have formed if H_2S had migrated into the formation and oxidized to make anhydrite. The sulfur isotopic composition of the anhydrite, however, does not allow such an explanation. Values for $\delta^{34}S_{CDT}$, which vary from +9.6 to +12.5 ‰ (Lee and Bethke, 1994), span the worldwide range for Permian evaporite minerals, but are much heavier than values associated with H_2S in sedimentary basins. Furthermore, the cements could not have precipitated from the original pore fluid of the Lyons, because the solubility of anhydrite (as well as gypsum, $CaSO_4 \cdot 2H_2O$) in aqueous solution is much too low to account for the amounts of cement observed. For these reasons, the cements of the gray facies must have formed after the Lyons was buried, in a system open to groundwater flow.

Lee and Bethke (1994) suggested that the gray Lyons facies formed as an alteration product of the red facies in a groundwater flow regime set up by uplift of the Front Range, which began to rise in the early Tertiary and reached its peak in Eocene time (McCoy, 1953). Groundwater in the basin today flows from west to east (Belitz and Bredehoeft, 1988) in response to the elevation of the Front Range. Past flow was more rapid than in the present day because erosion has reduced the elevation of the basin's western margin. Paleohydrologic models calculated for the basin (Lee and Bethke, 1994) suggest that in the Eocene groundwater flowed eastward through the Lyons at an estimated discharge of about 1 m/yr.

Flow in the Pennsylvanian Fountain formation, a sandstone aquifer that underlies the Lyons and is separated from it by an aquitard complex, was more restricted because the formation grades into less permeable dolomites and evaporites in the deep basin. Groundwater in the Fountain recharged along the Front Range and flowed eastward at an estimated discharge of about 0.1 m/yr. Where Fountain groundwater encountered less permeable sediments along the basin axis, it discharged upward and mixed by dispersion into the Lyons formation.

According to Lee and Bethke's (1994) interpretation, the gray facies formed in this zone of dispersive mixing when saline Fountain groundwater reacted with Lyons sediments and groundwater. To simulate the mixing reaction, we start by developing chemical models of the two groundwaters, assuming that the mixing occurred at 100°C.

To set the initial composition of the Lyons fluid, we use an analysis of modern Lyons groundwater sampled at 51°C (McConaghy et al., 1964), which we correct to the temperature of the simulation by heating it in the presence of calcite and quartz. In REACT, the commands

```
T = 51
swap Calcite for H+
swap Quartz for SiO2(aq)

Na+      = 108    mg/kg
Ca++     =  40    mg/kg
K+       =   4.6  mg/kg
Mg++     =   1    mg/kg
Cl-      =   9    mg/kg
SO4      =  36    mg/kg
HCO3-    = 340    mg/kg
balance on HCO3-

1 free cm3 Calcite
1 free cm3 Quartz
```

describe an initial system containing a kilogram of groundwater and excess amounts of calcite and quartz. The commands

```
(cont'd)
T final = 100
go
```

cause the program to equilibrate the fluid and heat it to the temperature of interest. The resulting fluid is a predominantly sodium-bicarbonate solution (Table 19.2).

Since we have no direct information about the chemistry of the Fountain fluid, we assume that its composition reflects reaction with minerals in the evaporite strata that lie beneath the Lyons. We take this fluid to be a three molal NaCl solution that has equilibrated with dolomite, anhydrite, magnesite

TABLE 19.2 Predicted compositions of Lyons groundwater and Fountain brine, before mixing (Lee and Bethke, 1994)

	Lyons	Fountain
Na^+ (mg/kg)	108	56 400
Ca^{++}	16	516
K^+	4.6	—
Mg^{++}	.5	3 450
SiO_2(aq)	49	23
HCO_3^-	419	28 780
Cl^-	9	87 000
SO_4^{--}	36	14 200
pH (100°C)	6.7	4.6

($MgCO_3$), and quartz. The choice of NaCl concentration reflects the upper correlation limit of the B-dot (modified Debye-Hückel) equations (see Chapter 7). To set pH, we assume a CO_2 fugacity of 50, which we will show leads to a reasonable interpretation of the isotopic composition of the dolomite cement.

In fact, the choice of CO_2 fugacity has little effect on the mineralogical results of the mixing calculation. In the model, the critical property of the Fountain fluid is that it is undersaturated with respect to calcite, so that calcite dissolves when the fluid mixes into the Lyons. Because we assume equilibrium with dolomite and magnesite, the saturation index (log Q/K) of calcite is fixed by the reaction

$$\underset{calcite}{CaCO_3} \rightleftarrows \underset{dolomite}{CaMg(CO_3)_2} - \underset{magnesite}{MgCO_3} \qquad (19.3)$$

to a value of -1.3, and hence, is independent of pH and CO_2 fugacity.

To calculate the composition of the Fountain brine, we start anew in REACT, enter the commands

```
T = 100
swap CO2(g) for H+
swap Magnesite for Mg++
swap Anhydrite for SO4--
swap Dolomite-ord for Ca++
swap Quartz for SiO2(aq)

Na+ = 3 molal
```

```
Cl- = 3 molal
f CO2(g) = 50
1 free mole Magnesite
1 free mole Anhydrite
1 free mole Dolomite-ord
1 free mole Quartz

balance on HCO3-
```

and type `go`. In contrast to the Lyons groundwater, the Fountain brine (Table 19.2) is a sodium chloride water.

To model the mixing of these fluids in contact with quartz and calcite of the red Lyons formation, we follow three steps. First, we calculate the composition of Lyons groundwater, as before, and save it to a file. Second, we compute the composition of the Fountain brine. Finally, we "pick up" the Fountain brine as a reactant, multiply its mass by 15 (giving 15 kg of solvent plus the solute mass), and titrate it into a system containing 1 kg of Lyons groundwater and excess amounts of quartz and calcite. Here, the factor 15 is largely arbitrary; we chose it to give a calculation endpoint with a high ratio of brine to Lyons groundwater. The REACT procedure, starting anew, is:

— Step 1 —

(constrain composition of Lyons groundwater, as before)
```
T initial = 51, final = 100
go

pickup fluid
save Lyons_100
```

— Step 2 —

```
reset
```
(constrain composition of Fountain brine, as before)
```
T = 100
go
```

— Step 3 —

```
pickup reactants = fluid
reactants times 15

read Lyons_100
swap Quartz for SiO2(aq)
swap Calcite for HCO3-
100 free mol Quartz
100 free mol Calcite
balance on Cl-
go
```

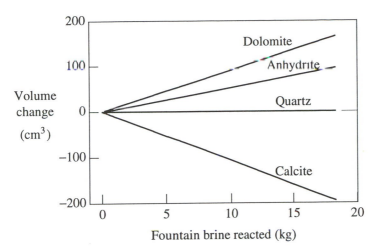

FIG. 19.3 Mineralogical consequences of mixing Fountain brine into the Lyons formation. Vertical axis shows changes in mineral volume, expressed per kg of Lyons groundwater; positive changes indicate precipitation and negative, dissolution.

In the resulting reaction (Fig. 19.3), calcite progressively dissolves as the Fountain brine mixes into the Lyons formation. The Ca^{++} and HCO_3^- added to solution drive precipitation of anhydrite and dolomite by a common ion effect. The overall reaction

$$5 \; CaCO_3 + 2 \; SO_4^- + \frac{5}{2} \; Mg^{++} \rightarrow$$
$$\underset{\text{calcite}}{}$$

$$2 \underset{\text{anhydrite}}{CaSO_4} + \frac{5}{2} \underset{\text{dolomite}}{CaMg(CO_3)_2} + \frac{1}{2} Ca^{++} \qquad (19.4)$$

explains the origin of the cements in the gray facies as well as the facies' lack of calcite.

We can predict the oxygen and carbon isotopic compositions of the dolomite produced by this reaction path, using the techniques described in Chapter 15. Figure 19.4 shows the compositions of calcite and dolomite cements in the Lyons, as determined by Levandowski et al. (1973). The calcite and dolomite show broad ranges in oxygen isotopic content. The dolomite, however, spans a much narrower range in carbon isotopic composition than does the calcite.

To set up the calculation, we specify initial isotopic compositions for the fluid and calcite. We choose a value of $-13‰$ for $\delta^{18}O_{SMOW}$ of the Lyons fluid, reflecting Tertiary rainfall in the region, and set the calcite composition to $+11‰$, the mean of the measured values (Fig. 19.4). We further set $\delta^{13}C_{PDB}$ for the fluid to $-12‰$. We do not specify an initial carbon composition for the

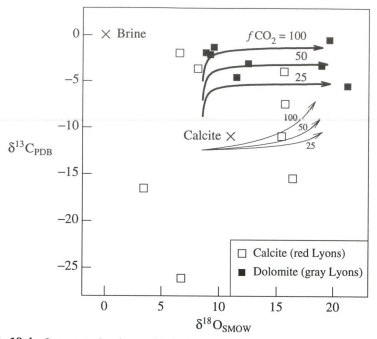

FIG. 19.4 Oxygen and carbon stable isotopic compositions of calcite (□) and dolomite (■) cements from Lyons sandstone (Levandowski et al., 1973), and isotopic trends (bold arrows) predicted for dolomite cements produced by mixing reaction shown in Fig. 19.3, assuming differing CO_2 fugacities (25, 50, and 100) for the Fountain brine. Fine arrows, for comparison, show isotopic trends predicted in calculations which assume (improperly) that fluid and minerals maintain isotopic equilibrium over the course of the simulation. Figure after Lee and Bethke (1996).

calcite, so the model sets this value to −11‰, in isotopic equilibrium with the fluid. Again, this value is near the mean of the measurements.

We then set the isotopic compositions of each oxygen and carbon-bearing component in the reactant, the Fountain brine, to $\delta^{18}O$ and $\delta^{13}C$ values of zero, as might be expected in a sedimentary brine. Finally, we segregate each mineral in the calculation from isotopic exchange, as discussed in Chapter 15. The procedure for small water/rock ratios is

```
(cont'd)
oxygen initial = -13, Calcite = +11
carbon initial = -12

carbon HCO3- = 0
oxygen H2O = 0, SiO2(aq) = 0, HCO3- = 0, SO4-- = 0
segregate Calcite, Quartz, Dolomite-ord, Anhydrite
```

```
reactants times 1
go
```

We then repeat the calculation to carry the model to high ratios with the commands

```
(cont'd)
reactants times 50
go
```

Figure 19.4 shows the predicted isotopic trends for the dolomite produced by the reaction path, calculated assuming several values for the CO_2 fugacity of the Fountain brine. These results suggest that the cement's carbon isotopic composition reflects the composition of a CO_2-rich brine more closely than that of the precursor calcite cement, explaining the narrow range observed in $\delta^{13}C$. The spread in oxygen composition results from mixing of fresh Lyons water and Fountain brine in varying proportions.

As noted by Lee and Bethke (1996), if we had calculated this model without holding minerals segregated from isotopic exchange, we would have predicted broadly different isotopic trends that are not in accord with the observed data. To verify this point, we enter the command

```
(cont'd)
unsegregate ALL
```

and type go. The resulting isotopic trends for the equilibrium case are shown in Fig. 19.4.

As we have demonstrated, reaction modeling provides an explanation of the observed diagenetic mineralogy as well as the isotopic compositions of the cements. The model also helps explain the association of the gray facies with oil reservoirs. Fractures that developed along the basin axis when the Front Range was uplifted provided pathways for oil to migrate by buoyancy upward into the Lyons. Continued uplift set up regional groundwater flow that drove brine from the Fountain formation upward along the same pathways as the oil. The petroleum or brine, or both, reduced iron oxides in the Lyons, changing its color from red to gray. As brine mixed into the Lyons, it dissolved the existing calcite cement and, by a common ion effect, precipitated anhydrite and dolomite on top of bitumen left behind by migrating oil.

Oil in the Cretaceous Dakota sandstone, a shallower aquifer than the Lyons, has migrated laterally as far as 150 km into present-day reservoirs (Clayton and Swetland, 1980). In contrast, oil has yet to be found in the Lyons outside the deep strata where it was generated. The formation of anhydrite and dolomite cements may have served to seal the oil into reservoirs, preventing it from migrating farther.

20

Kinetic Reaction Paths

In calculating most of the reaction paths in this book, we have measured reaction progress with respect to the dimensionless variable ξ. We showed in Chapter 14, however, that by incorporating kinetic rate laws into a reaction model, we can trace reaction paths using time as the reaction coordinate.

In this chapter we construct a variety of kinetic reaction paths to explore how this class of models behaves. Our calculations in each case are based on kinetic rate laws determined by laboratory experiment. In considering the calculation results, therefore, it is important to keep in mind the uncertainties entailed in applying laboratory measurements to model reaction processes in nature, as discussed in detail in Section 14.2.

20.1 Approach to Equilibrium and Steady State

In Chapter 14 we considered how quickly quartz dissolves into water at 100°C, using a kinetic rate law determined by Rimstidt and Barnes (1980). In this section we take up the reaction of silica (SiO_2) minerals in more detail, this time working at 25°C. We use kinetic data for quartz and cristobalite from the same study, as shown in Table 20.1.

Each silica mineral dissolves and precipitates in our calculations according to the rate law

$$r_{SiO_2} = A_S \, k_+ \left[\frac{Q}{K} - 1 \right] \tag{20.1}$$

(Eqn. 14.13) as discussed in Chapter 14. Here, r_{SiO_2} is the reaction rate (mol/sec; positive for precipitation), A_S and k_+ are the mineral's surface area (cm^2) and

TABLE 20.1 Rate constants k_+ (mol/cm^{2}sec) for the reaction
of silica minerals with water at various temperatures,
as determined by Rimstidt and Barnes (1980)

T(°C)	Quartz	α-Cristobalite	Amorphous silica
25	4.20×10^{-18}	1.71×10^{-17}	7.32×10^{-17}
70	2.30×10^{-16}	6.47×10^{-16}	2.19×10^{-15}
100	1.88×10^{-15}	4.48×10^{-15}	1.33×10^{-14}
150	3.09×10^{-14}	6.12×10^{-14}	1.49×10^{-13}
200	2.67×10^{-13}	4.81×10^{-13}	9.81×10^{-13}
250	1.46×10^{-12}	2.55×10^{-12}	4.43×10^{-12}
300	5.71×10^{-12}	1.01×10^{-11}	1.51×10^{-11}

rate constant (mol/cm^{2}sec), and Q and K are the activity product and equilibrium constant for the dissolution reaction. The reaction for quartz, for example, is

$$SiO_2 \rightleftarrows SiO_2(aq) \qquad\qquad (20.2)$$
$$\textit{quartz}$$

According to Knauss and Wolery (1988), this rate law is valid for neutral to acidic solutions; a distinct rate law applies in alkaline fluids, reflecting the dominance of a second reaction mechanism under conditions of high pH.

The procedure in REACT is similar to that used in the earlier calculation (Section 14.4)

```
time end = 1 year
1 mg/kg SiO2(aq)

react 5000 grams Quartz
kinetic Quartz   rate_con = 4.2e-18, surface = 1000
go
```

except that we work at 25°C and, since reaction proceeds more slowly at low temperature, set a longer time span. By including 5 kg of quartz sand in the calculation, we imply that the system's porosity is about 35%, since the density of quartz is 2.65 g/cm^3. At the specified silica concentration of 1 mg/kg, the initial fluid is undersaturated with respect to quartz, so we can expect quartz to dissolve over the reaction path.

In the calculation results (Fig. 20.1), the silica concentration gradually increases from the initial value, asymptotically approaching the equilibrium value of 6 mg/kg after about half a year of reaction. We repeat the calculation, this time starting with a supersaturated fluid

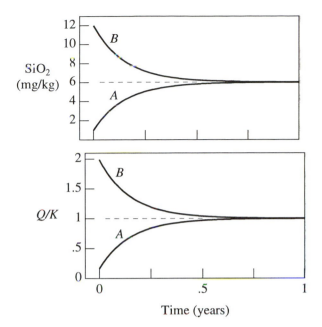

FIG. 20.1 Reaction of quartz with water at 25°C, showing approach to equilibrium (dashed lines) with time. Top diagram shows variation in SiO_2(aq) concentration and bottom plot shows change in quartz saturation. In calculation A, the fluid is initially undersaturated with respect to quartz; in B it is supersaturated.

(cont'd)
```
12 mg/kg SiO2(aq)
suppress Tridymite, Chalcedony
go
```

To keep our discussion simple for the moment, we suppress the silica polymorphs tridymite and chalcedony. In the calculation results (Fig. 20.1), the silica concentration gradually decreases from its initial value and, as in the previous calculation, approaches equilibrium with quartz after about half a year.

We could have anticipated the results in Fig. 20.1 from the form of the rate law (Eqn. 20.1). If we let m_{SiO_2} represent the molality of SiO_2(aq) and m_{eq} represent this value at equilibrium, we can rewrite the rate law as

$$\frac{dm_{SiO_2}}{dt} = -\frac{A_S \, k_+}{n_w \, m_{eq}} \, (m_{SiO_2} - m_{eq}) \qquad (20.3)$$

Here, we have assumed that the activity coefficient γ_{SiO_2} does not vary with silica concentration. As before, n_w is the mass of solvent water in the system.

Since we can take each variable except m_{SiO_2} to be constant, Eqn. 20.3 has the form of an ordinary differential equation in time. We can use standard techniques to solve the equation for $m_{SiO_2}(t)$. The solution corresponding to the initial condition $m_{SiO_2} = m_0$ at $t = 0$ is

$$m_{SiO_2} = (m_0 - m_{eq})\, e^{-(A_S\, k_+/n_w\, m_{eq})\, t} + m_{eq} \qquad (20.4)$$

The first term on the right side of the solution represents the extent to which the silica concentration deviates from equilibrium. Since the term appears as a negative exponential function in time, its value decays to zero (as can be seen in Fig. 20.1) at a rate that depends on the surface area and rate constant. As t becomes large, the first term disappears, leaving only the equilibrium concentration m_{eq}.

The calculation results in Fig. 20.1 suggest that groundwaters in quartz sand aquifers should approach equilibrium with quartz in less than a year, a short period compared to typical residence times in groundwater flow regimes. At low temperature, however, groundwaters in nature seldom appear to be in equilibrium with quartz, often appearing supersaturated. As discussed in Section 14.2, this discrepancy between calculation and observation might be accounted for if the surfaces of quartz grains in real aquifers were coated in a way that inhibited reaction. Alternatively, the discrepancy may arise from the effects of reactions with minerals other than quartz that consume and produce silica. If these reactions proceed rapidly compared to the dissolution and precipitation of quartz, they can control the fluid's silica content.

To see how a second kinetic reaction might affect the fluid's silica concentration, we add 250 grams of cristobalite to the system. The mass ratio of quartz to cristobalite, then, is twenty to one. Taking the fluid to be in equilibrium with quartz initially, the procedure in REACT is

```
time end = 1 year
swap Quartz for SiO2(aq)
5000 free grams Quartz

react 250 grams of Cristobalite
kinetic Cristobalite   rate_con = 1.7e-17   surface = 5000
kinetic Quartz         rate_con = 4.2e-18   surface = 1000

suppress Tridymite, Chalcedony
go
```

Here, we take a specific surface area for cristobalite five times greater than the value assumed for quartz. To calculate a second case in which the fluid starts the reaction path in equilibrium with cristobalite, we enter the commands

```
(cont'd)
swap Cristobalite for SiO2(aq)
250 free grams Cristobalite
```

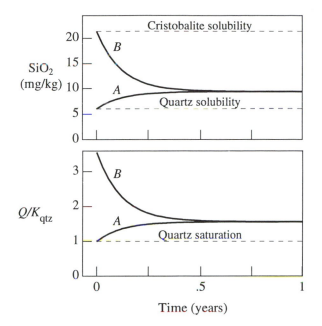

FIG. 20.2 Kinetic reaction of quartz and cristobalite with water at 25°C. In calculation *A* the fluid is originally in equilibrium with quartz, in *B* with cristobalite. The top diagram shows how the $SiO_2(aq)$ concentration varies with time, and the bottom plot shows the change in quartz saturation. The reaction paths approach a steady state in which the fluid is supersaturated with respect to quartz and undersaturated with respect to cristobalite.

```
react 0 grams Cristobalite
react 5000 grams Quartz
go
```

Figure 20.2 shows the results of the two calculations.

As in the previous example (Fig. 20.1), silica concentration in the calculations asymptotically approaches a single value, regardless of the initial concentration. The final silica concentration, however, does not represent a thermodynamic equilibrium, since (as we can see in Fig. 20.2) it is supersaturated with respect to quartz and undersaturated with respect to cristobalite.

Instead, this concentration marks a *steady state* (or *dynamic equilibrium*, a term that persists despite being an oxymoron) at which the rate of cristobalite dissolution matches the rate at which quartz precipitates. The steady state concentration will decay toward equilibrium with continued reaction only to the extent that the surface area of cristobalite decreases as the mineral dissolves. Because only a few tens of milligrams of cristobalite dissolve each year (as we can quickly compute using Eqn. 20.1), the steady state will persist for a long time, decaying only gradually over tens of thousands of years.

We can calculate the value of the steady state silica concentration m_{SS} directly from the rate law (Eqn. 20.1). Noting that the steady state is marked by the condition $r_{qtz} = -r_{cri}$, we can write

$$m_{SS} = \frac{1}{\gamma_{SiO_2}} \cdot \frac{(A_S\, k_+)_{qtz} + (A_S\, k_+)_{cri}}{(A_S\, k_+/K)_{qtz} + (A_S\, k_+/K)_{cri}} \tag{20.5}$$

To evaluate this equation, we use the values of the rate constants k_+ and surface areas A_S (the latter given as the product of specific surface area and mineral mass) for the two minerals and the equilibrium constants K for quartz (1.00×10^{-4}) and cristobalite (3.56×10^{-4}), and take γ_{SiO_2} to be one. The resulting steady-state concentration is 1.57×10^{-4} molal, or 9.4 mg/kg, which agrees with the simulation results in Fig. 20.2.

It is interesting to note that adding only a small amount of cristobalite to the system gives rise to a significant departure from equilibrium with the predominant mineral, quartz. The cristobalite plays a role in the calculation disproportionate to its abundance because the assumed values for its surface area and rate constant are considerably larger than those for quartz. In nature, therefore, we might expect highly reactive minerals to have significant effects on fluid chemistry, even when they are present in small quantities.

It is further interesting to observe that the behavior of a system approaching a thermodynamic equilibrium differs little from one approaching a steady state. According to the kinetic interpretation of equilibrium, as discussed in Chapter 14, a mineral is saturated in a fluid when it precipitates and dissolves at equal rates. At a steady state, similarly, the net rate at which a component is consumed by the precipitation reactions of two or more minerals balances with the net rate at which it is produced by the minerals' dissolution reactions. Thermodynamic equilibrium viewed from the perspective of kinetic theory, therefore, is a special case of the steady state.

In experimental studies, it is common practice to attempt to "bracket" a measured solubility by reacting a sample with undersaturated as well as supersaturated solutions. As is shown in Fig. 20.2, however, this technique might equally well identify a steady-state condition as an equilibrium state.

20.2 Quartz Deposition in a Fracture

In a second application, we consider the rate at which quartz might precipitate in an open fracture as a hydrothermal water flows along it, gradually cooling from 300°C to surface temperature. We assume that the fracture has an aperture δ and is lined with quartz. We further assume that the fluid is initially in equilibrium with quartz and that temperature varies linearly along the fracture. The latter condition imposes an important constraint on the calculation since at high rates of discharge the fluid can in reality control temperature along the fracture, causing temperature to deviate broadly from the assumed linear gradient. To investigate such conditions, we would want to construct a more sophisticated model in which we specifically account for advective heat transfer.

In our calculation, we need not be concerned with the dimensions of the fracture or the velocity of the fluid. Instead, we need specify only the length of time Δt it takes the fluid to travel from high-temperature conditions at depth to cool surface conditions.

To model the problem, we take a packet of water in contact with the fracture walls over a polythermal reaction path. The fact that the packet moves relative to the walls is of no concern, since the fracture surface area exposed to the packet is approximately constant. Since the system contains 1 kg of water, we can show from geometry that the surface area A_S (in cm^2) of the fracture lining is

$$A_S = \frac{2000\ \psi}{\rho\ \delta} \qquad (20.6)$$

Here, ψ is the surface roughness (surface area per unit area in cross section) of the fracture walls, ρ is fluid density in g/cm^3, and the aperture δ is taken in cm. For our purposes, it is sufficiently accurate to choose a value of 2 for ψ and set ρ to 1 g/cm^3. In a fracture with an aperture of 10 cm, for example, each kg of water is exposed to 400 cm^2 of quartz surface.

We use the Arrhenius equation

$$k_+ = A\ e^{-E_A/RT_K} \qquad (20.7)$$

(Eqn. 14.3) in our calculations to set the rate constant k_+ as a function of temperature along the fracture. A preexponential factor A of about 2.35×10^{-5} mol/cm^2 sec and an activation energy E_A of 72,800 J/mol fit the kinetic data for quartz in Table 20.1.

To assign the amount of quartz in the system, we arbitrarily specify a specific surface area of 1 cm^2/g. Then, we need only set the quartz mass to a value in grams equal to the desired surface area in cm^2. Finally, we set for each run the amount of time Δt it takes the packet of water to flow along the fracture.

To model the effects of flow through a 10-cm wide fracture, assuming a time span of one year, for example, the procedure in REACT is

```
time end = 1 year
T initial = 300, final = 25

swap Quartz for SiO2(aq)
400 free grams Quartz
kinetic Quartz   surface = 1
kinetic Quartz   pre-exp = 2.35e-5, act_eng = 72800

suppress Tridymite Chalcedony Cristobalite Amrph^silica
delxi = .001
go
```

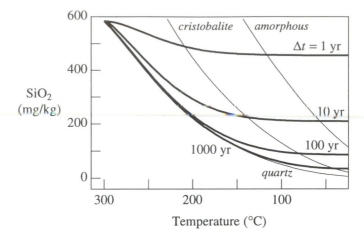

FIG. 20.3 Silica concentration (bold lines) in a fluid packet that cools from 300°C as it flows along a quartz-lined fracture of 10 cm aperture, calculated assuming differing traversal times Δt. Fine lines show solubilities of the silica polymorphs quartz, cristobalite, and amorphous silica.

By suppressing the other silica polymorphs, we prevent them from forming in the calculation. We will show, however, that except at large Δt these minerals tend to become supersaturated in the low-temperature end of the fracture. In reality, therefore, these minerals would be likely to form within the fracture under such conditions.

Figure 20.3 shows how SiO_2 concentration in the fluid varies along the fracture, calculated assuming different traversal times Δt. For slow flow rates (values of Δt of about 1000 years or longer), the fluid has enough reaction time to remain near equilibrium with quartz. When flow is more rapid, however, the fluid maintains much of its silica content, quickly becoming supersaturated with respect to quartz. Farther along the fracture, the fluid also becomes supersaturated with respect to cristobalite, and at traversal times of less than about 100 years, with respect to amorphous silica.

It is interesting to examine Fig. 20.3 in light of our discussion in Chapter 17 of the silica geothermometer. If we wish to derive an estimate for the fluid's original temperature of 300°C from its silica content under surface conditions, then our calculations suggest that under the modeled conditions the fluid must traverse the fracture (e.g., a fracture through the cap rock in a geothermal field, assuming that temperature in the cap rock varies linearly with depth) in less than one year.

We can use the calculated reaction rates (Fig. 20.3) to compute how rapidly quartz precipitation seals the fracture. The sealing rate, the negative rate at which fracture aperture changes, can be expressed

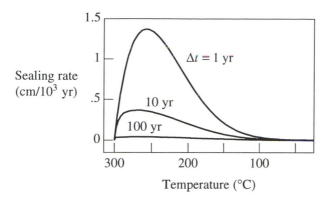

Sealing rate (cm/10^3 yr)

$\Delta t = 1$ yr

10 yr

100 yr

Temperature (°C)

FIG. 20.4 Fracture sealing rates (cm/10^3 yr) for the quartz precipitation calculations shown in Fig. 20.3.

$$-\frac{d\delta}{dt} = \frac{r_{qtz}\, M_V\, \rho\delta}{1000} \qquad (20.8)$$

Here, r_{qtz} is the rate of quartz precipitation (mol/sec from a kg of water) and M_V is the mineral's molar volume (22.7 cm^3/mol). Figure 20.4 shows the resulting sealing rates calculated for several traversal times. For a Δt of one year, for example, we expect the fracture to become occluded near its high-temperature end over a time scale of about 10,000 years.

20.3 Silica Transport in an Aquifer

Next we look at how temperature gradients along an aquifer might affect the silica content of flowing groundwater. We consider a symmetrical aquifer that descends from a recharge area at the surface to a depth of about 2 km and then ascends to a discharge area. Temperature in the calculation varies linearly from 20°C at the surface to 80°C at the aquifer's maximum depth.

To set up the calculation, we take a quartz sand of the same porosity as in the calculations in Section 20.1 and assume that the quartz reacts according to the same rate law (Eqn. 20.1). We let the rate constant vary with temperature according to the Arrhenius equation (Eqn. 20.7), using the values for the preexponential factor and activation energy given in Section 20.2. As in the previous section, we need only be concerned with the time available for water to react as it flows through the aquifer. We need not specify, therefore, either the aquifer length or the flow velocity.

To model reaction within the descending leg of the flow path, along which water warms with depth, the procedure in REACT is

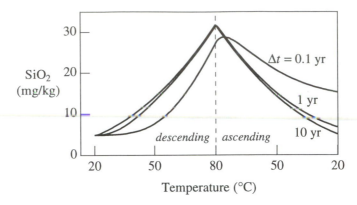

FIG. 20.5 Calculated silica concentration in a fluid packet flowing through a quartz sand aquifer. The fluid descends from the surface ($T = 20°C$) to a depth of about 2 km ($80°C$) and then returns to the surface ($20°C$). Results are shown for time spans Δt (representing half of the time the fluid takes to migrate through the aquifer) of 0.1, 1, and ten years. In the latter calculation, the fluid remains near equilibrium with quartz.

```
time end = 1 year
T initial = 20, final = 80
swap Quartz for SiO2(aq)
5000 free grams Quartz

kinetic Quartz   surface = 1000
kinetic Quartz   pre-exp = 2.35e-5   act_eng = 72800

suppress Tridymite, Chalcedony
go
```

Then, to model the ascending limb, we start with the final composition of the descending fluid and let it cool. The corresponding commands are

```
(cont'd)
pickup
T final 20
react 5000 grams Quartz
kinetic Quartz   surface = 1000
kinetic Quartz   pre-exp = 2.35e-5   act_eng = 72800
go
```

Here we take a reaction time Δt (the time it takes water to descend to maximum depth or to ascend back to the surface) of one year. We then repeat the calculation assuming other reaction times.

   In the calculation results (Fig. 20.5), silica concentration increases as the fluid flows downward, reflecting the increase in quartz solubility with depth, and then decreases as fluid moves back toward the surface. In calculations in which the reaction time exceeds about ten years, the fluid remains close to equilibrium with quartz. Given shorter reaction times, however, the fluid does not react quickly enough to respond to the changing quartz solubility along the flow path. Along the descending limb, the fluid appears undersaturated with silica, and it becomes supersaturated along the upflowing leg of the flow path. If we do not suppress the other silica polymorphs in runs assuming short reaction times, tridymite (the most stable) tends to precipitate as the fluid cools.

The results in Fig. 20.5 are interesting because they suggest that gradual temperature variations along quartz aquifers are unlikely to produce silica concentrations that deviate significantly from equilibrium, except for fluids that quickly traverse the flow system. The effects of other reactions that consume and produce silica, as discussed in Section 20.1, would appear more likely to cause disequilibrium between groundwater and quartz than would flow along temperature gradients.

20.4 Ostwald's Step Rule

Ostwald's step rule holds that a thermodynamically unstable mineral reacts over time to form a sequence of progressively more stable minerals (e.g., Morse and Casey, 1988; Steefel and Van Cappellen, 1990; Nordeng and Sibley, 1994). The step rule is observed to operate, especially at low temperature, in a number of mineralogic systems, including the carbonates, silica polymorphs, iron and manganese oxides, iron sulfides, phosphates, clay minerals, and zeolites.

Various theories, ranging from qualitative interpretations to those rooted in irreversible thermodynamics and geochemical kinetics, have been put forward to explain the step rule. A kinetic interpretation of the phenomenon, as proposed by Morse and Casey (1988), may provide the most insight. According to this interpretation, Ostwald's sequence results from the interplay of the differing reactivities of the various phases in the sequence, as represented by A_S and k_+ in Eqn. 20.1, and the thermodynamic drive for their dissolution and precipitation of each phase, represented by the $(Q/K - 1)$ term.

To investigate the kinetic explanation for the step rule, we model the reaction of three silica polymorphs — quartz, cristobalite, and amorphous silica — over time. We consider a system that initially contains 100 cm^3 of amorphous silica, the least stable of the polymorphs, in contact with 1 kg of water, and assume that the fluid is initially in equilibrium with this phase. We include in the system small amounts of cristobalite and quartz, thereby avoiding the question of how best to model nucleation. In reality, nucleation, crystal growth, or both of these factors might control the nature of the reaction; we will consider only the effect of crystal growth in our simple calculation.

Each mineral in the calculation dissolves and precipitates according to the kinetic rate law (Eqn. 20.1) used in the previous examples and the rate constants listed in Table 20.1. We take the same specific surface areas for quartz and cristobalite as we did in our calculations in Section 20.1, and assume a value of

20,000 cm^2/g for the amorphous silica, consistent with measurements of Leamnson et al. (1969). The procedure in REACT is

```
time end = 400000 years
swap Amrph^silica for SiO2(aq)
100 free cm3 Amrph^silica

react 0.1 cm3 of Cristobalite
react 0.1 cm3 of Quartz

kinetic Amrph^silica   rate_con = 7.3e-17   surface = 20000
kinetic Cristobalite   rate_con = 1.7e-17   surface = 5000
kinetic Quartz         rate_con = 4.2e-18   surface = 1000

suppress Tridymite, Chalcedony
delxi  = .001
dxplot = .001
go
```

In the calculation results (Fig. 20.6), the initial segment of the path is marked by the disappearance of the amorphous silica as it reacts to form cristobalite. The amorphous silica is consumed nearly completely after about 10,000 years of reaction. The mineral's mass approaches zero asymptotically because (as can be seen in Eqn. 20.1) as its surface area A_S decreases, the dissolution rate slows proportionately. During the initial period, only a small amount of quartz forms.

Once the amorphous silica has nearly disappeared, the cristobalite that formed early in the calculation begins to redissolve to form quartz. The cristobalite dissolves, however, much more slowly than it formed, reflecting the slow rate of quartz precipitation. After about 300,000 years of reaction, nearly all of the cristobalite has been transformed into quartz, the most stable silica polymorph, and the reaction has virtually ceased.

The step sequence in the calculation results arises from the fact that the values assumed for the specific surface areas A_{sp} (cm^2/g) and rate constants k_+ ($mol/cm^2 sec$), and hence the product of these two terms, decrease among the minerals with increasing thermodynamic stability:

Mineral	$\log K$	A_{sp}	$k_+ \times 10^{18}$	$A_{sp} \cdot k_+ \times 10^{15}$
Amorphous silica	−2.71	20 000	73	1460
Cristobalite	−3.45	5000	17	85
Quartz	−4.00	1000	4.2	4.2

The product $A_{sp} \cdot k_+$ for amorphous silica is about 17 times greater than it is for cristobalite, and this value in turn exceeds the value for quartz about twenty-fold. The least stable minerals, therefore, are the most reactive.

In the first segment of the calculation, the high reactivity of the amorphous silica assures that the fluid remains near equilibrium with this mineral, as shown in Fig. 20.7. Only after the mineral has nearly disappeared does the silica concentration begin to decrease. Since the surface area and rate constant for

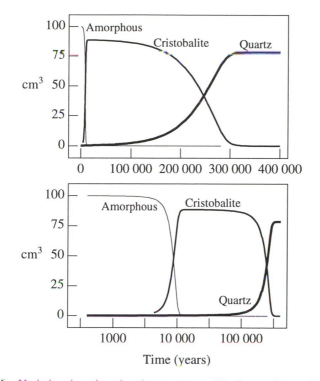

FIG. 20.6 Variation in mineral volumes over a kinetic reaction path designed to illustrate Ostwald's step sequence. The calculation traces the reaction at 25°C among the minerals amorphous silica (fine line), cristobalite (medium line), and quartz (bold line). The top diagram shows results plotted against time on a linear scale; the time scale on the bottom diagram is logarithmic. The decrease in total volume with time reflects the differing molar volumes of the three minerals.

cristobalite are considerably greater than those of quartz, the fluid remains near equilibrium with cristobalite until it in turn nearly disappears. Finally, after several hundred thousand years of reaction, the fluid approaches saturation with quartz and hence thermodynamic equilibrium.

20.5 Dissolution of Albite

In a final application of kinetic reaction modeling, we consider how sodium feldspar (albite, $NaAlSi_3O_8$) might dissolve into a subsurface fluid at 70°C. We consider a Na-Ca-Cl fluid initially in equilibrium with kaolinite $[Al_2Si_2O_5(OH)_4]$, quartz, muscovite $[KAl_3Si_3O_{10}(OH)_2$, a proxy for illite], and calcite ($CaCO_3$), and in contact with a small amount of albite. Feldspar cannot be in equilibrium with quartz and kaolinite, since the minerals will react to form

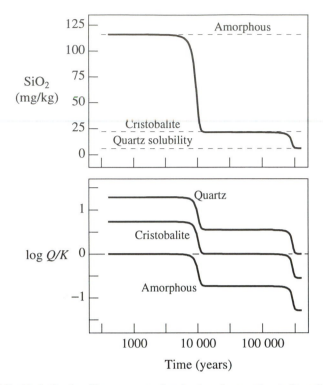

FIG. 20.7 Variation in silica concentration (top) and saturation indices (log Q/K) of the silica polymorphs (bottom) over the course of the reaction path shown in Fig. 20.6. The dashed lines in the top diagram show SiO_2(aq) concentrations in equilibrium with quartz, cristobalite, and amorphous silica.

a mica or a mica-like clay mineral such as illite. Hence, the initial fluid is necessarily undersaturated with respect to the albite.

We assume that albite and quartz react with the fluid according to kinetic rate laws. We take a rate law for albite

$$r_{alb} = A_S \, k_+ \left[\frac{Q}{K} - 1 \right] \tag{20.9}$$

of the form discussed in Chapter 14 (Eqn. 14.5). According to Knauss and Wolery (1986), this form is valid at 70°C over a range in pH of 2.9 to 8.9. The corresponding rate constant k_+ at this temperature is 7.9×10^{-16} mol/cm^{2}sec. The law for quartz (Eqn. 20.1) is the same one used in the previous examples in this chapter. The rate constant for quartz at 70°C (Table 20.1) is 2.3×10^{-16} mol/cm^{2}sec. For both minerals we assume a specific surface area of 1000 cm^2/g, which is reasonable for sand-size grains. All other minerals in the system remain in equilibrium with the fluid over the reaction path.

The procedure in REACT is

```
time end = 1500 years
T = 70

pH = 5.7
Na+    = .3  molal
Ca++   = .05 molal
Cl-    = .4  molal

swap Kaolinite for Al+++
swap Quartz     for SiO2(aq)
swap Muscovite for K+
swap Calcite    for HCO3-
 10 free grams Kaolinite
100 free grams Quartz
 10 free grams Muscovite
  1 free gram  Calcite

react 25 grams Albite
kinetic Albite   rate_con = 7.9e-16   surface = 1000
kinetic Quartz   rate_con = 2.3e-16   surface = 1000

delxi = .001
go
```

The predicted reaction path (Fig. 20.8) is interesting because the albite twice achieves saturation with the fluid. In the initial part of the calculation, the albite dissolves

$$NaAlSi_3O_8 + .4\, CO_2(aq) + .1\, HCO_3^- + \tfrac{1}{2}\, Ca^{++} + .1\, K^+ + .9\, H_2O \rightarrow Na^+ +$$
albite

$$.4\, Al_2Si_2O_5(OH)_4 + .1\, KAl_3Si_3O_{10}(OH)_2 + \tfrac{1}{2}\, CaCO_3 + 2\, SiO_2(aq) \quad (20.10)$$
   kaolinite   *muscovite*   *calcite*

to produce kaolinite, muscovite, and calcite and to add silica to the solution. Some of the silica precipitates to form quartz

$$\tfrac{1}{2}\, SiO_2(aq) \rightarrow \tfrac{1}{2}\, SiO_2 \quad (20.11)$$
         quartz

but the reaction rate is not sufficient to consume all of the silica produced by Reaction 20.10. The remaining silica accumulates in solution, quickly causing the silica polymorph tridymite to become saturated:

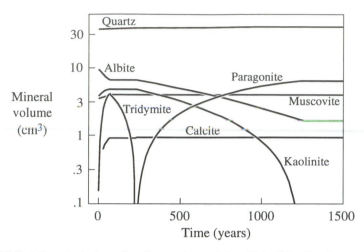

FIG. 20.8 Mineralogical results of a reaction path in which albite dissolves and quartz precipitates at 70°C according to kinetic rate laws.

$$1.5 \ SiO_2(aq) \rightarrow 1.5 \quad SiO_2 \qquad\qquad (20.12)$$
$$\textit{tridymite}$$

and precipitate.

When tridymite has formed, the $SiO_2(aq)$ activity and hence the saturation state of quartz is fixed at a constant value (Fig. 20.9). Albite continues to dissolve for about 100 years until, according to the reaction

$$NaAlSi_3O_8 + \tfrac{1}{2} \ H_2O + H^+ \rightleftarrows Na^+ + \tfrac{1}{2} \ Al_2Si_2O_5(OH)_4 + 2 \quad SiO_2 \qquad (20.13)$$
$$\textit{albite} \qquad\qquad\qquad\qquad \textit{kaolinite} \qquad \textit{tridymite}$$

it becomes saturated in fluid. Note that although albite and kaolinite cannot achieve equilibrium in the presence of quartz, they can coexist in a metastable equilibrium with tridymite.

From this point until about 225 years of reaction time, when the last tridymite disappears, the only reaction occurring in the system

$$SiO_2 \quad \rightarrow \quad SiO_2 \qquad\qquad (20.14)$$
$$\textit{tridymite} \qquad \textit{quartz}$$

is the conversion of tridymite to quartz. Once the conversion is complete, the silica activity decreases, causing the albite to become undersaturated once again (Fig. 20.9). It begins to dissolve

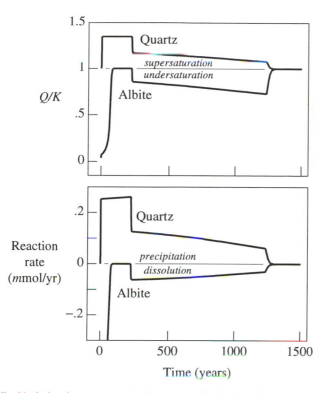

FIG. 20.9 Variation in quartz and albite saturation (top) and the kinetic reaction rates for these minerals (bottom) over the course of the reaction path shown in Fig. 20.8.

$$NaAlSi_3O_8 + Al_2Si_2O_5(OH)_4 \rightarrow NaAl_3Si_3O_{10}(OH)_2 + 2\ SiO_2 + H_2O \quad (20.15)$$
$$\text{\textit{albite} \quad\quad \textit{kaolinite} \quad\quad\quad \textit{paragonite} \quad\quad \textit{quartz}}$$

consuming kaolinite in the process while producing paragonite and quartz (Fig. 20.8). In this segment of the reaction path, the rate of quartz precipitation is sufficient to consume the silica produced as the albite dissolves. The reaction continues until all of the kaolinite in the system is consumed, after which point the albite and quartz quickly reach equilibrium with the fluid.

21

Waste Injection Wells

Increasingly since the 1930s, various industries around the world that generate large volumes of liquid byproducts have disposed of their wastes by injecting them into the subsurface of sedimentary basins. In the United States, according to a 1985 survey by Brower et al. (1989), 411 "Class I" wells were licensed to inject hazardous and nonhazardous waste into deep strata, and 48 more were proposed or under construction. Legal restrictions on the practice vary geographically, as does the suitability of geologic conditions. Nonetheless, the practice of deep-well injection had increased over time, partly in response to environmental laws that emphasize protection of surface water and shallow groundwater. More restrictive regulations introduced in the late 1980s and 1990s have begun to cause a decrease in the number of operating Class I wells.

Some injected wastes are persistent health hazards that need to be isolated from the biosphere indefinitely. For this reason, and because of the environmental and operational problems posed by loss of permeability or formation caving, well operators seek to avoid deterioration of the formation accepting the wastes and its confining layers. When wastes are injected, they are commonly far from chemical equilibrium with the minerals in the formation and, therefore, can be expected to react extensively with them (Boulding, 1990). The potential for subsurface damage by chemical reaction, nonetheless, has seldom been considered in the design of injection wells.

According to Brower et al. (1989; Fig. 21.1), nine wells at seven industrial sites throughout the state of Illinois were in use in the late 1980s for injecting industrial wastes into deeply buried formations; these wells accepted about 300 million gallons of liquid wastes per year. In this chapter, we look at difficulties stemming from reaction between waste water and rocks of the host formation at

309

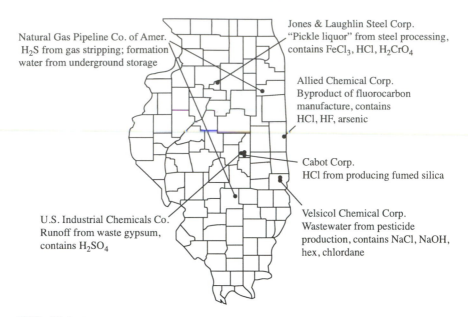

Natural Gas Pipeline Co. of Amer.
H$_2$S from gas stripping; formation
water from underground storage

Jones & Laughlin Steel Corp.
"Pickle liquor" from steel processing,
contains FeCl$_3$, HCl, H$_2$CrO$_4$

Allied Chemical Corp.
Byproduct of fluorocarbon
manufacture, contains
HCl, HF, arsenic

Cabot Corp.
HCl from producing fumed silica

U.S. Industrial Chemicals Co.
Runoff from waste gypsum,
contains H$_2$SO$_4$

Velsicol Chemical Corp.
Wastewater from pesticide
production, contains NaCl, NaOH,
hex, chlordane

FIG. 21.1   Locations of "Class I" injection wells in Illinois, from Brower et al. (1989).

several of these wells and consider how geochemical modeling might be used to
help predict deterioration and prevent blowouts.

21.1 Caustic Waste Injected in Dolomite

Velsicol Chemical Corporation maintained two injection wells at its plant near
Marshall, Illinois, to dispose of caustic wastes from pesticide production, as well
as contaminated surface runoff. In September 1965, the company began to inject
the wastes into Devonian dolomites of the Grand Tower Formation at a depth of
about 2600 feet. The wells accepted about 6 million gallons of waste monthly.

The waste contained about 3.5% dissolved solids, 1.7% chlorides, 0.4%
sodium hydroxide, and tens to hundreds of ppm of chlorinated hydrocarbons and
chlordane; its pH was generally greater than 13 (Brower et al., 1989). At the
time of drilling, analysis of formation samples indicated that the injection zone
was composed of nearly pure dolomite [CaMg(CO$_3$)$_2$]. The carbonate formation
was thought to be safe for accepting an alkaline waste water because carbonates
are considered stable at high pH.

With time, however, the company encountered problems, including caving of
the formation into the wellbore and the loss of permeability in zones that had
accepted fluid. In June 1987, a number of sidewall cores were taken from the
formation (Mehnert et al., 1990). Mineralogic analysis by x-ray diffraction

showed that significant amounts of calcite ($CaCO_3$) and brucite [$Mg(OH)_2$], as well as some amorphous matter, had formed from the original dolomite. In some samples, the dolomite was completely consumed and the rock was found to be composed entirely of a mixture of brucite and calcite.

The plant eventually closed for environmental reasons, including surface contamination unrelated to the injection wells, causing a loss of jobs in an economically depressed area. Could geochemical modeling techniques have predicted the wells' deterioration? Using REACT, we trace the irreversible reaction of dolomite into the NaOH-NaCl waste. To calculate the waste's initial state and then titrate dolomite into it, we enter the commands

```
T = 35
pH = 13
balance on Na+
Cl-   =  .5   molal
Ca++  = 1     mg/kg
HCO3- = 1     mg/kg
Mg++  =  .01 ug/kg

react 40 grams Dolomite
go
```

The fluid contains arbitrarily small amounts of Ca^{++}, Mg^{++}, and HCO_3^-, as is necessary in order for the program to be able to recognize dolomite. The initial magnesium content is set small to assure that the hydroxide mineral brucite is not supersaturated in the alkaline fluid.

Figure 21.2 shows the mineralogic results of the calculation. Dolomite dissolves, since it is quite undersaturated in the waste fluid. The dissolution adds calcium, magnesium, and carbonate to solution. Calcite and brucite precipitate from these components, as observations from the wells indicated. The fluid reaches equilibrium with dolomite after about 11.6 cm^3 of dolomite have dissolved per kg water. About 11 cm^3 of calcite and brucite form during the reaction. Since calculation predicts a net decrease in mineral volume, damage to the dolomite formation is evidently not due to loss of porosity. Instead, the reaction products likely formed a fine-grained, poorly coherent material capable of clogging pore openings and slumping into the well bore.

At the point of dolomite saturation, where reaction ceases, *p*H has decreased from 13 to about 11.9 (Fig. 21.3). The overall reaction between the fluid and dolomite

$$CaMg(CO_3)_2 + 2\,OH^- \rightarrow CaCO_3 + Mg(OH)_2 + CO_3^{--} \qquad (21.1)$$
$$\text{dolomite} \qquad\qquad \text{calcite} \quad \text{brucite}$$

is given by the slopes-of-the-lines method, as discussed in Section 11.2. For simplicity, we have lumped the ion pairs NaOH and $NaCO_3^-$ with the ions OH$^-$ and CO_3^- in writing the reaction.

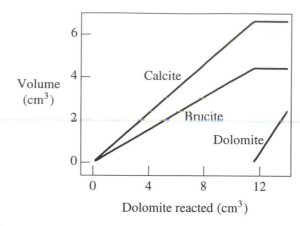

FIG. 21.2 Mineralogic results of reacting at 35°C alkaline waste water from the Velsicol plant with dolomite.

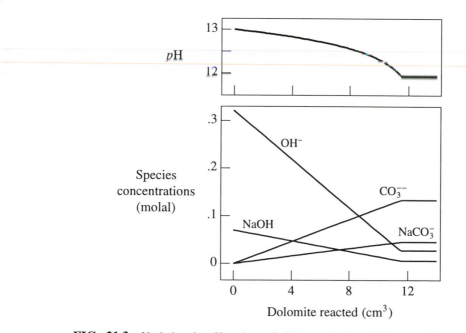

FIG. 21.3 Variation in *p*H and species' concentrations during reaction at 35°C of alkaline waste water with dolomite.

An interesting further experiment is to test the effects of letting the waste fluid equilibrate with the atmosphere before it is injected. Because the waste is so alkaline, its calculated CO_2 fugacity is $10^{-13.5}$, about ten orders of magnitude less than the atmospheric value of $10^{-3.5}$. To simulate the effects of leaving the waste in a lagoon long enough to equilibrate with atmospheric CO_2, we trace two reaction steps, which, for simplicity, we assume to occur at the same temperature.

```
(cont'd)
remove reactant Dolomite

slide log f CO2(g) to -3.5
go

pickup fluid
react 1 mg Dolomite
go
```

The first step adjusts the fluid's CO_2 fugacity to the atmospheric value, reducing pH to less than 10 by the reaction

$$CO_2(g) + H_2O \rightarrow CO_3^{--} + 2\,H^+ \tag{21.2}$$

After this step, dolomite appears only slightly undersaturated in the calculation results. In the second step, which simulates reaction of the formation into the equilibrated fluid, only about 2×10^{-5} cm^3 of dolomite dissolve per kilogram of waste water. This result suggests that the plant's waste stream could have been neutralized inexpensively by aeration.

21.2 Gas Blowouts

According to the Illinois Environmental Protection Agency (IEPA), a series of gas blowouts has occurred at two waste injection wells in the state (Brower et al., 1989). In each case, well operators were injecting concentrated hydrochloric acid into a dolomite bed. At its plant near Tuscola, the Cabot Corporation injects acid waste from the production of fumed silica into the Cambrian Eminence and Potosi Formations below 5,000 ft (1,500 m) depth. Allied Chemical Corporation injects acid into the Potosi formation below about 3,600 ft (1,100 m). The acid, which is contaminated with arsenic, is a by-product of the manufacture of refrigerant gas. Since some of the blowouts have caused damage such as fish kills, there is environmental interest as well as operational concern in preventing such accidents.

The blowouts seem to have occurred at times when especially acidic wastes were being injected. The acid apparently reacted with carbonate in the formations to produce a CO_2 gas cap at high pressure. In some cases, the injected waste was more than 31 wt.% HCl. As a temporary measure, the plants are now limited by the IEPA to injecting wastes containing no more than 6 wt.% HCl.

We can use reaction modeling techniques to test the conditions under which dolomite will react with hydrochloric acid to produce gas in the injection zones. Equivalent values of wt.% and molality for HCl are:

Wt.% HCl	molal HCl
2	.57
4	1.2
6	1.8
10	3.1
20	6.9
30	11.9

To configure a model of the reaction of dolomite with 30 wt.% hydrochloric acid, we start REACT and enter the commands

```
T = 50
H+    = 11.9  molal
Cl-   = 11.9  molal
Na+   =  .01  molal
Ca++  = 1     mg/kg
HCO3  = 1     mg/kg
Mg++  = 1     mg/kg

react 600 grams Dolomite
go
```

We then repeat the calculation for solutions of differing HCl contents, according to the chart above.

Reacting dolomite into the waste water increases the pH as well as the CO_2 fugacity. The predicted reaction is

$$CaMg(CO_3)_2 + 4\,H^+ \rightarrow 2\,CO_2(aq) + Ca^{++} + Mg^{++} + 2\,H_2O \qquad (21.3)$$
$$\text{\textit{dolomite}}$$

Here, we lump the ion pairs $CaCl^+$ and $MgCl^+$ with the free Ca^{++} and Mg^{++} ions. When the CO_2 fugacity exceeds the confining pressure in the formation, CO_2 exsolves as a free gas

$$CO_2(aq) \rightarrow CO_2(g) \qquad (21.4)$$

producing a gas cap in the subsurface.

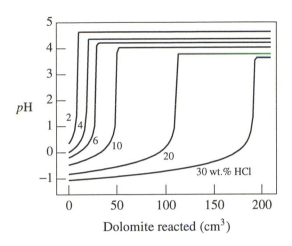

FIG. 21.4 Calculated variation in *p*H as HCl solutions of differing initial concentrations react at 50°C with dolomite.

Figure 21.4 shows how *p*H changes as hydrochloric acid in differing concentrations reacts at 50°C with dolomite, and Fig. 21.5 shows how CO_2 fugacity varies. Solutions of greater HCl contents dissolve more dolomite and, hence, produce more CO_2. At 30 wt.%, for example, each kilogram of waste water consumes almost 200 cm^3 of dolomite. At an injection rate of 25,000 kg/day at this concentration, about 200,000 m^3 of dolomite would dissolve each year.

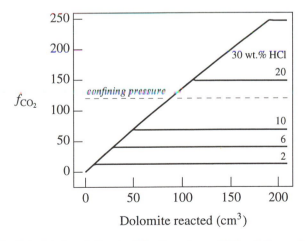

FIG. 21.5 Calculated variation in CO_2 fugacity as HCl solutions of differing initial concentrations react at 50°C with dolomite.

The most acidic solutions, as expected, produce the greatest CO_2 fugacities. For the 30 wt.% fluid, the partial pressure of CO_2 escaping from the fluid would approach 250 atm. Assuming a confining pressure of about 120 atm at the Allied well, solutions containing more than 15 wt.% HCl are likely to exsolve CO_2. The calculations indicate, on the other hand, that even the most acidic waste can be injected without fear of a gas blowout if it is first diluted by an equal amount of water.

22

Petroleum Reservoirs

In efforts to increase and extend production from oil and gas fields, as well as to keep wells operational, petroleum engineers pump a wide variety of fluids into the subsurface. Fluids are injected into petroleum reservoirs for a number of purposes, including:

- Waterflooding, where an available fresh or saline water is injected into the reservoir to displace oil toward producing wells.

- Improved Oil Recovery (IOR), where a range of more exotic fluids such as steam (hot water), caustic solutions, carbon dioxide, foams, polymers, surfactants, and so on are injected to improve recovery beyond what might be obtained by waterflooding alone.

- Near-well treatments, in which chemicals are injected into producing and sometimes injector wells, where they are intended to react with the reservoir rock. Well stimulation techniques such as acidization, for example, are intended to increase the formation's permeability. Alternatively, producing wells may receive "squeeze treatments" in which a mineral scale inhibitor is injected into the formation. In this case, the treatment is designed so that the inhibitor sorbs onto mineral surfaces, where it can gradually desorb into the formation water during production.

- Pressure management, where fluid is injected into oil fields in order to maintain adequate fluid pressure in reservoir rocks. Calcium carbonate may precipitate as mineral scale, for example, if pressure is allowed to deteriorate, especially in fields where formation fluids are rich in Ca^{++} and HCO_3^- and CO_2 fugacity is high.

317

In each of these procedures, the injected fluid can be expected to be far from equilibrium with sediments and formation waters. As such, it is likely to react extensively once it enters the formation, causing some minerals to dissolve and others to precipitate. Hutcheon (1984) appropriately refers to this process as "artificial diagenesis," drawing an analogy to the role of groundwater flow in the diagenesis of natural sediments (see Chapter 19). Further reaction is likely if the injected fluid breaks through to producing wells and mixes there with formation waters.

There is considerable potential, therefore, for mineral scale, such as barium sulfate (see the next section), to form during these procedures. The scale may be deposited in the formation, the wellbore, or in production tubing. Scale that forms in the formation near wells, known as "formation damage," can dramatically lower permeability and throttle production. When it forms in the wellbore and production tubing, mineral scale is costly to remove and may lead to safety problems if it blocks release valves.

In this chapter, in an attempt to devise methods for helping to foresee such unfavorable consequences, we construct models of the chemical interactions between injected fluids and the sediments and formation waters in petroleum reservoirs. We consider two cases: the effects of using seawater as a waterflood, taking oil fields of the North Sea as an example, and the potential consequences of using alkali flooding (i.e., the injection of a strong caustic solution) in order to to increase oil production from a clastic reservoir.

22.1 Sulfate Scaling in North Sea Oil Fields

A common problem in offshore petroleum production is that sulfate scale may form when seawater is injected into the formation during waterflooding operations. The scale forms when seawater, which is rich in sulfate but relatively poor in Ca^{++} and nearly depleted in Sr^{++} and Ba^{++}, mixes with formation fluids, many of which contain bivalent cations in relative abundance but little sulfate. The mixing causes minerals such as gypsum ($CaSO_4 \cdot 2H_2O$), anhydrite ($CaSO_4$), celestite ($SrSO_4$), and barite ($BaSO_4$, an almost insoluble salt) to become saturated and precipitate as scale.

Sulfate scaling poses a special problem in oil fields of the North Sea (e.g., Todd and Yuan, 1990, 1992; Yuan et al., 1994), where formation fluids are notably rich in barium and strontium. The scale can reduce permeability in the formation, clog the wellbore and production tubing, and cause safety equipment (such as pressure release valves) to malfunction. To try to prevent scale from forming, reservoir engineers use chemical inhibitors such as phosphonate (a family of organic phosphorus compounds) in "squeeze treatments," as described in the introduction to this chapter.

Table 22.1 shows the compositions of formation waters from three North Sea oil fields, and the composition of seawater (from Drever, 1988). The origin of

TABLE 22.1 Compositions of formation fluids from three North Sea oil fields (Edward Warren, personal communication) and seawater

	Miller	Forties	Amethyst	Seawater
Na$^+$ (mg/kg)	27 250	27 340	51 900	10 760
K$^+$	1 730	346	1 100	399
Mg^{++}	110	469	2 640	1 290
Ca^{++}	995	2 615	16 320	411
Sr^{++}	105	534	1 000	8
Ba^{++}	995	235	10	.01
Cl$^-$	45 150	48 753	121 550	19 350
HCO$_3^-$	1 980	462	85	142
SO$_4^{--}$	0	10	0	2 710
T (°C)	121	96	88	20
pH (20°C)	7	7	6.7	8.1
TDS (mg/kg)	78 300	80 800	194 600	35 100

the scaling problem is clear. Seawater contains more than 2500 mg/kg of sulfate but little strontium and almost no barium. The formation waters, however, are depleted in sulfate, but they contain strontium and barium in concentrations up to about 1000 mg/kg and significant amounts of calcium. A mixture of seawater and formation fluid, therefore, will contain high concentrations of both sulfate and the cations, and hence, will probably be supersaturated with respect to sulfate minerals.

To model the results of the mixing process, we calculate the effects of titrating seawater into the formation waters. Two aspects of the modeling results are of interest. First, the volume of mineral scale produced during mixing provides a measure of the potential severity of a scaling problem. Second, in simulations in which scale is prevented from forming (simulating the case of a completely successful inhibition treatment), the saturation states of the sulfate minerals give information about the thermodynamic driving force for precipitation. This information is of value because it provides a measure of the difficulty of inhibiting scale formation (Sorbie et al., 1994). In general, to be effective against scaling, the inhibitor must be present at greater concentration where minerals are highly supersaturated than where they are less supersaturated (e.g., He et al., 1994).

To start the simulation, we equilibrate seawater (as we did in Chapter 6), using REACT to carry it to formation temperature, and then "pick up" the resulting fluid as a reactant in the mixing calculation. The procedure (taking the temperature of 121°C reported for the Miller field) is

```
pH = 8.1
Na+    =  10760     mg/kg
K+     =    399     mg/kg
Mg++   =   1290     mg/kg
Ca++   =    411     mg/kg
Sr++   =      8     mg/kg
Ba++   =        .01 mg/kg
Cl-    -  19350     mg/kg
HCO3-  =    142     mg/kg
SO4--  =   2710     mg/kg

TDS = 35100
T initial = 20, final = 121
go

pickup reactants = fluid
```

We then equilibrate the formation fluid, using data from Table 22.1. Since *pH* measurements from saline solutions are not reliable, we assume that *pH* in the reservoir is controlled by equilibrium with the most saturated carbonate mineral, which turns out to be witherite ($BaCO_3$) or, for the Amethyst field, strontianite ($SrCO_3$). Using the Miller analysis, the procedure for completing the calculation is

```
(cont'd)
T = 121
Na+    = 27250 mg/kg
K+     =  1730 mg/kg
Mg++   =   110 mg/kg
Ca++   =   995 mg/kg
Sr++   =   105 mg/kg
Ba++   =   995 mg/kg
Cl-    = 45150 mg/kg
HCO3-  =  1980 mg/kg
SO4--  =    10 ug/kg

swap Witherite for H+
1 free cm3 Witherite

TDS = 78300
reactants times 10
delxi = .001
dump
go
```

To model the case in which scale is prevented from forming, we repeat the calculation

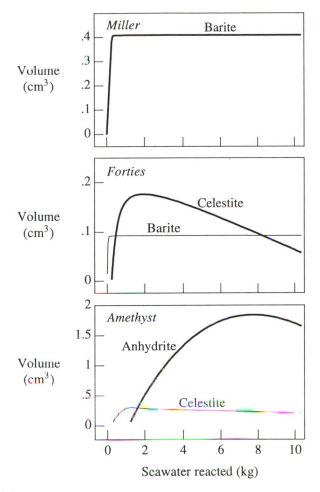

FIG. 22.1 Volumes of minerals precipitated during a reaction model simulating the mixing at reservoir temperature of seawater into formation fluids from the Miller, Forties, and Amethyst oil fields in the North Sea. The reservoir temperatures and compositions of the formation fluids are given in Table 22.1. The initial extent of the system in each case is 1 kg of solvent water. Not shown for the Amethyst results are small volumes of strontianite, barite, and dolomite that form during mixing.

(cont'd)

```
precip = off
go
```

with mineral precipitation disabled. Figure 22.1 shows the volumes of minerals precipitated during the mixing reactions for fluids from the three fields; Fig. 22.2

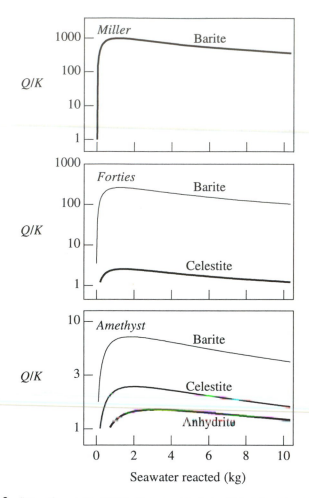

FIG. 22.2 Saturation states (Q/K) of supersaturated sulfate minerals over the courses of simulations in which seawater mixes at reservoir temperature with formation fluids from three North Sea oil fields. Reaction paths are the same as shown in Fig. 22.1, except that minerals are not allowed to precipitate.

shows the saturation states of sulfate minerals during mixing in the absence of precipitation.

For the Miller fluid, barite precipitates

$$Ba^{++} + SO_4^{--} \rightarrow BaSO_4 \qquad (22.1)$$
$$\text{barite}$$

early in the mixing reaction. Because of the mineral's low solubility, virtually all of the seawater sulfate added to the system is consumed by the precipitation reaction until the barium is depleted from the fluid. Barite is the only mineral to form in the simulation, reaching a maximum volume of 0.4 cm^3 (per kg of water in the formation fluid) after reaction with just a small amount of seawater.

The Forties fluid contains less barium but is richer in strontium. Barite forms initially (according to Reaction 22.1), but in a lesser amount than from the Miller fluid. Celestite forms

$$Sr^{++} + SO_4^{--} \rightarrow SrSO_4 \qquad (22.2)$$
$$celestite$$

shortly thereafter. In reality, the precipitate would occur as a barium-strontium solid solution rather than as separate phases. Later in the mixing reaction, as seawater comes to dominate the mixed fluid, the celestite starts to redissolve, reflecting the small amount of strontium in seawater. Less than about 0.3 cm^3 of scale forms during the mixing reaction.

The Amethyst fluid is richer in strontium and calcium than the other fluids, but nearly depleted in barium. Celestite becomes saturated first, and more of this mineral forms from this fluid than from the Forties fluid. Anhydrite becomes saturated later in the mixing process and precipitates

$$Ca^{++} + SO_4^{--} \rightarrow CaSO_4 \qquad (22.3)$$
$$anhydrite$$

More than 2 cm^3 of scale form during the simulation, making this fluid more potentially damaging than the Miller or Forties formation waters by a factor of five or greater.

In the three simulations, the sulfate minerals form at mixing ratios related to their solubilities. Barite, the least soluble, forms early, when small amounts of seawater are added. The more soluble celestite forms only after the addition of somewhat larger quantities of seawater. Anhydrite, the most soluble of the minerals, forms from the Amethyst fluid at still higher ratios of seawater to formation fluid.

Comparing the volumes of scale produced in simulations in which minerals are allowed to form (Fig. 22.1) with the minerals' saturation states when precipitation is disabled (Fig. 22.2), it is clear that no direct relationship exists between these values. In the Forties and Amethyst simulations, the saturation state of barite is greater than those of the other minerals that form, although the other minerals form in greater volumes. Whereas the Amethyst fluid is capable of producing about five times as much scale as fluids from the Miller and Forties fields, furthermore, the saturation states (Q/K) predicted for it are 1.5 to 2 orders of magnitude lower than for the other fluids. This result indicates that although the Amethyst fluid is the most potentially damaging of the three, it is also the fluid for which scale formation might be most readily inhibited by chemical treatment.

We could, of course, attempt more sophisticated simulations of scale formation. Since the fluid mixture is quite concentrated early in the mixing, we might use a virial model to calculate activity coefficients (see Chapter 7). The Harvie-Møller-Weare (1984) activity model is limited to 25°C and does not consider barium or strontium, but Yuan and Todd (1991) suggested a similar model for the Na-Ca-Ba-Sr-Mg-SO$_4$-Cl system in which the virial coefficients can be extrapolated to typical reservoir temperatures in the North Sea.

Given a specific application, we might also include precipitation kinetics in our calculations, as described in Chapter 14. Wat et al. (1992) present a brief study of the kinetics of barite formation, including the effects of scale inhibitors on precipitation rates. For a variety of reasons (see Section 14.2), however, it remains difficult to construct reliable models of the kinetics of scale precipitation.

22.2 Alkali Flooding

Some of the most radical changes to the geochemistry of a petroleum reservoir are induced by the highly reactive fluids injected in well stimulation and improved oil recovery (IOR) procedures. Stimulation (e.g., acidization) is generally a near-well treatment designed to improve the productivity of a formation, sometimes by reversing previous formation damage. The fluids used in IOR may react with the formation water, the mineral assemblage in the formation, and with the crude oil itself. Alkali flooding is an example, considered in this section, of an IOR procedure employing an extremely reactive fluid.

The purpose of alkali flooding (Jennings et al., 1974) is to introduce alkali into a reservoir where it can react with organic acids in the oil to produce organic salts, which act as surfactants. The surfactants (or "petroleum soaps") generated reduce the surface tension between the oil and water and this in turn reduces the level of capillary trapping of the oil. Thus, more oil is recovered because less of it remains trapped in the formation's pore spaces.

This type of flood can be successful only if, as the fluid moves through the reservoir, a sufficient amount of the alkali remains in solution to react with the oil. Reaction of the flood with minerals and fluid in the reservoir, however, can consume the flood's alkali content. Worse, the reactions may precipitate minerals in the formation's pore space, decreasing permeability near the wellbore where free flow is most critical. A special problem for this type of flood is the reaction of clay minerals to form zeolites (Sydansk, 1982).

The effectiveness of alkali flooding, and, in fact, most reservoir treatments, varies widely from formation to formation in a manner that is often difficult to predict. Quantitative techniques have been applied to model the migration and consumption of alkali as it moves through a reservoir (e.g., Bunge and Radke, 1982; Zabala et al., 1982; Dria et al., 1988). There have been fewer attempts, however, to predict the specific chemical reactions that might occur in a reservoir or the effects of the initial mineralogy of the reservoir and the composition of the flood on those reactions (Bethke et al., 1992).

To consider how such predictions might be made, we model how three types of alkali floods might affect a hypothetical sandstone reservoir. The floods, which are marketed commercially for this purpose, are NaOH, Na_2CO_3, and Na_2SiO_3. We take each flood at 0.5 N strength and assume that reaction occurs at a temperature of 70°C.

The reservoir rock in our model is composed of quartz grains, carbonate cement, and clay minerals in the following proportions, by volume:

Quartz	SiO_2	85%
Calcite	$CaCO_3$	6
Dolomite	$CaMg(CO_3)_2$	4
Muscovite (illite)	$KAl_3Si_3O_{10}(OH)_2$	3
Kaolinite	$Al_2Si_2O_5(OH)_4$	2

The initial formation fluid is a solution at *p*H 5 of 1 molal NaCl and 0.2 molal $CaCl_2$ which is in chemical equilibrium with the minerals in the reservoir. The extent of the system is 1 kg water, which at 15% porosity corresponds to a 1 cm-thick slice of the formation extending about 20 cm from the wellbore.

Using the "flush" configuration (Chapter 2), we continuously displace the pore fluid with the flooding solution. In this way, we replace the pore fluid in the system a total of ten times over the course of the simulated flood, which lasts twenty days. Because of the short interval selected for the flood, we assume that the pore fluid does not remain in equilibrium with quartz or framework silicates such as feldspar.

We set quartz dissolution and precipitation according to a kinetic rate law (Knauss and Wolery, 1988; see Chapter 14)

$$r_{qtz} = A_S\, k_+ \, a_{H^+}^{-1/2} \left[\frac{Q}{K} - 1 \right] \qquad (22.4)$$

valid at 70°C for *p*H values greater than about 6, which are quickly reached in the simulation; the corresponding rate constant k_+ is 1.6×10^{-18} mol/cm^{2}sec. As in our calculations in Chapter 20, we assume a specific surface area for quartz of 1000 cm^2/g. Feldspars are suppressed so they will not precipitate if they become supersaturated, since these minerals presumably do not have time to nucleate and grow. We assume, however, that the carbonate and clay minerals in the simulation maintain equilibrium with the fluid as it moves through the system. In a more sophisticated simulation, we might also set kinetic rate laws for these minerals.

For the NaOH flood, the complete procedure in REACT is

```
T = 70

pH = 5
Na+  = 1    molal
Ca++ = 0.2 molal
Cl-  = 1    molal

swap Dolomite-ord for Mg++
swap Muscovite for K+
swap Kaolinite for Al+++
swap Quartz for SiO2(aq)
swap Calcite for HCO3-

 365 free cm3 Calcite
 235 free cm3 Dolomite-ord
 180 free cm3 Muscovite
 120 free cm3 Kaolinite
5150 free cm3 Quartz

react 1     kg H2O
react 0.5   moles Na+
react 0.5   moles OH-

suppress Albite, "Albite high", "Albite low"
suppress "Maximum Microcline", K-feldspar, "Sanidine high"
kinetic Quartz   rate_con = 1.8e-18,   power(H+) = -1/2
kinetic Quartz   surface = 1000

reactants times 10
time end = 20 days
flush
go
```

To run the Na_2CO_3 and Na_2SiO_3 simulations, we need only alter the composition of the reactant fluid:

```
(cont'd)
remove reactant OH-
react 0.25 moles CO3--
go
```

and

```
(cont'd)
remove reactant CO3--
remove reactant Na+
react 0.25 moles Na2O
react 0.25 moles SiO2(aq)
go
```

Figures 22.3–22.4 show the results of the simulations.

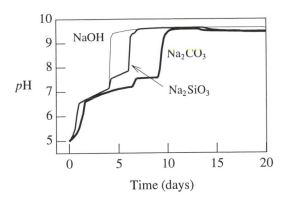

FIG. 22.3 Variation in *pH* during simulated alkali floods of a clastic petroleum reservoir at 70°C, using 0.5 N NaOH, Na_2CO_3, and Na_2SiO_3 solutions. Pore fluid is displaced by unreacted flooding solution at a rate of one-half of the system's pore volume per day.

As the alkaline solution enters the system, *pH* increases in each of the simulations, eventually reaching a value of about 9.5. In the early portions of the simulations, however, some of the alkali is consumed in the conversion of kaolinite to paragonite [$NaAl_3Si_3O_{10}(OH)_2$]. In the NaOH flood, for example, the reaction

$$2\ NaOH + 3\ Al_2Si_2O_5(OH)_4 \rightarrow 2\ NaAl_3Si_3O_{10}(OH)_2 + 5\ H_2O \quad (22.5)$$
$$\text{kaolinite} \qquad\qquad\qquad \text{paragonite}$$

consumes most of the sodium hydroxide and helps maintain a low *pH* until the kaolinite has been exhausted.

Later in the simulations, the zeolite analcime ($NaAlSi_2O_6 \cdot H_2O$) begins to form, largely at the expense of micas, which serve as proxies for clay minerals. In the NaOH flood, the overall reaction (expressed per formula unit of analcime) is

$$1.75\ NaOH + CaMg(CO_3)_2 + .9\ SiO_2 + .33\ KAl_3Si_3O_{10}(OH)_2$$
$$\qquad\qquad \text{dolomite} \qquad\quad \text{quartz} \qquad\quad \text{muscovite}$$

$$+ .36\ NaAl_3Si_3O_{10}(OH)_2 + .6\ H_2O \rightarrow CaCO_3 + NaAlSi_2O_6 \cdot H_2O$$
$$\qquad \text{paragonite} \qquad\qquad\qquad\qquad\quad \text{calcite} \qquad \text{analcime}$$

$$+ .75\ NaAlCO_3(OH)_2 + .33\ KAlMg_3Si_3O_{10}(OH)_2$$
$$\qquad\quad \text{dawsonite} \qquad\qquad\qquad \text{phlogopite}$$

$$+ .4\ Na^+ + .15\ HCO_3^- + .1\ CO_3^{--} \qquad (22.6)$$

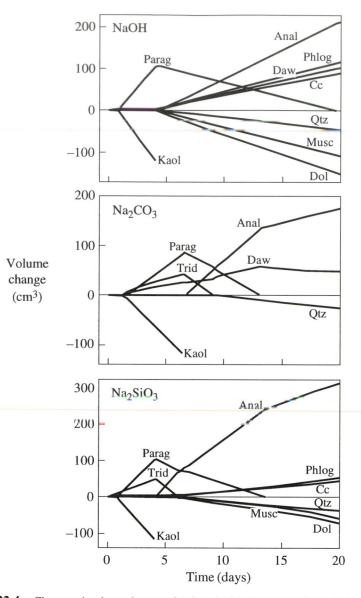

FIG. 22.4 Changes in the volumes of minerals in the reservoir rock during the simulated alkali floods (Fig. 22.3) of a clastic petroleum reservoir using NaOH, Na_2CO_3, and Na_2SiO_3 solutions. Minerals that react in small volumes are omitted from the plots. Abbreviations: Anal = analcime, Cc = calcite, Daw = dawsonite, Dol = dolomite, Kaol = kaolinite, Musc = muscovite, Parag = paragonite, Phlog = phlogopite, Qtz= quartz, Trid = tridymite.

TABLE 22.2 Comparison of the alkali flood simulations*

	NaOH	Na_2CO_3	Na_2SiO_3
Net change in mineral volume (cm^3)			
zeolite	+213	+175	+316
carbonate	+42	+53	−27
quartz	−46	−25	−36
clay (& mica)	−111	−126	−118
	+98	+77	+135
Change in pore volume (%)	−9	−7	−13
Alkali consumed (%)	78	53	65

*Extent of system is 1 kg water.

In part because of the open crystal structure and resulting low density characteristic of zeolite minerals, analcime is the most voluminous reaction product in the simulations.

Table 22.2 summarizes the simulation results. In each case, the flood dissolves clay minerals and quartz. The simulations, however, predict the production of significant volumes of analcime, in accord with the observation that zeolites are prone to form during alkali floods. The volume of analcime produced in each case is sufficient to offset the volumes of the dissolved minerals and lead to a net decrease in porosity. Of the simulations, the Na_2SiO_3 flood leads to the production of the most analcime and to the greatest loss in porosity.

In the simulations, a significant fraction (about 50% to 80%) of the alkali present in solution is consumed by reactions near the wellbore with the reservoir minerals (as shown in Reaction 22.6 for the NaOH flood), mostly by the production of analcime, paragonite, and dawsonite [$NaAlCO_3(OH)_2$]. In the clastic reservoir considered, therefore, alkali floods might be expected to cause formation damage (mostly due to the precipitation of zeolites) and to be less effective at increasing oil mobility than they would in a reservoir where they do not react extensively with the formation.

23

Acid Drainage

Acid drainage is a persistent environmental problem in many mineralized areas. The problem is especially pronounced in areas that host or have hosted mining activity (e.g., Lind and Hem, 1993), but it also occurs naturally in unmined areas. The acid drainage results from the weathering of sulfide minerals that oxidize to produce hydrogen ions and contribute dissolved metals to solution.

These acidic waters are toxic to plant and animal life, including fish and aquatic insects. Streams affected by acid drainage may be rendered nearly lifeless, their stream beds coated with unsightly yellow and red precipitates of oxy-hydroxide minerals. In some cases, the heavy metals in acid drainage threaten water supplies and irrigation projects.

Where acid drainage is well developed and extensive, the costs of remediation can be high. In the Summitville, Colorado district (USA), for example, efforts to limit the contamination of fertile irrigated farmlands in the nearby San Luis Valley and protect aquatic life in the Alamosa River will cost an estimated $100 million or more (Plumlee, 1994a).

Not all mine drainage, however, is acidic or rich in dissolved metals (e.g., Ficklin et al., 1992; Mayo et al., 1992; Plumlee et al., 1992). Drainage from mining districts in the Colorado Mineral Belt ranges in pH from 1.7 to greater than 8 and contains total metal concentrations ranging from as low as about 0.1 mg/kg to more than 1000 mg/kg. The primary controls on drainage pH and metal content seem to be (1) the exposure of sulfide minerals to weathering, (2) the availability of atmospheric oxygen, and (3) the ability of nonsulfide minerals to buffer acidity.

In this chapter we construct geochemical models to consider how the availability of oxygen and the buffering of host rocks affect the pH and

331

composition of acid drainage. We then look at processes that can attenuate the dissolved metal content of drainage waters.

23.1 Role of Atmospheric Oxygen

Acid drainage results from the reaction of sulfide minerals with oxygen in the presence of water. As we show in this section, water in the absence of a supply of oxygen gas becomes saturated with respect to a sulfide mineral after only a small amount of the mineral has dissolved. The dissolution reaction in this case (when oxygen gas is not available) causes little change in the water's pH or composition. In a separate effect, it is likely that atmospheric oxygen further promotes acid drainage because of its role in the metabolism of bacteria that catalyze both the dissolution of sulfide minerals and the oxidation of dissolved iron (Nordstrom, 1982).

For these reasons, there is a clear connection between the chemistry of mine drainage and the availability of oxygen. Plumlee et al. (1992) found that the most acidic, metal-rich drainage waters in Colorado tend to develop in mine dumps, which are highly permeable and open to air circulation, and in mineral districts like Summitville (see King, 1995), where abandoned workings and extensive fracturing give atmospheric oxygen access to the ores. Drainage waters that are depleted in dissolved oxygen, on the other hand, tend to be less acidic and have lower heavy metal concentrations.

To investigate how the presence of atmospheric oxygen affects the reaction of pyrite (FeS_2) with oxidizing groundwater, we construct a simple model in REACT. First we take a hypothetical groundwater at $25°C$ that has equilibrated with atmospheric oxygen but is no longer in contact with it. We suppress hematite (Fe_2O_3), which does not form directly at low temperature, and goethite (FeOOH); each of these minerals is more stable thermodynamically than the ferric precipitate observed to form in acid drainage. To set the initial fluid, we type

```
swap O2(g) for O2(aq)
f O2(g) = 0.2
pH = 6.8

Ca++   = 10 mg/kg
Mg++   =  2 mg/kg
Na+    =  6 mg/kg
K+     =  2 mg/kg
HCO3-  = 75 mg/kg
SO4--  =  2 mg/kg
Cl-    =  1 mg/kg
Fe++   =  1 ug/kg
balance on HCO3-
```

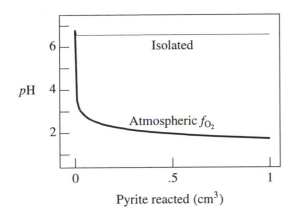

FIG. 23.1 Calculated variation in *p*H during reaction of pyrite with a hypothetical groundwater at 25°C, assuming that the fluid is isolated from (fine line) and in contact with (bold line) atmospheric oxygen.

```
suppress Hematite, Goethite
go
```

The resulting fluid contains about 8 mg/kg (.25 *m*mol) of dissolved oxygen.
We then add pyrite to the system

```
(cont'd)
react 1 cm3 Pyrite
go
```

letting it react to equilibrium with the fluid. The reaction proceeds until the $O_2(aq)$ has been consumed, dissolving a small amount of pyrite (about .08 *m*mol) according to

$$FeS_2 + \frac{7}{2} O_2(aq) + 2\ HCO_3^- \rightarrow Fe^{++} + 2\ SO_4^{--} + 2\ CO_2(aq) + H_2O \quad (23.1)$$
pyrite

The fluid's *p*H in the model changes slightly (Fig. 23.1), decreasing from 6.8 to about 6.6.

To see how contact with atmospheric oxygen might affect the reaction, we repeat the calculation, assuming this time that oxygen fugacity is fixed at its atmospheric level

```
(cont'd)
fix f O2(g)
go
```

In this case, the reaction proceeds without exhausting the oxygen supply, which in the calculation is limitless, driving *p*H to a value of about 1.7 (Fig. 23.1). We

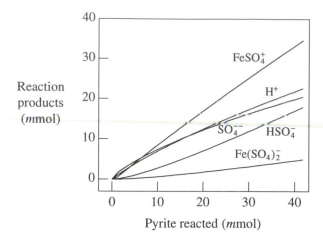

FIG. 23.2 Masses of species produced by reacting pyrite with a hypothetical groundwater that is held in equilibrium with atmospheric oxygen, according to the reaction path calculation shown in Fig. 23.1.

could, in fact, continue to dissolve pyrite into the fluid indefinitely, thereby reaching even lower pH values.

Initially, pyrite oxidation in the model proceeds according to the reaction

$$FeS_2 + \frac{15}{4} O_2(g) + \frac{1}{2} H_2O \rightarrow FeSO_4^+ + SO_4^{--} + H^+ \qquad (23.2)$$
$$pyrite$$

as shown in Fig. 23.2, producing H^+ and thus driving the fluid acidic. According to the model, the pyrite dissolution produces ferric iron in an ion pair with sulfate. As the pH decreases, HSO_4^- comes to dominate SO_4^{--} and a second reaction

$$FeS_2 + \frac{15}{4} O_2(g) + \frac{1}{2} H_2O \rightarrow FeSO_4^+ + HSO_4^- \qquad (23.3)$$
$$pyrite$$

becomes important. This reaction produces no free hydrogen ions and hence does not contribute to the fluid's acidity.

Our calculated reaction path may reasonably well represent the overall reaction of pyrite as it oxidizes, but it does little to illustrate the steps that make up the reaction process. Reaction 23.2, for example, involves the transfer of 16 electrons to oxygen, the electron acceptor in the reaction, from the iron and sulfur in each FeS_2 molecule. Elementary reactions (those that proceed on a molecular level), however, seldom transfer more than one or two electrons. Reaction 23.2, therefore, would of necessity represent a composite of elementary reactions.

In nature, at least two aqueous species, $O_2(aq)$ and Fe^{+++}, can serve as electron acceptors during the pyrite oxidation (Moses et al., 1987). In the case of Fe^{+++}, the oxidation reaction proceeds as

$$FeS_2 + 14\,Fe^{+++} + 8\,H_2O \rightarrow 15\,Fe^{++} + 2\,SO_4^{--} + 16\,H^+ \qquad (23.4)$$
pyrite

This reaction, while still not an elementary reaction, more closely describes how the oxidation proceeds on a molecular level. Even where Reaction 23.4 operates, however, our model may still represent the overall process occurring in nature, since O_2 is needed to produce the Fe^{+++} that drives the reaction forward and is, therefore, the ultimate oxidant in the system.

Even neglecting the question of the precise steps that make up the overall reaction, our calculations are a considerable simplification of reality. The implicit assumption that iron in the fluid maintains redox equilibrium with the dissolved oxygen, as described in Section 6.4, is especially vulnerable. In reality, the ferrous iron added to solution by the dissolving pyrite must react with dissolved oxygen to produce ferric species, a process that may proceed slowly. To construct a more realistic model, we could treat the dissolution in two steps by disenabling the Fe^{++}/Fe^{+++} redox couple. In the first step we would let pyrite dissolve, and in the second, let the ferrous species oxidize.

23.2 Buffering by Wall Rocks

The most important control on the chemistry of drainage from mineralized areas (once we assume access of oxygen to the sulfide minerals) is the nature of the nonsulfide minerals available to react with the drainage before it discharges to the surface (e.g., Sherlock et al., 1995). These minerals include gangue minerals in the ore, the minerals making up the country rock, and the minerals found in mine dumps. The drainage chemistry of areas in which these minerals have the ability to neutralize acid differs sharply from that of areas in which they do not.

In the Colorado Mineral Belt (USA), for example, deposits hosted by argillically altered wallrocks (as in Summitville) tend to produce highly acidic and metal-rich drainage, because the wallrocks have a negligible capacity for buffering acid (Plumlee, 1994b). In contrast, ores that contain carbonate minerals or are found in carbonate terrains, as well as those with propylitized wallrocks, can be predicted to produce drainages of near-neutral pH. These waters may be rich in zinc but are generally not highly enriched in other metals.

In the historic silver mining districts of the Wasatch Range (Utah, USA), Mayo et al. (1992) found that few springs in the area discharge acid drainage. Acid drainage occurs only where groundwater flows through aquifers found in rocks nearly devoid of carbonate minerals. Where carbonate minerals are abundant in the country rock, the discharge invariably has a near-neutral pH.

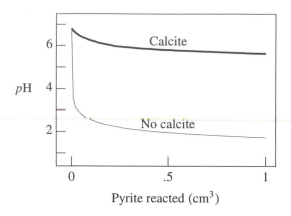

FIG. 23.3 Variation in pH as pyrite reacts at 25°C with a groundwater held in equilibrium with atmospheric O_2, calculated assuming that the reaction occurs in the absence of buffering minerals (fine line, from Fig. 23.1) and in the presence of calcite (bold line).

To model the effect of carbonate minerals on drainage chemistry, we continue our calculations from the previous section (in which we reacted pyrite with a hypothetical groundwater in contact with atmospheric oxygen). This time, we include calcite ($CaCO_3$) in the initial system

(cont'd)
```
swap Calcite for HCO3-
10 free cm3 Calcite
balance on Ca++
go
```

In contrast to the previous calculation, the fluid maintains a near-neutral pH (Fig. 23.3), reflecting the acid-buffering capacity of the calcite.

As the pyrite dissolves by oxidation, calcite is consumed and ferric hydroxide precipitates (Fig. 23.4) according to the reaction

$$\underset{pyrite}{FeS_2} + 2\,\underset{calcite}{CaCO_3} + \frac{3}{2}\,H_2O + \frac{15}{4}\,O_2(g) \rightarrow$$

$$\underset{ferric\ hydroxide}{Fe(OH)_3} \quad + 2\,Ca^{++} + 2\,SO_4^{--} + 2\,CO_2(aq) \quad (23.5)$$

Ca^{++} and SO_4^{--} accumulate in the fluid, eventually causing gypsum ($CaSO_4{\cdot}2H_2O$) to saturate and precipitate. At this point, the overall reaction becomes

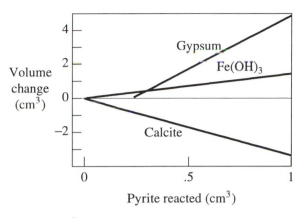

FIG. 23.4 Volumes (cm³) of minerals consumed (negative values) and precipitated (positive values) as pyrite reacts at 25°C with a groundwater in equilibrium with atmospheric O_2 in the presence of calcite.

$$\underset{pyrite}{FeS_2} + \underset{calcite}{2\,CaCO_3} + \frac{11}{2}\,H_2O + \frac{15}{4}\,O_2(g) \;\rightarrow$$

$$\underset{ferric\ hydroxide}{Fe(OH)_3} + \underset{gypsum}{2\,CaSO_4 \cdot 2H_2O} + 2\,CO_2(aq) \qquad (23.6)$$

In the calculation, reaction of 1 cm³ of pyrite consumes about 3.4 cm³ of calcite, demonstrating that considerable quantities of buffering minerals may be required in mineralized areas to neutralize drainage waters.

The accumulation of $CO_2(aq)$ over the reaction path (according to Reactions 23.5–23.6) raises the fluid's CO_2 fugacity to a value greater than atmospheric pressure. In nature, $CO_2(g)$ would begin to effervesce

$$CO_2(aq) \;\rightarrow\; CO_2(g) \qquad (23.7)$$

at about this point in the reaction, causing pH to increase to values somewhat greater than predicted by the reaction path, which did not account for degassing. (As a variation of the calculation, we could account for the degassing by reacting enough pyrite to bring f_{CO_2} to one, "picking up" the calculation results, fixing the fugacity, and then reacting the remaining pyrite.)

In the two calculations (one including, the other excluding calcite), the resulting fluids differ considerably in composition. After reaction of 1 cm³ of pyrite at atmospheric oxygen fugacity, the compositions are

	No calcite	Calcite
pH	1.7	5.6
Fe (mg/kg)	2300.	.03
SO_4	8000.	1800.
Ca	9.9	1100.
HCO_3	74.	6000.

Whereas pyrite oxidation in the absence of calcite produces H-Fe-SO_4 drainage, the reaction in the presence of calcite yields a Ca-HCO_3-SO_4 drainage.

23.3 Fate of Dissolved Metals

The metal concentrations in acid drainage can be alarmingly high, but in many cases they are attenuated to much lower levels by natural processes (and sometimes by remediation schemes) near the drainage discharge area. Attenuation results from a variety of processes. The drainage, of course, may be diluted by base flow and by mixing with other surface waters. More significantly, drainage waters become less acidic after discharge as they react with various minerals, mix with other waters (especially those rich in HCO_3^-), and in some cases lose CO_2 to the atmosphere.

As pH rises, the metal content of drainage water tends to decrease. Some metals precipitate directly from solution to form oxide, hydroxide, and oxy-hydroxide phases. Iron and aluminum are notable is this regard. They initially form colloidal and suspended phases known as hydrous ferric oxide (HFO, FeOOH·nH_2O) and hydrous aluminum oxide (HAO, AlOOH·nH_2O), both of which are highly soluble under acidic conditions but nearly insoluble at near-neutral pH.

The concentrations of other metals attenuate when the metals sorb onto the surfaces of precipitating minerals (see Chapter 8). HFO, the behavior of which is well studied (Dzombak and Morel, 1990), has a large specific surface area and is capable of sorbing metals from solution in considerable amounts, especially at moderate to high pH; HAO may behave similarly. The process by which HFO or HAO forms and then adsorbs metals from solution, known as *coprecipitation*, represents an important control on the mobility of heavy metals in acid drainages (e.g., Chapman et al., 1983; Johnson, 1986; Davis et al., 1991; Smith et al., 1992).

To see how this process works, we construct a model in which reaction of a hypothetical drainage water with calcite leads to the precipitation of ferric hydroxide [Fe(OH)$_3$, which we use to represent HFO] and the sorption of dissolved species onto this phase. We assume that the precipitate remains suspended in solution with its surface in equilibrium with the changing fluid chemistry, using the surface complexation model described in Chapter 8. In our model, we envisage the precipitate eventually settling to the stream bed and hence removing the sorbed metals from the drainage.

We do not concern ourselves with the precipitate that lines sediments in the stream bed, since it formed earlier while in contact with the drainage, and hence would not be expected to continue to sorb from solution. Smith et al. (1992), for example, found that in an acid drainage from Colorado (USA), sorption on the suspended solids, rather than the sediments along the stream bed, controls the dissolved metal concentrations.

As the first step in the coprecipitation process, ferric hydroxide precipitates either from the effect of the changing pH on the solubility of ferric iron

$$Fe^{+++} + 3\,H_2O \rightarrow \quad Fe(OH)_3 \quad + 3\,H^+ \tag{23.8}$$
$$\textit{ferric hydroxide}$$

or by the oxidation of ferrous species

$$Fe^{++} + \frac{1}{4}\,O_2(aq) + \frac{5}{2}\,H_2O \rightarrow \quad Fe(OH)_3 \quad + 2\,H^+ \tag{23.9}$$
$$\textit{ferric hydroxide}$$

in solution. For simplicity, we assume in our calculations that the dissolved iron has already oxidized, so that Reaction 23.8 is responsible for forming the sorbing phase.

In REACT, we prepare the calculation by disenabling the redox couple between trivalent and pentavalent arsenic (arsenite and arsenate, respectively). As well, we disenable the couples for ferric iron and cupric copper, since we will not consider either ferrous or cupric species, and set the Dzombak and Morel (1990) surface complexation model

```
decouple AsO4---
decouple Cu++
decouple Fe+++
surface_data = FeOH.data
```

We then define an initial fluid representing the unreacted drainage. We set the fluid's iron content by assuming equilibrium with jarosite [$NaFe_3(SO_4)_2(OH)_6$] and prescribe a high content of dissolved arsenite, arsenate, cupric copper, lead, and zinc. Finally, we suppress hematite and goethite (which are more stable than ferric hydroxide) and two ferrite minerals (e.g., $ZnFe_2O_4$), which we consider unlikely to form. The procedure is

```
(cont'd)
swap Jarosite-Na for Fe+++
1 free ug Jarosite-Na

pH = 3
Na+      =     10 mg/kg
Ca++     =     10 mg/kg
```

```
Cl-        =     20 mg/kg
HCO3-      =    100 mg/kg
balance on SO4--

As(OH)4- =    200 ug/kg
AsO4--- =    1000 ug/kg
Cu++     =     500 ug/kg
Pb++     =     200 ug/kg
Zn++     =   18000 ug/kg

suppress Hematite, Goethite, Ferrite-Zn, Ferrite-Cu
delxi = .002
```

To calculate the effect of reacting calcite with the drainage, we enter the command

(cont'd)
```
react 4 mmol Calcite
```

and type `go` to trigger the calculation.

In the calculation, calcite dissolves into the drainage according to

$$CaCO_3 + 2\,H^+ \rightarrow Ca^{++} + CO_2(aq) + H_2O \qquad (23.10)$$
$$\text{\textit{calcite}}$$

and

$$CaCO_3 + H^+ \rightarrow Ca^{++} + HCO_3^- \qquad (23.11)$$
$$\text{\textit{calcite}}$$

consuming H^+ and causing pH to increase. The shift in pH drives the ferric iron in solution to precipitate as ferric hydroxide, according to Reaction 23.8. A total of about .89 $m$mol of precipitate forms over the reaction path.

In the calculation, the distribution of metals between solution and the ferric hydroxide surface varies strongly with pH (Fig. 23.5). As discussed in Sections 8.4 and 12.3, pH exerts an important control over the sorption of metal ions for two reasons. First, the electrical charge on the sorbing surface tends to decrease as pH increases, lessening the electrical repulsion between surface and ions. Second, because hydrogen ions are involved in the sorption reactions, pH affects ion sorption by mass action. The sorption of bivalent cations such as Cu^{++}

$$>(s)FeOH_2^+ + Cu^{++} \rightarrow >(s)FeOCu^+ + 2\,H^+ \qquad (23.12)$$

is favored as pH increases, and there is a similar effect for Pb^{++} and Zn^{++}. Sorption of arsenite is also favored by increasing pH, according to the reaction

$$>(w)FeOH_2^+ + As(OH)_3 \rightarrow >(w)FeH_2AsO_3 + H_2O + H^+ \qquad (23.13)$$

As a result, the metal ions are progressively partitioned onto the ferric hydroxide surface as pH increases.

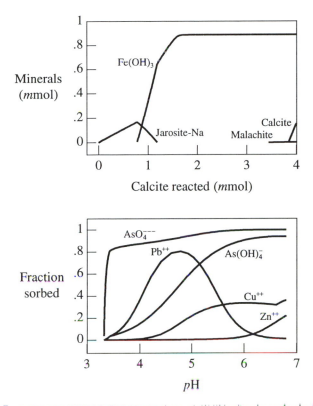

FIG. 23.5 Minerals formed during reaction at 25°C of a hypothetical acid drainage water with calcite (top), and fractions of the amounts of arsenite, arsenate, copper, lead, and zinc present initially in solution that sorb onto ferric hydroxide over the course of the reaction path (bottom). Bottom figure is plotted against *p*H, which increases as the water reacts with calcite.

Arsenate sorbs onto the ferric hydroxide surface over the *p*H range of the calculation because of its ability to bond tightly with the surface's weakly sorbing sites. Initially, the arsenate sorbs

$$>(w)FeOH_2^+ + H_2AsO_4^- \rightarrow >(w)FeH_2AsO_4 + H_2O \qquad (23.14)$$

by complexing with the protonated weak sites. As *p*H increases, the surface species $>(w)FeHAsO_4^-$ and $>(w)FeOHAsO_4^{---}$ come to predominate, and the arsenate is partitioned even more strongly onto the sorbing surface.

Interestingly, Pb^{++} sorbs in the initial stages of the reaction but then desorbs in the calculation as *p*H continues to rise. The desorption reaction

$$>(s)FeOPb^+ + HCO_3^- + Cu^{++} \rightarrow >(s)FeOCu^+ + H^+ + PbCO_3 \qquad (23.15)$$

is driven by the formation of the $PbCO_3$ ion pair from the bicarbonate contributed to solution by the dissolving calcite, and by competition with Cu^{++} (and Zn^{++}) for the strongly sorbing sites.

It is useful to compare the sorption limit for each metal (the amount of each that could sorb if it occupied each surface site) with the metal concentrations in solution. To calculate the sorption limits, we take into account the amount of ferric precipitate formed in the calculation (.89 $mmol$), the number of moles of strongly and weakly binding surface sites per mole of precipitate (.005 and .2, respectively, according to the surface complexation model), and the site types that accept each metal [Pb^{++} and Cu^{++} sorb on strong sites, $As(OH)_4^-$ and AsO_4^{---} sorb on weak sites, and Zn^{++} sorbs on both].

Comparing the initial metal concentrations to the resulting sorption limits

Component	Sorption limit (mg/kg)	In solution (mg/kg)
$As(OH)_4^-$	25.4	.2
AsO_3^{---}	24.7	1.0
Cu^{++}	.28	.5
Pb^{++}	.92	.2
Zn^{++}	11.9	18.0

we see that the precipitate formed in sufficient quantity to sorb the arsenic initially present in solution as well as the lead, but only a fraction of the copper and zinc. An initial consideration in evaluating the ability of coprecipitation to attenuate metal concentrations in a drainage water, therefore, is whether the sorbing mineral forms in a quantity sufficient to sorb the water's metal content.

Appendix 1
Sources of Modeling Software

The following is a list, current at the time of publication, of sources of some of the most popular geochemical modeling software programs and packages. Most sources require that the requesting party pay a licensing or distribution fee or provide magnetic media for distribution. Software authors do not necessarily provide user assistance.

For more information, consult the references listed for each software package, as well as overviews of these and other programs by Mangold and Tsang (1991) and Appel and Reilly (1994).

CHESS

inquiries:	Jan van der Lee
	Centre d'Informatique Géologique
	Ecole des Mines de Paris
	35, rue Saint-Honoré
	77305 Fontainebleau France
internet:	jan@cig.ensmp.fr
reference:	van der Lee (1993)

CHILLER

inquiries:	Mark H. Reed
	Department of Geological Sciences
	1272 University of Oregon
	Eugene, Oregon 97404-1272 USA
internet:	mhreed@oregon.uoregon.edu
reference:	Reed (1982)

DISSOL, THERMAL, EQSEL
      inquiries: Bertrand Fritz
                      Centre de Géochimie de la Surface
                      CNRS
                      1, rue Blessig
                      67084 Strasbourg CEDEX France
      internet: `bfritz@illite.u-strasbg.fr`
   references: Fritz (1981), Fritz et al. (1987), Zins-Pawlas (1988)

EQ3/EQ6
      inquiries: Thomas J. Wolery
                      Mail Stop L–219
                      Lawrence Livermore National Laboratory
                      P.O. Box 808
                      Livermore, California 94550 USA
      internet: `wolery1@llnl.gov`
    reference: Wolery (1992a)

GEOCHEM-PC
      inquiries: Dr. David R. Parker
                      Department of Soil and Environmental Sciences
                      University of California
                      Riverside, California 92521 USA
      internet: `dparker@mail.ucr.edu`
    reference: Parker et al. (1995)

The Geochemist's Workbench®
      inquiries: Hydrogeology Program
                      Department of Geology
                      University of Illinois
                      1301 West Green Street
                      Urbana, Illinois 61801 USA
      internet: `help@aquifer.geology.uiuc.edu`
    reference: Bethke (1994)

KINDISP

      inquiries: Benoit Madé
                 Centre d'Informatique Géologique
                 Ecole des Mines de Paris
                 35, rue Saint-Honoré
                 77305 Fontainebleau France
      internet: made@cig.ensmp.fr
     reference: Madé et al. (1994)

MINEQL+

      inquiries: Dr. William Schecher
                 Environmental Research Software
                 16 Middle Street
                 Hallowell, Maine 04347 USA
      internet: (none)
    references: Westall et al. (1976), Schecher and McAvoy (1994)

MINTEQA2

      inquiries: Center for Exposure Assessment Modeling
                 Attn: Mr. David W. Disney
                 U.S. Environmental Protection Agency
                 Environmental Research Laboratory
                 960 College Station Road
                 Athens, Georgia 30605-2720 USA
      internet: ceam@athens.ath.epa.gov
     reference: Allison et al. (1991)

PHREEQC

      inquiries: David Parkhurst
                 U.S. Geological Survey
                 Box 25046, Mail Stop 418
                 Denver Federal Center
                 Lakewood, Colorado 80225 USA
      internet: brrcrftp.cr.usgs.gov (anonymous ftp)
     reference: Parkhurst (1995)

PHREEQE

    inquiries: WATSTORE Program Office
                  U.S. Geological Survey
                  437 National Center
                  12201 Sunrise Valley Drive
                  Reston, Virginia 22092 USA
    internet: brrcrftp.cr.usgs.gov (anonymous ftp)
    reference: Parkhurst et al. (1980)

PHRQPITZ

    inquiries: WATSTORE Program Office
                  U.S. Geological Survey
                  437 National Center
                  12201 Sunrise Valley Drive
                  Reston, Virginia 22092 USA
    internet: brrcrftp.cr.usgs.gov (anonymous ftp)
    reference: Plummer et al. (1988)

REACT

    inquiries: Ronald Stoessell
                  Department of Geology and Geophysics
                  University of New Orleans
                  New Orleans, Louisiana 71048 USA
    internet: rkses@uno.edu
    reference: Stoessell (1988)

SOILCHEM

    inquiries: Steven Santiago
                  Office of Technology Licensing, #1620
                  University of California
                  Berkeley, California 94720-1620 USA
    internet: (none)
    reference: Sposito and Coves (1995)

SOLMNEQ.88

  inquiries: Yousif Kharaka
        U.S. Geological Survey
        345 Middlefield Road
        Mail Stop 27
        Menlo Park, California 94025 USA
  internet: ykharaka@rcamnl.wr.usgs.gov
  reference: Kharaka et al. (1988)

SOLMNEQ.88 pc/shell

  inquiries: E.H. Perkins
        Alberta Research Council
        P.O. Box 8330
        Edmonton, Alberta T6H 5X2 Canada
  internet: perkins@arc.ab.ca
  reference: Perkins (1992)

WATEQ4F

  inquiries: WATSTORE Program Office
        U.S. Geological Survey
        437 National Center
        12201 Sunrise Valley Drive
        Reston, Virginia 22092 USA
  internet: brrcrftp.cr.usgs.gov (anonymous ftp)
  reference: Ball and Nordstrom (1991)

The following programs are time-space continuum models, as discussed in Section 2.2.

MPATH, GEM

  inquiries: Peter C. Lichtner
        Southwest Research Institute
        Center for Nuclear Waste Regulatory Analyses
        6220 Culebra Road
        San Antonio, Texas 78238-5166 USA
  internet: lichtner@swri.edu
  references: Lichtner (1992), Steinmann et al. (1994),
        Lichtner (1995)

1DREACT

    inquiries: Carl Steefel
                Department of Geology
                University of South Florida
                4202 East Fowler Avenue
                Tampa, Florida 33620 USA
    internet: `csteefel@chuma.cas.usf.edu`
    references: Steefel and Lasaga (1994),
                      Steefel and Lichtner (1994)

Appendix 2
Evaluating the HMW Activity Model

The best way to fully understand the calculation procedure for the Harvie-Mφller-Weare activity model (Harvie et al., 1984) is to carry through a simple example by hand. In this appendix, we follow the steps in the procedure outlined in Tables 7.1–7.3, using the model coefficients given in Tables 7.4–7.7.

We take as an example a 6 molal NaCl solution containing .01 molal CaSO$_4$. Since the only species considered for this chemical system are Na$^+$, Cl$^-$, Ca^{++}, and SO$_4^{--}$, we can immediately write down the species molalities m_i along with their charges z_i:

	m_i	z_i
Na$^+$	6	+1
Cl$^-$	6	−1
Ca^{++}	.01	+2
SO$_4^{--}$	.01	−2

The only task left to us is to calculate the activity coefficients.

GIVEN DATA. The following data are model parameters from Tables 7.4–7.7.

$\beta_{MX}^{(0)}$	Cl$^-$	Ca^{++}	SO$_4^{--}$
Na$^+$	.0765	—	.01958
Cl$^-$	×	.3159	—
Ca^{++}	×	×	.2

$\beta_{MX}^{(1)}$	Cl$^-$	Ca^{++}	SO$_4^{--}$
Na$^+$	.2664	—	1.113
Cl$^-$	×	1.614	—
Ca^{++}	×	×	3.1973

$\beta_{MX}^{(2)}$	Cl$^-$	Ca^{++}	SO$_4^{--}$
Na$^+$	0	—	0
Cl$^-$	×	0	—
Ca^{++}	×	×	−54.24

C_{MX}^{ϕ}	Cl$^-$	Ca^{++}	SO$_4^{--}$
Na$^+$	.00127	—	.00497
Cl$^-$	×	−.00034	—
Ca^{++}	×	×	0

α_{MX}	Cl^-	Ca^{++}	SO_4^{--}
Na^+	2	—	2
Cl^-	×	2	—
Ca^{++}	×	×	1.4

θ_{ij}	Cl^-	Ca^{++}	SO_4^{--}
Na^+	—	.07	—
Cl^-	×	—	.02
Ca^{++}	×	×	—

$\psi_{aa'Na^+}$	Cl^-	Ca^{++}	SO_4^{--}
Na^+	—	—	—
Cl^-	×	—	.0014
Ca^{++}	×	×	—

$\psi_{aa'Ca^{++}}$	Cl^-	Ca^{++}	SO_4^{--}
Na^+	—	—	—
Cl^-	×	—	-.018
Ca^{++}	×	×	—

$\psi_{cc'Cl^-}$	Cl^-	Ca^{++}	SO_4^{--}
Na^+	—	-.007	—
Cl^-	×	—	—
Ca^{++}	×	×	—

$\psi_{cc'SO_4^-}$	Cl^-	Ca^{++}	SO_4^{--}
Na^+	—	-.055	—
Cl^-	×	—	—
Ca^{++}	×	×	—

There are no neutral species; hence, no λ_{ni}.

STEP 1. The solution ionic strength I and total molal charge Z are

$$I = \frac{1}{2}\left[1\times6 + 1\times6 + 4\times.01 + 4\times.01\right] = 6.04 \text{ molal}$$

$$Z = 1\times6 + 1\times6 + 2\times.01 + 2\times.01 = 12.04 \text{ molal}$$

STEP 2. Using the program in Table A2.1, we calculate values for ${}^E\theta_{ij}(I)$ and ${}^E\theta'_{ij}(I)$:

${}^E\theta_{ij}(I)$	Cl^-	Ca^{++}	SO_4^{--}
Na^+	—	-0.05933	—
Cl^-	×	—	-0.05933
Ca^{++}	×	×	—

${}^E\theta'_{ij}(I)$	Cl^-	Ca^{++}	SO_4^{--}
Na^+	—	.004861	—
Cl^-	×	—	.004861
Ca^{++}	×	×	—

STEP 3. Values for functions $g(x)$ and $g'(x)$, taking $x = \alpha_{MX}\sqrt{I}$, are

$g(\alpha_{MX}\sqrt{I})$	Cl^-	Ca^{++}	SO_4^{--}
Na^+	.07919	—	.07919
Cl^-	×	.07919	—
Ca^{++}	×	×	.1449

$g'(\alpha_{MX}\sqrt{I})$	Cl^-	Ca^{++}	SO_4^{--}
Na^+	-.07186	—	-.07186
Cl^-	×	-.07186	—
Ca^{++}	×	×	-.1129

TABLE A2.1 Program (ANSI C) for calculating $^E\theta_{ij}(I)$ and $^E\theta'_{ij}(I)$ at 25°C

```c
#include <stdio.h>
#include <stdlib.h>
#include <math.h>

void calc_lambdas(double is);
void calc_thetas(double is, int z1, int z2,
                 double *etheta, double *etheta_prime);
double elambda[17], elambda1[17];

void main() {
    double etheta, etheta_prime, is;
    int z1, z2;
    while (printf("Enter I, z1, z2: ")
           && scanf("%lf %i %i", &is, &z1, &z2) == 3) {
        if (abs(z1) <= 4 && abs(z2) <= 4 && is > 0) {
            calc_lambdas(is);
            calc_thetas(is, z1, z2, &etheta, &etheta_prime);
            printf("E-theta(I) = %f, E-theta'(I) = %f\n\n", etheta, etheta_prime);
        }
        else
            printf("Input data out of range\n");
    }
}
void calc_lambdas(double is) {
    double aphi, dj, jfunc, jprime, t, x, zprod;
    int i, ij, j;

    /* Coefficients c1-c4 are used to approximate the integral function "J";
       aphi is the Debye-Huckel constant at 25 C */

    double c1 = 4.581, c2 = 0.7237, c3 = 0.0120, c4 = 0.528;

    aphi = 0.392;    /* Value at 25 C */

    /* Calculate E-lambda terms for charge combinations of like sign,
       using method of Pitzer (1975). */

    for (i=1; i<=4; i++) {
        for (j=i; j<=4; j++) {
            ij = i*j;
            zprod = (double)ij;
            x = 6.0* zprod * aphi * sqrt(is);                    /* eqn 23 */
            jfunc = x / (4.0 + c1*pow(x,-c2)*exp(-c3*pow(x,c4)));  /* eqn 47 */

            t = c3 + c4 * pow(x,c4);
            dj = c1* pow(x,(-c2-1.0)) * (c2+t) * exp(-c3*pow(x,c4));
            jprime = (jfunc/x)*(1.0 + jfunc*dj);

            elambda[ij] = zprod*jfunc / (4.0*is);                /* eqn 14 */
            elambda1[ij] = (3.0*zprod*zprod*aphi*jprime/(4.0*sqrt(is))
                           - elambda[ij])/is;
        }
    }
}
void calc_thetas(double is, int z1, int z2,
                 double *etheta, double *etheta_prime) {
    int i, j;
    double f1, f2;

    /* Calculate E-theta(I) and E-theta'(I) using method of Pitzer (1987) */

    i = abs(z1);
    j = abs(z2);
    if (z1*z2 < 0) {
        *etheta = 0.0;
        *etheta_prime = 0.0;
    }
    else {
        f1 = (double)i/(double)(2*j);
        f2 = (double)j/(double)(2*i);                            /* eqn A14 */
        *etheta = elambda[i*j] - f1*elambda[j*j] - f2*elambda[i*i];
        *etheta_prime = elambda1[i*j] - f1*elambda1[j*j] - f2*elambda1[i*i];
    }
}
```

STEP 4. The second virial coefficients for cation-anion pairs are

B_{MX}	Cl^-	Ca^{++}	SO_4^{--}
Na^+	.09744	—	.1077
Cl^-	×	.4437	—
Ca^{++}	×	×	.5386

B'_{MX}	Cl^-	Ca^{++}	SO_4^{--}
Na^+	−.003169	—	−.01324
Cl^-	×	−.0192	—
Ca^{++}	×	×	−.03909

B^{ϕ}_{MX}	Cl^-	Ca^{++}	SO_4^{--}
Na^+	.07845	—	.02774
Cl^-	×	.3277	—
Ca^{++}	×	×	.3024

STEP 5. The third virial coefficients for cation-anion pairs are

C_{MX}	Cl^-	Ca^{++}	SO_4^{--}
Na^+	.000635	—	.001757
Cl^-	×	−.0001202	—
Ca^{++}	×	×	0

STEP 6. The second virial coefficients for cation-cation and anion-anion pairs are

Φ_{ij}	Cl^-	Ca^{++}	SO_4^{--}
Na^+	—	.01067	—
Cl^-	×	—	−.03933
Ca^{++}	×	×	—

Φ'_{ij}	Cl^-	Ca^{++}	SO_4^{--}
Na^+	—	.004861	—
Cl^-	×	—	.004861
Ca^{++}	×	×	—

Φ^{ϕ}_{ij}	Cl^-	Ca^{++}	SO_4^{--}
Na^+	—	.04003	—
Cl^-	×	—	−.00997
Ca^{++}	×	×	—

STEP 7. From the above results, the value of F is

$F = -1.2568$

STEP 8. The ion activity coefficients are calculated

	Na^+	Ca^{++}	Cl^-	SO_4^{--}
$z_i^2 F$	−1.2568	−5.0271	−1.2568	−5.0271
First sum	1.2194	5.3266	1.2259	1.4303
Second sum	−.0002	−.1273	−.0007	−.4226
Third sum	.0001	−.0011	−.0004	−.0033
Fourth sum	.0230	.0459	.0230	.0459
Fifth sum	0.0	0.0	0.0	0.0
$\ln \gamma_i$	−.0145	.2170	−.0090	−3.9768
γ_i	.986	1.242	.991	.019

STEP 9. The quantity $\sum_i m_i (\phi - 1)$ is calculated

First term	−1.4735
First sum	3.1211
Second sum	−0.0002
Third sum	−0.0001
Fourth sum	0.0
Fifth sum	0.0
	$1.6474 \times 2 = 3.2948$

STEP 10. The activity of water is given

$\sum_i m_i$	12.02
$(\sum_i m_i) \phi$	15.3168
$\ln a_w$	−.2759
a_w	.759

We can use program REACT to quickly verify our calculations:

```
hmw
Na+   = 6 molal
Cl-   = 6 molal
Ca++  = .01 molal
SO4-- = .01 molal
go
```

The values for γ_i and a_w in the program output can be compared to the results obtained in Steps 8 and 10.

Appendix 3

Band Filtering

Reaction models commonly produce data points that when plotted fall very closely together. Whether using a program like GTPLOT to render the results, importing them into a word processor, or (horrors!) plotting them by hand, the data density can slow things down and get in the way. Wouldn't it be nice to filter out the unnecessary points, leaving just those needed to define a smooth and accurate curve?

Band filtering does just that. The program listed below takes a stream of x-y coordinates, one data pair per line, and returns just the points needed to plot the curve. You control the smoothness of the curve by setting the -filter (or -f) option. By default, the filter is 0.001, but you can set it smaller to get a smoother curve with more data points, or larger to get a rougher curve. By default, the program takes input from the standard input stream, but you can specify an input file with the -input (or -i) option. The results are sent to the standard output stream, which you can redirect as usual.

```
/*
 *   Program band_filter
 *
 *   An ANSI C program to filter from a series of x-y pairs the data
 *   points not needed to define a smooth curve.
 *
 *   To compile under Unix (note: an ANSI-compatible C compiler is required):
 *           cc band_filter.c -o band_filter -lm
 *
 *   Use:
 *           band_filter data.long > data.short
 *           cat data.long | band_filter > data.short
 *           band_filter -filter .005 -input input_file
 *
 *   Options:
 *           -input or -i:   specify input dataset (default = stdin)
 *           -filter or -f:  specify filter width (default = 0.001)
 *
 *   Note: set filter width to a smaller value for a smoother curve with
 *   more data points, or to a larger value for a rougher curve with fewer
 *   points.
 *
 *   Input dataset must be a listing of numerical x-y pairs, one pair
 *   per line.
 *
 *   Prepared by Craig M. Bethke, December 23, 1994.
 */
```

```c
#include <stdlib.h>
#include <stdio.h>
#include <math.h>
#include <string.h>

#define IN_BAND              0
#define OUT_OF_BAND          1
#define max(a,b)             ((a)<(b)? (b)  :  (a))
#define min(a,b)             ((a)>(b)? (b)  :  (a))
#define equal(str1,str2)     !strcmp(str1,str2)
#define abs(a)               ((a)>=0 ? (a)  : -(a))

/* Global references. */

double *xarray, *yarray;
double min_xval, min_yval, max_xval, max_yval, xrange, yrange;
double filter = 0.001, filter2;
int npts;
char input_file[256] = "";
FILE *input;

/* Function prototypes. */

void get_opts(int argc, char *argv[]);
int scan_data(void);
int check_band (double x1, double y1,
                double x2, double y2,
                double x,  double y);

/* Main program. */

int main(int argc, char *argv[])
{
    int i, j, k;

    /* Get command line options. */

    get_opts(argc, argv);
    filter = abs(filter);
    filter2 = filter*filter;

    /* Scan in the data points. */

    if ((npts = scan_data()) == 0) {
        fprintf(stderr, "Empty input stream\\n");
        exit(-1);
    }

    /* Check for points out of band. */

    printf("%g %g   (%i)\\n", xarray[0], yarray[0], 1);
    k = 0;
    for (j=1; j<npts; j++) {
        for (i=j-1; i>k; i--) {
            if (check_band(xarray[k], yarray[k],
                           xarray[j], yarray[j],
                           xarray[i], yarray[i]) == OUT_OF_BAND) {
                printf("%g %g   (%i)\\n", xarray[i], yarray[i], i+1);
                k = i;
            }
        }
    }
    printf("%g %g   (%i)\\n", xarray[npts-1], yarray[npts-1], npts);

    exit(0);
}

void get_opts(int argc, char *argv[])
{
    int n;

    n = 1;
    while (n < argc) {
        if (equal(argv[n], "-f") || equal(argv[n], "-filter")) {
            if (sscanf(argv[++n], "%lf", &filter) != 1) {
                fprintf(stderr, "Bad filter value\\n");
```

```
                exit(-1);
            }
        }
        else if (equal(argv[n], "-i") || equal(argv[n], "-input"))
            strcpy(input_file, argv[++n]);
        else
            strcpy(input_file, argv[n]);

        n++;
    }
}

/* Function to read data from input stream. */

int scan_data()
{
    char uline[256];
    double xval, yval;
    int n;
    size_t size = 1000;

    /* Allocate main arrays. */

    xarray = (double *)malloc(size*sizeof(double));
    yarray = (double *)malloc(size*sizeof(double));

    n = 0;
    min_xval = 1.e30;
    min_yval = 1.e30;
    max_xval = -1.e30;
    max_yval = -1.e30;

    /* Open input file. */

    if (equal(input_file, ""))
        input = stdin;
    else {
        input = fopen(input_file, "r");
        if (input == NULL) return 0;
    }

    /* Scan in the data. */

    while (fgets(uline, 255, input) != NULL) {
        if (sscanf(uline, "%lf %lf", &xval, &yval) != 2) return n;
        if (n >= size-1) {
            size += 1000;
            xarray = (double *)realloc(xarray, size*sizeof(double));
            yarray = (double *)realloc(yarray, size*sizeof(double));
        }
        xarray[n] = xval;
        yarray[n] = yval;
        n++;

        min_xval = min(min_xval, xval);
        min_yval = min(min_yval, yval);
        max_xval = max(max_xval, xval);
        max_yval = max(max_yval, yval);
    }

    xrange = max_xval - min_xval;
    yrange = max_yval - min_yval;
    return n;
}

/* Function reports whether (x,y) is within current band. */

int check_band (double x1, double y1,
                double x2, double y2,
                double x,  double y)
{

    double a, d, d1, dx, dx12, dy, dy12;

    /* Check for points very close together. */
```

```
    dx = (x-x1)/xrange;
    dy = (y-y1)/yrange;
    d = (dx*dx + dy*dy);

    dx12 = (x2-x1)/xrange;
    dy12 = (y2-y1)/yrange;
    d1 = sqrt(dx12*dx12 + dy12*dy12);

    if (d1 <= filter) {
        if (d > filter2)
            return OUT_OF_BAND;
        else
            return IN_BAND,
    }

    /* Project (p-p1) onto (p2-p1). */

    a = (dx12*dx + dy12*dy)/d1;
    if (a < 0.0 || a > d1) return OUT_OF_BAND;

    /* Is (x,y) is close enough to count? */

    if ((d-(a*a)) > filter2) return OUT_OF_BAND;

    return IN_BAND;
}
```

Appendix 4

Minerals in the LLNL Database

Mineral	Chemical formula	General type
Acanthite	Ag_2S	sulfide
Akermanite	$Ca_2MgSi_2O_7$	
Alabandite	MnS	sulfide
Albite	$NaAlSi_3O_8$	feldspar
Albite high	$NaAlSi_3O_8$	feldspar
Albite low	$NaAlSi_3O_8$	feldspar
Alstonite	$BaCa(CO_3)_2$	carbonate
Alunite	$KAl_3(OH)_6(SO_4)_2$	sulfate
Amesite-14A	$Mg_4Al_4Si_2O_{10}(OH)_8$	serpentine
Amrph°silica	SiO_2	silica
Analc-dehydr	$NaAlSi_2O_6$	zeolite
Analcime	$NaAlSi_2O_6 \cdot H_2O$	zeolite
Andalusite	Al_2SiO_5	
Andradite	$Ca_3Fe_2(SiO_4)_3$	garnet
Anglesite	$PbSO_4$	sulfate
Anhydrite	$CaSO_4$	sulfate
Annite	$KFe_3AlSi_3O_{10}(OH)_2$	mica
Anorthite	$CaAl_2Si_2O_8$	feldspar
Antarcticite	$CaCl_2 \cdot 6H_2O$	halide
Anthophyllite	$Mg_7Si_8O_{22}(OH)_2$	amphibole
Antigorite	$Mg_{24}Si_{17}O_{42.5}(OH)_{31}$	serpentine
Aragonite	$CaCO_3$	carbonate
Arcanite	K_2SO_4	sulfate
Arsenolite	As_2O_3	oxide
Arsenopyrite	$AsFeS$	sulfide
Artinite	$Mg_2CO_3(OH)_2 \cdot 3H_2O$	carbonate
Azurite	$Cu_3(CO_3)_2(OH)_2$	carbonate

(continues)

Minerals in LLNL Database

Mineral	Chemical formula	General type
Barite	$BaSO_4$	sulfate
Barytocalcite	$BaCa(CO_3)_2$	carbonate
Bassanite	$CaSO_4 \cdot .5H_2O$	sulfate
Bassetite	$Fe(UO_2)_2(PO_4)_2$	phosphate
Beidellit-Ca	$Ca_{.165}Al_{2.33}Si_{3.67}O_{10}(OH)_2$	smectite
Beidellit-H	$H_{.33}Al_{2.33}Si_{3.67}O_{10}(OH)_2$	smectite
Beidellit-K	$K_{.33}Al_{2.33}Si_{3.67}O_{10}(OH)_2$	smectite
Beidellit-Mg	$Mg_{.165}Al_{2.33}Si_{3.67}O_{10}(OH)_2$	smectite
Beidellit-Na	$Na_{.33}Al_{2.33}Si_{3.67}O_{10}(OH)_2$	smectite
Berlinite	$AlPO_4$	phosphate
Bieberite	$CoSO_4 \cdot 7H_2O$	sulfate
Birnessite	$Mn_8O_{19}H_{10}$	
Bischofite	$MgCl_2 \cdot 6H_2O$	chloride
Bixbyite	Mn_2O_3	oxide
Bloedite	$Na_2Mg(SO_4)_2 \cdot 4H_2O$	sulfate
Boehmite	$AlOOH$	hydroxide
Boltwood-Na	$Na_{.7}K_{.3}H_3OUO_2SiO_4 \cdot H_2O$	
Boltwoodite	$K(H_3O)UO_2(SiO_4)$	
Borax	$Na_2B_4O_5(OH)_4 \cdot 8H_2O$	borate
Boric acid	$B_3(c)$	
Bornite	Cu_5FeS_4	sulfide
Brezinaite	Cr_3S_4	sulfide
Brucite	$Mg(OH)_2$	hydroxide
Burkeite	$Na_6CO_3(SO_4)_2$	sulfide
Ca-Al Pyroxene	$CaAl_2SiO_6$	pyroxene
Calcite	$CaCO_3$	carbonate
Carnallite	$KMgCl_3 \cdot 6H_2O$	halide
Carnotite	$K_2(UO_2)_2(VO_4)_2$	vanadate
Cattierite	CoS_2	sulfide
Celestite	$SrSO_4$	sulfate
Cerussite	$PbCO_3$	carbonate
Chalcedony	SiO_2	silica
Chalcocite	Cu_2S	sulfide
Chalcopyrite	$CuFeS_2$	sulfide
Chamosite-7A	$Fe_2Al_2SiO_5(OH)_4$	7 Å clay
Chloromagnesite	$MgCl_2$	halide
Chloropyromorphite	$Pb_5(PO_4)_3Cl$	phosphate
Chrysotile	$Mg_3Si_2O_5(OH)_4$	serpentine
Cinnabar	HgS	sulfide
Claudetite	As_2O_3	oxide

(continues)

Mineral	Chemical formula	General type
Clinochl-14A	$Mg_5Al_2Si_3O_{10}(OH)_8$	chlorite
Clinochl-7A	$Mg_5Al_2Si_3O_{10}(OH)_8$	
Clinoptil-Ca	$CaAl_2Si_{10}O_{24} \cdot 8H_2O$	zeolite
Clinoptil-K	$K_2Al_2Si_{10}O_{24} \cdot 8H_2O$	zeolite
Clinoptil-Mg	$MgAl_2Si_{10}O_{24} \cdot 8H_2O$	zeolite
Clinoptil-Na	$Na_2Al_2Si_{10}O_{24} \cdot 8H_2O$	zeolite
Clinozoisite	$Ca_2Al_3Si_3O_{12}(OH)$	epidote
Coffinite	$USiO_4$	epidote
Colemanite	$Ca_2B_6O_8(OH)_6 \cdot 2H_2O$	borate
Copper	Cu	native element
Cordier^anhy	$Mg_2Al_4Si_5O_{18}$	
Cordier^hydr	$Mg_2Al_4Si_5O_{18} \cdot H_2O$	
Corundum	Al_2O_3	oxide
Covellite	CuS	sulfide
Cristobalite	SiO_2	silica
Cronstedt-7A	$Fe_4SiO_5(OH)_4$	serpentine
Cuprite	Cu_2O	oxide
Daphnite-14A	$Fe_5Al_2Si_3O_{10}(OH)_8$	chlorite
Daphnite-7A	$Fe_5Al_2Si_3O_{10}(OH)_8$	
Dawsonite	$NaAlCO_3(OH)_2$	carbonate
Diaspore	$AlHO_2$	hydroxide
Diopside	$CaMgSi_2O_6$	pyroxene
Dolomite	$CaMg(CO_3)_2$	carbonate
Dolomite-dis	$CaMg(CO_3)_2$	carbonate
Dolomite-ord	$CaMg(CO_3)_2$	carbonate
Enstatite	$MgSiO_3$	pyroxene
Epidote	$Ca_2FeAl_2Si_3O_{12}OH$	epidote
Epidote-ord	$Ca_2FeAl_2Si_3O_{12}OH$	epidote
Epsomite	$MgSO_4 \cdot 7H_2O$	sulfate
Eu	Eu	native element
Eucryptite	$LiAlSiO_4$	
Fayalite	Fe_2SiO_4	olivine
Ferrite-Ca	$CaFe_2O_4(c)$	
Ferrite-Cu	$CuFe_2O_4(c)$	
Ferrite-Mg	$MgFe_2O_4(c)$	
Ferrite-Zn	$ZnFe_2O_4$	
Ferrosilite	$FeSiO_3$	pyroxene
Fluorapatite	$Ca_5(PO_4)_3F$	phosphate
Fluorite	CaF_2	fluoride
Forsterite	Mg_2SiO_4	olivine

(continues)

Mineral	Chemical formula	General type
Galena	PbS	sulfide
Gaylussite	$CaNa_2(CO_3)_2 \cdot 5H_2O$	carbonate
Gehlenite	$Ca_2Al_2SiO_7$	
Gibbsite	$Al(OH)_3$	hydroxide
Goethite	$FeOOH$	hydroxide
Gold	Au	native element
Graphite	C	native element
Greenalite	$Fe_3Si_2O_5(OH)_4$	serpentine
Grossular	$Ca_3Al_2Si_3O_{12}$	garnet
Gummite	UO_3	oxide
Gypsum	$CaSO_4 \cdot 2H_2O$	sulfate
Haiweeite	$Ca(UO_2)_2(Si_2O_5)_3 \cdot 5H_2O$	
Halite	$NaCl$	halide
Hausmannite	Mn_3O_4	oxide
Hedenbergite	$CaFe(SiO_3)_2$	pyroxene
Hematite	Fe_2O_3	oxide
Hercynite	$FeAl_2O_4$	oxide
Heulandite	$CaAl_2Si_7O_{18} \cdot 6H_2O$	zeolite
Hexahydrite	$MgSO_4 \cdot 6H_2O$	sulfate
Hinsdalite	$PbAl_3(PO_4)(SO_4)(OH)_6$	phosphate
Huntite	$CaMg_3(CO_3)_4$	carbonate
Hydroboracite	$MgCa(B_6O_{11}) \cdot 6H_2O$	borate
Hydromagnesite	$Mg_5(CO_3)_4(OH)_2 \cdot 4H_2O$	carbonate
Hydrophilite	$CaCl_2$	halide
Hydroxyapatite	$Ca_5(PO_4)_3OH$	phosphate
Hydroxypyromorphite	$Pb_5(PO_4)_3OH$	phosphate
Illite	$K_{.6}Mg_{.25}Al_{2.3}Si_{3.5}O_{10}(OH)_2$	10 Å clay
Jadeite	$NaAl(SiO_3)_2$	pyroxene
Jarosite-K	$KFe_3(SO_4)_2(OH)_6$	sulfate
Jarosite-Na	$NaFe_3(SO_4)_2(OH)_6$	sulfate
K-feldspar	$KAlSi_3O_8$	feldspar
Kainite	$KMgClSO_4 \cdot 3H_2O$	sulfate
Kalicinite	$KHCO_3$	carbonate
Kalsilite	$KAlSiO_4$	feldspathoid
Kaolinite	$Al_2Si_2O_5(OH)_4$	7 Å clay
Kasolite	$PbUO_2SiO_4 \cdot H_2O$	
Kieserite	$MgSO_4 \cdot H_2O$	sulfate
Kyanite	Al_2SiO_5	
Larnite	Ca_2SiO_4	olivine
Laumontite	$CaAl_2Si_4O_{12} \cdot 4H_2O$	zeolite

(continues)

Mineral	Chemical formula	General type
Lawrencite	$FeCl_2$	halide
Lawsonite	$CaAl_2Si_2O_7(OH)_2 \cdot H_2O$	epidote
Leonhardtite	$MgSO_4 \cdot 4H_2O$	sulfate
Lime	CaO	oxide
Linnaeite	Co_3S_4	sulfide
Magnesite	$MgCO_3$	carbonate
Magnetite	Fe_3O_4	oxide
Malachite	$Cu_2CO_3(OH)_2$	carbonate
Manganite	$MnOOH$	hydroxide
Manganosite	MnO	oxide
Margarite	$CaAl_4Si_2O_{10}(OH)_2$	mica
Maximum Microcline	$KAlSi_3O_8$	feldspar
Melanterite	$FeSO_4 \cdot 7H_2O$	sulfate
Mercallite	$KHSO_4$	sulfate
Merwinite	$Ca_3Mg(SiO_4)_2$	olivine
Metacinnabar	HgS	sulfide
Minnesotaite	$Fe_3Si_4O_{10}(OH)_2$	mica
Mirabilite	$Na_2SO_4 \cdot 10H_2O$	sulfate
Misenite	$K_8H_6(SO_4)_7$	sulfate
Modderite	$CoAs$	arsenide
Molysite	$FeCl_3$	halide
Monohydrocalcite	$CaCO_3 \cdot H_2O$	carbonate
Monticellite	$CaMgSiO_4$	olivine
Mordenite-K	$KAlSi_5O_{12} \cdot 3H_2O$	zeolite
Mordenite-Na	$NaAlSi_5O_{12} \cdot 3H_2O$	zeolite
Muscovite	$KAl_3Si_3O_{10}(OH)_2$	mica
Nepheline	$NaAlSiO_4$	feldspathoid
Nesquehonite	$Mg(HCO_3)(OH) \cdot 2H_2O$	carbonate
Ningyoite	$CaU(PO_4)_2 \cdot 2H_2O$	phosphate
Nontronit-Ca	$Ca_{.165}Fe_2Al_{.33}Si_{3.67}O_{10}(OH)_2$	smectite
Nontronit-K	$K_{.33}Fe_2Al_{.33}Si_{3.67}O_{10}(OH)_2$	smectite
Nontronit-Mg	$Mg_{.165}Fe_2Al_{.33}Si_{3.67}O_{10}(OH)_2$	smectite
Nontronit-Na	$Na_{.33}Fe_2Al_{.33}Si_{3.67}O_{10}(OH)_2$	smectite
Orpiment	As_2S_3	sulfide
Paragonite	$NaAl_3Si_3O_{10}(OH)_2$	mica
Pargasite	$NaCa_2Al_3Mg_4Si_6O_{22}(OH)_2$	amphibole
Pentahydrite	$MgSO_4 \cdot 5H_2O$	sulfate
Petalite	$Li_2Al_2Si_8O_{20}$	feldspathoid
Phengite	$KAlMgSi_4O_{10}(OH)_2$	mica
Phlogopite	$KAlMg_3Si_3O_{10}(OH)_2$	mica

(continues)

Mineral	Chemical formula	General type
Pirssonite	$Na_2Ca(CO_3)_2 \cdot 2H_2O$	carbonate
Plumbogummite	$PbAl_3(PO_4)_2(OH)_5 \cdot H_2O$	phosphate
Portlandite	$Ca(OH)_2$	hydroxide
Prehnite	$Ca_2Al_2Si_3O_{10}(OH)_2$	mica
Przhevalskite	$Pb(UO_2)_2(PO_4)_2$	phosphate
Pseudowollastonite	$CaSiO_3$	
Pyrite	FeS_2	sulfide
Pyrolusite	MnO_2	oxide
Pyrophyllite	$Al_2Si_4O_{10}(OH)_2$	mica
Pyrrhotite	$Fe_{.875}S$	sulfide
Quartz	SiO_2	silica
Quicksilver	Hg	native element
Rankinite	$Ca_3Si_2O_7$	
Realgar	AsS	sulfide
Rhodochrosite	$MnCO_3$	carbonate
Rhodonite	$MnSiO_3$	pyroxene
Ripidolit-14A	$Fe_2Mg_3Al_2Si_3O_{10}(OH)_8$	chlorite
Ripidolit-7A	$Fe_2Mg_3Al_2Si_3O_{10}(OH)_8$	
Rutherfordine	UO_2CO_3	carbonate
Safflorite	$CoAs_2$	arsenide
Saleeite	$Mg(UO_2)_2(PO_4)_2$	phosphate
Sanidine high	$KAlSi_3O_8$	feldspar
Saponite-Ca	$Ca_{.165}Mg_3Al_{.33}Si_{3.67}O_{10}(OH)_2$	smectite
Saponite-H	$H_{.33}Mg_3Al_{.33}Si_{3.67}O_{10}(OH)_2$	smectite
Saponite-K	$K_{.33}Mg_3Al_{.33}Si_{3.67}O_{10}(OH)_2$	smectite
Saponite-Mg	$Mg_{3.165}Al_{.33}Si_{3.67}O_{10}(OH)_2$	smectite
Saponite-Na	$Na_{.33}Mg_3Al_{.33}Si_{3.67}O_{10}(OH)_2$	smectite
Scacchite	$MnCl_2$	halide
Schoepite	$UO_2(OH)_2 \cdot H_2O$	hydroxide
Scorodite	$FeAsO_4 \cdot 2H_2O$	arsenide
Sepiolite	$Mg_4Si_6O_{15}(OH)_2 \cdot 6H_2O$	
Siderite	$FeCO_3$	carbonate
Sillimanite	Al_2SiO_5	
Silver	Ag	native element
Sklodowskite	$Mg(UO_2)_2(SiO_4)_2O_6H_{14}$	
Smectite-Reykjanes	$Na_{.33}Ca_{.66}K_{.03}Mg_{1.29}Fe_{.68}Mn_{.01} \cdot Al_{1.11}Si_{3.167}O_{10}(OH)_2$	smectite
Smectite-high-Fe-Mg	$Na_{.1}Ca_{.025}K_{.2}Mg_{1.15}Fe_{.7} \cdot Al_{1.25}Si_{3.5}O_{10}(OH)_2$	smectite

(continues)

Mineral	Chemical formula	General type
Smectite-low-Fe-Mg	$Na_{.15}Ca_{.02}K_{.2}Mg_{.9}Fe_{.45}$ · $Al_{1.25}Si_{3.75}O_{10}(OH)_2$	smectite
Smithsonite	$ZnCO_3$	carbonate
Soddyite	$(UO_2)_2(SiO_4)\cdot 2H_2O$	
Sphalerite	ZnS	sulfide
Spinel	Al_2MgO_4	oxide
Spodumene-a	$LiAlSi_2O_6$	pyroxene
Strengite	$FePO_4\cdot 2H_2O$	phosphate
Strontianite	$SrCO_3$	carbonate
Sulfur-Rhmb	S	native element
Sylvite	KCl	halide
Tachyhydrite	$Mg_2CaCl_6\cdot 12H_2O$	halide
Talc	$Mg_3Si_4O_{10}(OH)_2$	mica
Tenorite	CuO	oxide
Tephroite	Mn_2SiO_4	olivine
Thenardite	Na_2SO_4	sulfate
Thorianite	ThO_2	oxide
Todorokite	$Mn_7O_{12}\cdot 3H_2O$	oxide
Torbernite	$Cu(UO_2)_2(PO_4)_2$	phosphate
Tremolite	$Ca_2Mg_5Si_8O_{22}(OH)_2$	amphibole
Tridymite	SiO_2	silica
Troilite	FeS	sulfide
Tsumebite	$Pb_2Cu(PO_4)(OH)_3\cdot 3H_2O$	phosphate
Tyuyamunite	$Ca(UO_2)_2(VO_4)_2$	vanadate
Uraninite	UO_2	oxide
Uranocircite	$Ba(UO_2)_2(PO_4)_2$	phosphate
Uranophane	$Ca(H_2O)_2(UO_2)_2(SiO_2)_2(OH)_6$	hydroxide
Vivianite	$Fe_3(PO_4)_2\cdot 8H_2O$	phosphate
Wairakite	$CaAl_2Si_4O_{10}(OH)_4$	zeolite
Weeksite	$K_2(UO_2)_2(Si_2O_5)_3\cdot 4H_2O$	
Whitlockite	$Ca_3(PO_4)_2$	phosphate
Witherite	$BaCO_3$	carbonate
Wollastonite	$CaSiO_3$	
Wurtzite	ZnS	sulfide
Wustite	$Fe_{.947}O$	oxide
Zoisite	$Ca_2Al_3Si_3O_{12}(OH)$	epidote

Appendix 5
Nonlinear Rate Laws

As noted in Chapter 14, transition state theory does not require that kinetic rate laws take a linear form, although most kinetic studies have assumed that they do. The rate law for reaction of a mineral $A_{\vec{k}}$ can be expressed in the general nonlinear form

$$r_{\vec{k}} = sgn\,(Q_{\vec{k}}/K_{\vec{k}} - 1)\,(A_S\,k_+)_{\vec{k}} \prod^i (a_i)^{P_{i\vec{k}}} \prod^j (a_j)^{P_{j\vec{k}}} \left| (Q_{\vec{k}}/K_{\vec{k}})^\omega - 1 \right|^\Omega$$

where ω and Ω are arbitrary exponents that are determined empirically (e.g., Steefel and Lasaga, 1994). Here, *sgn* is a function that borrows the sign of its argument; it equals positive one when the fluid is supersaturated and negative one when it is undersaturated. This equation resembles the linear form of the rate law (Eqn. 14.2) except for the presence of the exponents ω and Ω. When the values of ω and Ω are set to one, the rate law reduces to its linear form.

To incorporate nonlinear rate laws into the solution procedure for tracing kinetic reaction paths (Section 14.3), we need to find the derivative of the reaction rate $r_{\vec{k}}$ with respect to the molalities m_i of the basis species A_i. The derivatives are given

$$\frac{dr_{\vec{k}}}{dm_i} = sgn\,(Q_{\vec{k}}/K_{\vec{k}} - 1)\,\frac{(A_S\,k_+)_{\vec{k}}}{m_i} \prod^{i'} (a_{i'})^{P_{i'\vec{k}}} \prod^j (a_j)^{P_{j\vec{k}}} \times$$

$$\left\{ sgn\,(Q_{\vec{k}}/K_{\vec{k}} - 1)\,\omega\,\Omega\,\nu_{i\vec{k}}\,(Q_{\vec{k}}/K_{\vec{k}})^\omega \left| (Q_{\vec{k}}/K_{\vec{k}})^\omega - 1 \right|^{\Omega-1} + \right.$$

$$\left. \left[P_{i\vec{k}} + \sum_j \nu_{ij}P_{j\vec{k}} \right] \left| (Q_{\vec{k}}/K_{\vec{k}})^\omega - 1 \right|^\Omega \right\}$$

This formula replaces Eqn. 14.11 in the calculation procedure.

To illustrate the effects of ω and Ω on reaction rates, we consider the reaction of quartz with dilute water, from Chapter 14. As before, we begin in REACT

```
time begin = 0 days, end = 5 days
T = 100
```

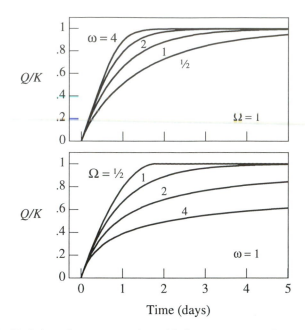

FIG. A5.1   Variation of quartz saturation with time as quartz sand reacts at 100°C with deionized water, calculated according to nonlinear forms of a kinetic rate law using various values of ω and Ω.

```
pH = 7
Cl-      = 10 umolal
Na+      = 10 umolal
SiO2(aq) =  1 umolal

react 5000 g Quartz
kinetic Quartz rate_con = 2.e-15   surface = 1000
```

and then set ω and Ω (keywords `order1` and `order2`, respectively) to differing values. For example,

```
(cont'd)
kinetic Quartz order1 = 1/2, order2 = 1
go
```

Figure A5.1 shows the calculation results.

References

Aagaard, P. and H.C. Helgeson, 1982, Thermodynamic and kinetic constraints on reaction rates among minerals and aqueous solutions, I. Theoretical considerations. *American Journal of Science* **282**, 237–285.

Adamson, A.W., 1976, *Physical Chemistry of Surfaces*. Wiley, New York.

Allison, J.D., D.S. Brown, and K.J. Novo-Gradac, 1991, MINTEQA2/PRODEFA2, a geochemical assessment model for environmental systems, version 3.0 user's manual. U.S. Environmental Protection Agency Report EPA/600/3–91/021.

Anderson, G.M. and D.A. Crerar, 1993, *Thermodynamics in Geochemistry, The Equilibrium Model*. Oxford University Press, New York.

Anderson, G.M. and G. Garven, 1987, Sulfate-sulfide-carbonate associations in Mississippi Valley-type lead-zinc deposits. *Economic Geology* **82**, 482–488.

Aplin, A.C. and E.A. Warren, 1994, Oxygen isotopic indications of the mechanisms of silica transport and quartz cementation in deeply buried sandstones. *Geology* **22**, 847–850.

Appel, C.A. and T.E. Reilly, 1994, Summary of selected computer programs produced by the U.S. Geological Survey for simulation of ground-water flow and quality. *U.S. Geological Survey Circular* **1104**.

Appelo, C.A.J. and D. Postma, 1993, *Geochemistry, Groundwater, and Pollution*. Balkema, Rotterdam.

Arnorsson, S., E. Gunnlaugsson, and H. Svavarsson, 1983, The chemistry of geothermal waters in Iceland, II. Mineral equilibria and independent variables controlling water compositions. *Geochimica et Cosmochimica Acta* **47**, 547–566.

Baccar, M.B. and B. Fritz, 1993, Geochemical modelling of sandstone diagenesis and its consequences on the evolution of porosity. *Applied Geochemistry* **8**, 285–295.

Baes, C.F., Jr. and R.E. Mesmer, 1976, *The Hydrolysis of Cations*. Wiley, New York.

Bahr, J.M. and J. Rubin, 1987, Direct comparison of kinetic and local equilibrium formulations for solute transport affected by surface reactions. *Water Resources Research* **23**, 438–452.

Ball, J.W. and D.K. Nordstrom, 1991, User's manual for WATEQ4F, with revised thermodynamic data base and test cases for calculating speciation of major, trace, and redox elements in natural waters. U.S. Geological Survey Open File Report 91–183.

Ball, J.W, E.A. Jenne, and D.K. Nordstrom, 1979, WATEQ2 — a computerized chemical model for trace and major element speciation and mineral equilibria of natural waters. In E.A. Jenne (ed.), *Chemical Modeling in Aqueous Systems*, American Chemical Society, Washington DC, pp. 815–835.

Barton, P.B., Jr., P.M. Bethke, and P. Toulmin, 3rd, 1963, Equilibrium in ore deposits. *Mineralogical Society of America Special Paper* **1**, 171–185.

Bear, J., 1972, *Dynamics of Fluids in Porous Media.* Elsevier, New York.

Belitz, K. and J. D. Bredehoeft, 1988, Hydrodynamics of Denver basin, explanation of subnormal fluid pressures. *American Association of Petroleum Geologists Bulletin* **72**, 1334–1359.

Berner, R.A., 1980, *Early Diagenesis, A Theoretical Approach.* Princeton University Press, Princeton, NJ.

Bethke, C.M., 1992, The question of uniqueness in geochemical modeling. *Geochimica et Cosmochimica Acta* **56**, 4315–4320.

Bethke, C.M., 1994, The Geochemist's Workbench™, version 2.0, A Users Guide to Rxn, Act2, Tact, React, and Gtplot. Hydrogeology Program, University of Illinois.

Bethke, C.M. and S. Marshak, 1990, Brine migrations across North America — The plate tectonics of groundwater. *Annual Review Earth and Planetary Sciences* **18**, 287–315.

Bethke, C.M., W.J. Harrison, C. Upson, and S.P. Altaner, 1988, Supercomputer analysis of sedimentary basins. *Science* **239**, 261–267.

Bethke, C.M., M.-K. Lee, and R.F. Wendlandt, 1992, Mass transport and chemical reaction in sedimentary basins, natural and artificial diagenesis. In M. Quintard and M.S. Todorovic (eds.), *Heat and Mass Transfer in Porous Media.* Elsevier, Amsterdam, pp. 421–434.

Bird, R.B., W.E. Stewart, and E.N. Lightfoot, 1960, *Transport Phenomena.* Wiley, New York.

Bischoff, J.L., D.B. Herbst, and R.J. Rosenbauer, 1991, Gaylussite formation at Mono Lake, California. *Geochimica et Cosmochimica Acta* **55**, 1743–1747.

Bischoff, J.L., J.A. Fitzpatrick, and R.J. Rosenbauer, 1993, The solubility and stabilization of ikaite ($CaCO_3 \cdot 6H_2O$) from 0°C to 25°C, environmental and paleoclimatic implications for thinalite tufa. *Journal of Geology* **101**, 21–33.

Bjorlykke, K. and P.K. Egeberg, 1993, Quartz cementation in sedimentary basins. *American Association of Petroleum Geologists Bulletin* **77**, 1538–1548.

Block, J. and O.B. Waters, Jr., 1968, The $CaSO_4\text{-}Na_2SO_4\text{-}NaCl\text{-}H_2O$ system at 25° to 100°C. *Journal of Chemical and Engineering Data* **13**, 336–344.

Bourcier, W.L., 1985, Improvements to the solid solution modeling capabilities of the EQ3/6 geochemical code. Lawrence Livermore National Laboratory Report UCID-20587.

Boulding, J.R., 1990, Assessing the geochemical fate of deep-well-injected hazardous waste. U.S. Environmental Protection Agency Report EPA/625/6–89/025a.

Bowers, T.S. and H.P. Taylor, Jr., 1985, An integrated chemical and stable isotope model of the origin of midocean ridge hot spring systems. *Journal of Geophysical Research* **90**, 12583–12606.

Bowers, T.S., K.L. Von Damm, and J.H. Edmond, 1985, Chemical evolution of mid-ocean ridge hot springs. *Geochimica et Cosmochimica Acta* **49**, 2239–2252.

Boynton, F.P., 1960, Chemical equilibrium in multicomponent polyphase systems. *Journal of Chemical Physics* **32**, 1880–1881.

Brady, J.B., 1975, Chemical components and diffusion. *American Journal of Science* **275**, 1073–1088.

Brady, P.V. and J.V. Walther, 1989, Controls on silicate dissolution rates in neutral and basic pH solutions at 25°C. *Geochimica et Cosmochimica Acta* **53**, 2823–2830.

Brantley, S.L., 1992, Kinetics of dissolution and precipitation — experimental and field results. In Y.K Kharaka and A.S Maest (eds.), *Water-Rock Interaction.* Balkema, Rotterdam, pp. 3–6.

Brantley, S.L., D.A. Crerar, N.E. Møller, and J.H. Weare, 1984, Geochemistry of a modern marine evaporite, Bocana de Virrilá, Peru. *Journal of Sedimentary Petrology* **54**, 447–462.

Brezonik, P.L., 1994, *Chemical Kinetics and Process Dynamics in Aquatic Systems.* Lewis Publishers, Boca Raton, FL.

Brinkley, S.R., Jr., 1947, Calculation of the equilibrium composition of systems of many components. *Journal of Chemical Physics* **15**, 107–110.

Brinkley, S.R., Jr., 1960, Discussion of "A brief survey of past and curent methods of solution for equilibrium composition" by H.E. Brandmaier and J.J. Harnett. In G.S. Bahn and E.E. Zukoski (eds.), *Kinetics, Equilibria and Performance of High Temperature Systems.* Butterworths, Washington DC, p. 73.

Brower, R. D., A. P. Visocky, I. G. Krapac, B. R. Hensel, G. R. Peyton, J. S. Nealon, and M. Guthrie, 1989, Evaluation of underground injection of industrial waste in Illinois. Illinois Scientific Surveys Joint Report 2.

Brown, T.H. and B.J. Skinner, 1974, Theoretical prediction of equilibrium phase assemblages in multicomponent systems. *American Journal of Science* **274**, 961–986.

Bunge, A.L. and C.J. Radke, 1982, Migration of alkaline pulses in reservoir sands. *Society of Petroleum Engineers Journal* **22**, 998–1012.

Butler, G.P., 1969, Modern evaporite deposition and geochemistry of coexisting brines, the sabkha, Trucial Coast, Arabian Gulf. *Journal of Sedimentary Petrology* **39**, 70–89.

Carnahan, B., H.A. Luther, and J.O. Wilkes, 1969, *Applied Numerical Methods.* Wiley, New York.

Carpenter, A.B., 1980, The chemistry of dolomite formation I: the stability of dolomite. *Society of Economic Paleontologists and Mineralogists Special Publication* **28**, 111–121.

Cederberg, G.A., R.L. Street, and J.O. Leckie, 1985, A groundwater mass transport and equilibrium chemistry model for multicomponent systems. *Water Resources Research* **21**, 1095–1104.

Chapelle, F.H., 1993, *Ground-water Microbiology and Geochemistry.* Wiley, New York.

Chapman, B.M., D.R. Jones, and R.F. Jung, 1983, Processes controlling metal ion attenuation in acid mine drainage streams. *Geochimica et Cosmochimica Acta* **47**, 1957–1973.

Clayton, J.L. and P.J. Swetland, 1980, Petroleum generation and migration in Denver basin. *American Association of Petroleum Geologists Bulletin* **64**, 1613–1633.

Cole, D.R. and H. Ohmoto, 1986, Kinetics of isotopic exchange at elevated temperatures and pressures. In J.W. Valley, H.P. Taylor, Jr., and J.R. O'Neil (eds.), Stable Isotopes in High Temperature Geological Processes. *Reviews in Mineralogy* **16**, 41–90.

Coudrain-Ribstein, A. and P. Jamet, 1989, Choix des composantes et spéciation d'une solution. *Comptes Rendus de l'Academie des Sciences* **309**-**II**, 239–244.

Crerar, D.A., 1975, A method for computing multicomponent chemical equilibria based on equilibrium constants. *Geochimica et Cosmochimica Acta* **39**, 1375–1384.

Criaud, A., C. Fouillac, and B. Marty, 1989, Low enthalpy geothermal fluids from the Paris basin, 2 — oxidation-reduction state and consequences for the prediction of corrosion and sulfide scaling. *Geothermics* **18**, 711–727.

Davies, C.W., 1962, *Ion Association.* Butterworths, Washington DC.

Davis, A., R.L. Olsen, and D.R. Walker, 1991, Distribution of metals between water and entrained sediment in streams impacted by acid mine drainage, Clear Creek, Colorado, U.S.A. *Applied Geochemistry* **6**, 333–348.

Davis, J.A. and D.B. Kent, 1990, Surface complexation modeling in aqueous geochemistry. In M.F. Hochella and A.F. White (eds.), Mineral-Water Interface Geochemistry. *Reviews in Mineralogy* **23**, 177–260.

Degens, E.T. and D.A. Ross (eds.), 1969, *Hot Brines and Recent Heavy Metal Deposits in the Red Sea.* Springer-Verlag, New York.

Delany, J.M. and S.R. Lundeen, 1989, The LLNL thermochemical database. Lawrence Livermore National Laboratory Report UCRL-21658.

Delany, J.M. and T.J. Wolery, 1984, Fixed-fugacity option for the EQ6 geochemical reaction path code. Lawrence Livermore National Laboratory Report UCRL-53598.

Deloule, E., 1982, The genesis of fluorspar hydrothermal deposits at Montroc and Le Burc, The Tarn, as deduced from fluid inclusion analysis. *Economic Geology,* **77**, 1867–1874.

Denbigh, K., 1971, *The Principles of Chemical Equilibrium,* 3rd ed. Cambridge University Press, London.

Domenico, P.A. and F.W. Schwartz, 1990, *Physical and Chemical Hydrogeology.* Wiley, New York.

Dongarra, J.J., C.B. Moler, J.R. Bunch and G.W. Stewart, 1979, *Linpack Users' Guide.* Society for Industrial and Applied Mathematics, Philadelphia.

Dove, P.M. and D.A. Crerar, 1990, Kinetics of quartz dissolution in electrolyte solutions using a hydrothermal mixed flow reactor. *Geochimica et Cosmochimica Acta* **54**, 955–969.

Drever, J.I., 1988, *The Geochemistry of Natural Waters,* 2nd ed. Prentice-Hall, Englewood Cliffs, NJ.

Dria, M.A., R.S. Schedchter, and L.W. Lake, 1988, An analysis of reservoir chemical treatments. *SPE Production Engineering* **3**, 52–62.

Drummond, S.E. and H. Ohmoto, 1985, Chemical evolution and mineral deposition in boiling hydrothermal systems. *Economic Geology* **80**, 126–147.

Dzombak, D.A. and F.M.M. Morel, 1987, Adsorption of inorganic pollutants in aquatic systems. *Journal of Hydraulic Engineering* **113**, 430–475.

Dzombak, D.A. and F.M.M. Morel, 1990, *Surface Complexation Modeling.* Wiley, New York.

Eugster, H.P. and B.F. Jones, 1979, Behavior of major solutes during closed-basin brine evolution. *American Journal of Science* **279**, 609–631.

Eugster, H.P., C.E. Harvie and J.H. Weare, 1980, Mineral equilibria in the six-component seawater system, Na-K-Mg-Ca-SO$_4$-Cl-H$_2$O, at 25°C. *Geochimica et Cosmochimica Acta* **44**, 1335–1347.

Faure, G., 1986, *Principles of Isotope Geology,* 2nd ed. Wiley, New York.

Felmy, A.R. and J.H. Weare, 1986, The prediction of borate mineral equilibria in natural waters, application to Searles Lake, California. *Geochimica et Cosmochimica Acta* **50**, 2771–2783.

Ficklin, W.H., G.S. Plumlee, K.S. Smith and J.B. McHugh, 1992, Geochemical classification of mine drainages and natural drainages in mineralized areas. In Y.K Kharaka and A.S Maest (eds.), *Water-Rock Interaction.* Balkema, Rotterdam, pp. 381–384.

Fournier, R.O., 1977, Chemical geothermometers and mixing models for geothermal systems. *Geothermics* **5**, 41–50.

Fournier, R.O and R.W. Potter II, 1979, Magnesium correction to the Na-K-Ca chemical geothermometer. *Geochimica et Cosmochimica Acta* **43**, 1543–1550.

Fournier, R.O. and J.J. Rowe, 1966, Estimation of underground temperatures from the silica content of water from hot springs and wet-steam wells. *American Journal of Science* **264**, 685–697.

Fournier, R.O. and A.H. Truesdell, 1973, An empirical Na-K-Ca geothermometer for natural waters. *Geochimica et Cosmochimica Acta* **37**, 1255–1275.

Freeze, R.A. and J.A. Cherry, 1979, *Groundwater*. Prentice Hall, Englewood Cliffs, NJ.

Fritz, B., 1981, Etude thermodynamique et modélisation des réactions hydrothermales et diagénétiques. Mémoires Sciences Géologiques number 65, Université Louis-Pasteur, Strasbourg, France.

Fritz, B., M.-P. Zins-Pawlas, and M. Gueddari, 1987, Geochemistry of silica-rich brines from Lake Natron (Tanzania). *Sciences Géologiques Bulletin* **40**, 97–110.

Garrels, R.M. and F.T. Mackenzie, 1967, Origin of the chemical compositions of some springs and lakes. *Equilibrium Concepts in Natural Waters*, Advances in Chemistry Series **67**, American Chemical Society, Washington, DC, pp. 222–242.

Garrels, R.M. and M.E. Thompson, 1962, A chemical model for sea water at 25°C and one atmosphere total pressure. *American Journal of Science* **260**, 57–66.

Garven, G. and R.A. Freeze, 1984, Theoretical analysis of the role of groundwater flow in the genesis of stratabound ore deposits, 2, quantitative results. *American Journal of Science* **284**, 1125–1174.

Giggenbach, W.F., 1988, Geothermal solute equilibria, derivation of Na-K-Mg-Ca geoindicators. *Geochimica et Cosmochimica Acta* **52**, 2749–2765.

Glynn, P.D., E.J. Reardon, L.N. Plummer, and E. Busenberg, 1990, Reaction paths and equilibrium end-points in solid-solution aqueous-solution systems. *Geochimica et Cosmochimica Acta* **54**, 267–282.

Greenberg, J.P. and N. Møller, 1989, The prediction of mineral solubilities in natural waters, a chemical equilibrium model for the Na-K-Ca-Cl-SO$_4$-H$_2$O system to high concentration from 0 to 250°C. *Geochimica Cosmochimica Acta* **53**, 2503–2518.

Greenwood, H.J., 1975, Thermodynamically valid projections of extensive phase relationships. *American Mineralogist* **60**, 1–8.

Guggenheim, E.A., 1967, *Thermodynamics, an Advanced Treatment for Chemists and Physicists*. 5th ed. North-Holland, Amsterdam.

Haas, J.L., Jr. and J.R. Fisher, 1976, Simultaneous evaluation and correlation of thermodynamic data. *American Journal of Science* **276**, 525–545.

Hardie, L.A., 1987, Dolomitization, a critical view of some current views. *Journal of Sedimentary Petrology* **57**, 166–183.

Hardie, L.A., 1991, On the significance of evaporites. *Annual Review Earth and Planetary Sciences* **19**, 131–168.

Hardie, L.A. and H.P. Eugster, 1970, The evolution of closed-basin brines. *Mineralogical Society of America Special Paper* **3**, 273–290.

Harrison, W.J., 1990, Modeling fluid/rock interactions in sedimentary basins. In T.A. Cross (ed.), *Quantitative Dynamic Stratigraphy*. Prentice Hall, Englewood Cliffs, NJ, pp. 195–231.

Harvie, C.E. and J.H. Weare, 1980, The prediction of mineral solubilities in natural waters, the Na-K-Mg-Ca-Cl-SO_4-H_2O system from zero to high concentration at 25°C. *Geochimica et Cosmochimica Acta* **44**, 981–997.

Harvie, C.E., J.H. Weare, L.A. Hardie, and H.P. Eugster, 1980, Evaporation of seawater, calculated mineral sequences. *Science* **208**, 498–500.

Harvie, C.E., N. Møller, and J.H. Weare, 1984, The prediction of mineral solubilities in natural waters, the Na-K-Mg-Ca-H-Cl-SO_4-OH-HCO_3-CO_3-CO_2-H_2O system to high ionic strengths at 25°C. *Geochimica et Cosmochimica Acta* **48**, 723–751.

Harvie, C.E., J.P. Greenberg, and J.H. Weare, 1987, A chemical equilibrium algorithm for highly non-ideal multiphase systems: free energy minimization. *Geochimica et Cosmochimica Acta* **51**, 1045–1057.

Hay, R.L., 1963, Stratigraphy and zeolitic diagenesis of the John Day formation of Oregon. *University of California Publications in Geological Sciences*, Berkeley, pp. 199–261.

Hay, R.L., 1966, Zeolites and zeolitic reactions in sedimentary rocks. *Geological Society of America Special Paper* **85**.

Hayes, J.B., 1979, Sandstone diagenesis — the hole truth. *Society of Economic Paleontologists and Mineralogists Special Publication* **26**, 127–139.

He, S., J.E. Oddo, and M.B. Tomson, 1994, The inhibition of gypsum and barite nucleation in NaCl brines at temperatures from 25 to 90°C. *Applied Geochemistry* **9**, 561–567.

Helgeson, H.C., 1968, Evaluation of irreversible reactions in geochemical processes involving minerals and aqueous solutions, I. Thermodynamic relations. *Geochimica et Cosmochimica Acta* **32**, 853–877.

Helgeson, H.C., 1969, Thermodynamics of hydrothermal systems at elevated temperatures and pressures. *American Journal of Science* **267**, 729–804.

Helgeson, H.C., 1970, A chemical and thermodynamic model of ore deposition in hydrothermal systems. *Mineralogical Society of America Special Paper* **3**, 155–186.

Helgeson, H.C. and D.H. Kirkham, 1974, Theoretical prediction of the thermodynamic behavior of aqueous electrolytes at high pressures and temperatures, II. Debye-Hückel parameters for activity coefficients and relative partial molal properties. *American Journal of Science* **274**, 1199–1261.

Helgeson, H.C., R.M. Garrels, and F.T. Mackenzie, 1969, Evaluation of irreversible reactions in geochemical processes involving minerals and aqueous solutions, II. Applications. *Geochimica et Cosmochimica Acta* **33**, 455–481.

Helgeson, H.C., J.M. Delany, H.W. Nesbitt, and D.K. Bird, 1978, Summary and critique of the thermodynamic properties of rock-forming minerals. *American Journal of Science* **278-A**, 1–229.

Helgeson, H.C., T.H. Brown, A. Nigrini, and T.A. Jones, 1970, Calculation of mass transfer in geochemical processes involving aqueous solutions. *Geochimica et Cosmochimica Acta* **34**, 569–592.

Helgeson, H.C., D.H. Kirkham and G.C. Flowers, 1981, Theoretical prediction of the thermodynamic behavior of aqueous electrolytes at high temperatures and pressures, IV. Calculation of activity coefficients, osmotic coefficients, and apparent molal and standard and relative partial molal properties to 600°C and 5 kB. *American Journal of Science* **281**, 1249–1516.

Hem, J.D., 1985, Study and interpretation of the chemical characteristics of natural water. *U.S. Geological Survey Water-Supply Paper* **2254**.

Hemley, J.J., G.L. Cygan, and W.M. d'Angelo, 1986, Effect of pressure on ore mineral solubilities under hydrothermal conditions. *Geology* **14**, 377–379.

Henley, R.W., 1984, Chemical structure of geothermal systems. In J.M. Robertson (ed.), Fluid-Mineral Equilibria in Hydrothermal Systems. *Reviews in Economic Geology* **1**, 9–28.

Hill, C.G., Jr., 1977, *An Introduction to Chemical Engineering Kinetics and Reactor Design*. Wiley, New York.

Hoffmann, R., 1991, Hot brines in the Red Sea. *American Scientist* **79**, 298–299.

Holland, H.D., 1978, *The Chemistry of the Atmosphere and Oceans*. Wiley, New York.

Hostettler, J.D., 1984, Electrode electrons, aqueous electrons, and redox potentials in natural waters. *American Journal of Science* **284**, 734–759.

Hubbert, M. K., 1940, The theory of ground-water motion. *Journal of Geology* **48**, 785–944.

Hubert, J.F., 1960, Petrology of the Fountain and Lyons formations, Front Range, Colorado. *Colorado School of Mines Quarterly* **55**.

Hunt, J.M., 1990, Generation and migration of petroleum from abnormally pressured fluid compartments. *American Association of Petroleum Geologists Bulletin* **74**, 1–12.

Hutcheon, I., 1984, A review of artificial diagenesis during thermally enhanced recovery. In D.A. MacDonald and R.C. Surdam (eds.), *Clastic Diagenesis*. American Association of Petroleum Geologists, Tulsa, pp. 413–429.

Interscience Publishers, 1954, *The Collected Papers of P.J.W. Debye*. Interscience Publishers, Inc., New York.

Jackson, K.J. and T.J. Wolery, 1985, Externson of the EQ3/6 computer codes to geochemical modeling of brines. *Materials Research Society Symposium Proceedings* **44**, 507–514.

Janecky, D.R. and W.E. Seyfried, Jr., 1984, Formation of massive sulfide deposits on oceanic ridge crests, incremental reaction models for mixing between hydrothermal solutions and seawater. *Geochimica et Cosmochimica Acta* **48**, 2723–2738.

Janecky, D.R. and W.C. Shanks, III, 1988, Computational modeling of chemical and sulfur isotopic reaction processes in seafloor hydrothermal systems, chimneys, massive sulfides, and subjacent alteration zones. *Canadian Mineralogist* **26**, 805–825.

Jankowski, J. and G. Jacobson, 1989, Hydrochemical evolution of regional groundwaters to playa brines in central Australia. *Journal of Hydrology* **108**, 123–173.

Jarraya, F. and M. El Mansar, 1987, *Modelisation Simplifieé du Gisement de Saumare à Sebkhat el Melah à Zarzis.* Projet de fin d'etudes, Ecole Nationale d'Ingénieurs de Tunis, Tunis, Tunisia.

Jennings, H.Y., Jr., C.E. Johnson, Jr., and C.D. McAuliffe, 1974, A caustic waterflooding process for heavy oils. *Journal of Petroleum Technology* **26**, 1344–1352.

Johnson, C.A., 1986, The regulation of trace element concentrations in river and estuarine waters with acid mine drainage, the adsorption of Cu and Zn on amorphous Fe oxyhydroxides. *Geochimica et Cosmochimica Acta* **50**, 2433–2438.

Johnson, J.W., E.H. Oelkers, and H.C. Helgeson, 1991, SUPCRT92: a software package for calculating the standard molal thermodynamic properties of minerals, gases, aqueous species, and reactions from 1 to 5000 bars and 0° to 1000°C. Earth Sciences Department, Lawrence Livermore Laboratory.

Karpov, I.K. and L.A. Kaz'min, 1972, Calculation of geochemical equilibria in heterogeneous multicomponent systems. *Geochemistry International* **9**, 252–262.

Karpov, I.K., L.A. Kaz'min, and S.A. Kashik, 1973, Optimal programming for computer calculation of irreversible evolution in geochemical systems. *Geochemistry International* **10**, 464–470.

Kastner, M., 1984, Control of dolomite formation. *Nature* **311**, 410–411.

Keenan, J.H., F.G. Keyes, P.G. Hill, and J.G. Moore, 1969, *Steam Tables, Thermodynamic Properties of Water Including Vapor, Liquid, and Solid Phases.* Wiley, New York.

Kennedy, V.C., G.W. Zellweger, and B.F. Jones, 1974, Filter pore-size effects on the analysis of Al, Fe, Mn, and Ti in water. *Water Resources Research* **10**, 785–790.

Kharaka, Y.K. and I. Barnes, 1973, SOLMNEQ: solution-mineral equilibrium computations. U.S. Geological Survey Computer Contributions Report PB-215-899.

Kharaka, Y.K., W.D. Gunter, P.K. Aggarwal, E.H. Perkins and J.D. DeBraal, 1988, SOLMNEQ.88, a computer program for geochemical modeling of water-rock interactions. *U.S. Geological Survey Water Resources Investigation Report 88–4227.*

King, T.V.V. (ed.), 1995, Environmental considerations of active and abandoned mine lands, lessons from Summitville, Colorado. *U.S. Geological Survey Bulletin* **2220**.

Knapp, R.B., 1989, Spatial and temporal scales of local equilibrium in dynamic fluid-rock systems. *Geochimica et Cosmochimica Acta* **53**, 1955–1964.

Knauss, K.G. and T.J. Wolery, 1986, Dependence of albite dissolution kinetics on pH and time at 25°C and 70°C. *Geochimica et Cosmochimica Acta* **50**, 2481–2497.

Knauss, K.G. and T.J. Wolery, 1988, The dissolution kinetics of quartz as a function of pH and time at 70 °C. *Geochimica et Cosmochimica Acta* **52**, 43–53.

Lafon, G.M., G.A. Otten, and A.M. Bishop, 1992, Experimental determination of the calcite-dolomite equilibrium below 200°C; revised stabilities for dolomite and magnesite support near-equilibrium dolomitization models. *Geological Society of America Abstracts with Programs* **24**, A210-A211.

Land, L.S. and G.L. Macpherson, 1992, Geothermometry from brine analyses, lessons from the Gulf Coast, U.S.A. *Applied Geochemistry* **7**, 333–340.

Lasaga, A.C., 1981a, Rate laws of chemical reactions. In A.C. Lasaga and R.J. Kirkpatrick (eds.), *Kinetics of Geochemical Processes.* Mineralogical Society of America, Washington DC, pp. 1–68.

Lasaga, A.C., 1981b, Transition state theory. In A.C. Lasaga and R.J. Kirkpatrick (eds.), *Kinetics of Geochemical Processes.* Mineralogical Society of America, Washington DC, pp. 135–169.

Lasaga, A.C., 1984, Chemical kinetics of water-rock interactions. *Journal of Geophysical Research* **89**, 4009–4025.

Lasaga, A.C., J.M. Soler, J. Ganor, T.E. Burch, and K.L. Nagy, 1994, Chemical weathering rate laws and global geochemical cycles. *Geochimica et Cosmochimica Acta* **58**, 2361–2386.

Leach, D.L., G.S. Plumlee, A.H. Hofstra, G.P. Landis, E.L. Rowan, and J.G. Viets, 1991, Origin of late dolomite cement by CO_2-saturated deep basin brines: evidence from the Ozark region, central United States. *Geology* **19**, 348–351.

Leamnson, R.N., J. Thomas, Jr., and H.P. Ehrlinger, III, 1969, A study of the surface areas of particulate microcrystalline silica and silica sand. *Illinois State Geological Survey Circular* **444**.

Lee, M.-K. and C.M. Bethke, 1994, Groundwater flow, late cementation, and petroleum accumulation in the Permian Lyons sandstone, Denver basin. *American Association of Petroleum Geologists Bulletin* **78**, 217–237.

Lee, M.-K. and C.M. Bethke, 1996, A model of isotope fractionation in reacting geochemical systems. *American Journal of Science*, submitted.

Levandowski, D.W., M.E. Kaley, S.R. Silverman, and R.G. Smalley, 1973, Cementation in Lyons sandstone and its role in oil accumulation, Denver basin, Colorado. *American Association of Petroleum Geologists Bulletin* **57**, 2217–2244.

Levenspeil, O., 1972, *Chemical Reaction Engineering*, 2nd ed. Wiley, New York.

Lichtner, P.C., 1985, Continuum model for simultaneous chemical reactions and mass transport in hydrothermal systems. *Geochimica et Cosmochimica Acta* **49**, 779–800.

Lichtner, P.C., 1988, The quasi-stationary state approximation to coupled mass transport and fluid-rock interaction in a porous medium. *Geochimica et Cosmochimica Acta* **52**, 143–165.

Lichtner, P.C., 1992, Time-space continuum description of fluid/rock interaction in permeable media. *Water Resources Research* **28**, 3135–3155.

Lichtner, P.C., 1995, Principles and practice of reactive transport modeling. In T. Murakami and R.C. Ewing (eds.), *Scientific Basis for Nuclear Waste Management XVIII*, vol. 353. Materials Research Society Proceedings, Pittsburgh, PA, pp. 117–130.

Lichtner, P.C., E.H. Oelkers, and H.C. Helgeson, 1986, Interdiffusion with multiple precipitation/dissolution reactions: transient model and the steady-state limit. *Geochimica et Cosmochimica Acta* **50**, 1951–1966.

Lico, M.S., Y.K. Kharaka, W.W. Carothers, and V.A. Wright, 1982, Methods for collection and analysis of geopressured geothermal and oil field waters. *U.S. Geological Survey Water Supply Paper* **2194**.

Lind, C.J. and J.D. Hem, 1993, Manganese minerals and associated fine particulates in the streambed of Pinal Creek, Arizona, U.S.A., a mining-related acid drainage problem. *Applied Geochemistry* **8**, 67–80.

Lindberg, R.D. and D.D. Runnells, 1984, Groundwater redox reactions: an analysis of equilibrium state applied to Eh measurements and geochemical modeling. *Science* **225**, 925–927.

Liu, C.W. and T.N. Narasimhan, 1989a, Redox-controlled multiple-species reactive chemical transport, 1. Model development. *Water Resources Research* **25**, 869–882.

Liu, C.W. and T.N. Narasimhan, 1989b, Redox-controlled multiple-species reactive chemical transport, 2. Verification and application. *Water Resources Research* **25**, 883–910.

Madé B., A. Clément and B. Fritz, 1994, Modeling mineral/solution interactions, the thermodynamic and kinetic code KINDISP. *Computers and Geosciences* **20**, 1347–1363.

Mangold, D.C. and C.-F. Tsang, 1991, A summary of subsurface hydrological and hydrochemical models. *Reviews of Geophysics* **29**, 51–79.

Marshall, W.L. and R. Slusher, 1966, Thermodynamics of calcium sulfate dihydrate in aqueous sodium chloride solutions, 0–110°. *The Journal of Physical Chemistry* **70**, 4015–4027.

Martin, C.A., 1965, Denver basin. *American Association of Petroleum Geologists Bulletin* **49**, 1908–1925.

Mattes, B.W. and E.W. Mountjoy, 1980, Burial dolomitization of the Upper Devonian Miette Buildup, Jasper National Park, Alberta. In D.H. Zenger, J.B. Dunham, and R.L. Effington (eds.), Concepts and Models of Dolomitization. *SEPM Special Publication* **28**, 259–297.

May, H., 1992, The hydrolysis of aluminum, conflicting models and the interpretation of aluminum geochemistry. In Y.K Kharaka and A.S Maest (eds.), *Water-Rock Interaction*. Balkema, Rotterdam, pp. 13–21.

Mayo, A.L., P.J. Nielsen, M. Loucks, and W.H. Brimhall, 1992, The use of solute and isotopic chemistry to identify flow patterns and factors which limit acid mine drainage in the Wasatch Range, Utah. *Ground Water* **30**, 243–249.

McConaghy, J.A., G.H. Chase, A.J. Boettcher, and T.J. Major, 1964, Hydrogeologic data of the Denver basin, Colorado. *Colorado Groundwater Basic Data Report* **15**.

McCoy, A.W., III, 1953, Tectonic history of Denver basin. *American Association of Petroleum Geologists Bulletin* **37**, 1873–1893.

McDuff, R.E. and F.M.M. Morel, 1980, The geochemical control of seawater (Sillen revisited). *Environmental Science and Technology* **14**, 1182–1186.

Mehnert, E., C.R. Gendron and R.D. Brower, 1990, Investigation of the hydraulic effects of deep-well injection of industrial wastes. *Illinois State Geological Survey Environmental Geology* **135**, 100 p.

Merino, E., D. Nahon, and Y. Wang, 1993, Kinetics and mass transfer of pseudomorphic replacement, application to replacement of parent minerals and kaolinite by Al, Fe, and Mn oxides during weathering. *American Journal of Science* **293**, 135–155.

Meyers, W.J. and K.C. Lohmann, 1985, Isotope geochemistry of regional extensive calcite cement zones and marine components in Mississippian limestones, New Mexico. In N. Schneidermann and P.M. Harris (eds.), Carbonate Cements. *SEPM Special Publication* **36**, 223–239.

Michard, G. and E. Roekens, 1983, Modelling of the chemical composition of alkaline hot waters. *Geothermics* **12**, 161–169.

Michard, G., C. Fouillac, D. Grimaud, and J. Denis, 1981, Une méthode globale d'estimation des températures des réservoirs alimentant les sources thermales, exemple du Massif Centrale Français. *Geochimica Cosmochimica Acta* **45**, 1199–1207.

Møller, N., 1988, The prediction of mineral solubilities in natural waters, a chemical equilibrium model for the Na-Ca-Cl-SO$_4$-H$_2$O system, to high temperature and concentration. *Geochimica Cosmochimica Acta* **52**, 821–837.

Morel, F.M.M., 1983, *Principles of Aquatic Chemistry*. Wiley, New York.

Morel, F. and J. Morgan, 1972, A numerical method for computing equilibria in aqueous chemical systems. *Environmental Science and Technology* **6**, 58–67.

Morse, J.W. and W.H. Casey, 1988, Ostwald processes and mineral paragenesis in sediments. *American Journal of Science* **288**, 537–560.

Moses, C.O., D.K. Nordstrom, J.S. Herman, and A.L. Mills, 1987, Aqueous pyrite oxidation by dissolved oxygen and by ferric iron. *Geochimica et Cosmochimica Acta* **51**, 1561–1571.

Mottl, M.J. and T.F. McConachy, 1990, Chemical processes in buoyant hydrothermal plumes on the East Pacific Rise near 21°N. *Geochimica et Cosmochimica Acta* **54**, 1911–1927.

Nagy, K.L., A.E. Blum, and A.C. Lasaga, 1991, Dissolution and precipitation kinetics of kaolinite at 80°C and *p*H 3, the dependence on solution saturation state. *American Journal of Science* **291**, 649–686.

Nordeng, S.H. and D.F. Sibley, 1994, Dolomite stoichiometry and Ostwald's step rule. *Geochimica et Cosmochimica Acta* **58**, 191–196.

Nordstrom, D.K., 1982, Aqueous pyrite oxidation and the consequent formation of secondary iron minerals. In Acid Sulfate Weathering. *Soil Science Society of America Special Publication* **10**, 37–56.

Nordstrom, D.K. and J.L. Munoz, 1994, *Geochemical Thermodyanmics*, 2nd ed. Blackwell, Boston.

Nordstrom, D.K., E.A. Jenne, and J.W. Ball, 1979, Redox equilibria of iron in acid mine waters. In E.A. Jenne (ed.), *Chemical Modeling in Aqueous Systems*, American Chemical Society, Washington DC, pp. 51–79.

Nordstrom, D.K., R.H. McNutt, I Puigdomènech, J.A.T. Smellie, and M. Wolf, 1992, Ground water chemistry and geochemical modeling of water-rock interactions at the Osamu Utsumi mine and the Morro do Ferro analogue study sites, Poços de Caldas, Minas Gerais, Brazil. *Journal of Geochemical Exploration* **45**, 249–287.

Okereke, A. and S.E. Stevens, Jr., 1991, Kinetics of iron oxidation by Thiobacillus ferrooxidans. *Applied and Environmental Microbiology* **57**, 1052–1056.

O'Neil, J.R., 1987, Preservation of H, C, and O isotopic ratios in the low temperature environment. In T.K. Kyser (ed.), Stable Isotope Geochemistry of Low Temperature Processes. *Mineralogical Society of Canada Short Course* **13**, 85–128.

Oreskes, N., K. Shrader-Frechette, and K. Belitz, 1994, Verification, validation, and confirmation of numerical models in the Earth sciences. *Science* **263**, 641–646.

Ortoleva, P.J., E. Merino, C. Moore, and J. Chadam, 1987, Geochemical self-organization, I., Reaction-transport feedbacks and modeling approach. *American Journal of Science* **287**, 979–1007.

Paces, T., 1975, A systematic deviation from Na-K-Ca geothermometer below 75°C and above 10^{-4} atm P_{CO_2}. *Geochimica Cosmochimica Acta* **39**, 541–544.

Paces, T., 1983, Rate constants of dissolution derived from the measurements of mass balance in hydrological catchments. *Geochimica et Cosmochimica Acta* **47**, 1855–1863.

Parker, D.R., W.A. Norvell, and R.L. Chaney, 1995, GEOCHEM-PC, a chemical speciation program for IBM and compatible personal computers. In R.H. Loeppert, A.P. Schwab, and S. Goldberg (eds.), Chemical Equilibrium and Reaction Models. *Soil Science Society of America Special Publication* **42**, 253–269.

Parkhurst, D.L., 1995, User's guide to PHREEQC, a computer model for speciation, reaction-path, advective-transport and inverse geochemical calculations. U.S. Geological Survey Water-Resources Investigations Report 95–4227, 143 p.

Parkhurst, D.L., D.C. Thorstenson, and L.N. Plummer, 1980, PHREEQE — a computer program for geochemical calculations. U.S. Geological Survey Water-Resources Investigations Report 80–96.

Peaceman, D.W., 1977, *Fundamentals of Numerical Reservoir Simulation*. Elsevier, New York.

Perkins, E.H., 1992, Integration of intensive variable diagrams and fluid phase equilibrium with SOLMNEQ.88 pc/shell. In Y.K Kharaka and A.S Maest (eds.), *Water-Rock Interaction*, Balkema, Rotterdam, p. 1079–1081.

Perkins, E.H. and T.H. Brown, 1982, Program PATH, calculation of isothermal and isobaric mass transfer. University of British Columbia, unpublished manuscript.

Perthuisot, J.P., 1980, Sebkha el Melah near Zarzis: a recent paralic salt basin (Tunisia). In G. Busson (ed.), *Evaporite Deposits, Illustration and Interpretation of Some Environmental Sequences*, Editions Technip, Paris, pp. 11–17, 92–95.

Phillips, S.L., A. Igbene, J.A. Fair, and H. Ozbek, 1981, A technical databook for geothermal energy utilization. Lawrence Berkeley Laboratory Report LBL-12810.

Pitzer, K.S., 1975, Thermodynamics of electrolytes, V, effects of higher order electrostatic terms. *Journal of Solution Chemistry* **4**, 249-265.

Pitzer, K.S., 1979, Theory: ion interaction approach. In R.M. Pytkowitz (ed.), *Activity Coefficients in Electrolyte Solutions*, vol. 1. CRC Press, Boca Raton, pp. 157–208.

Pitzer, K.S., 1987, A thermodynamic model for aqueous solutions of liquid-like density. In I.S.E. Carmichael and H.P. Eugster (eds.), Thermodynamic Modeling of Geological Materials: Minerals, Fluids and Melts. *Reviews in Mineralogy* **17**, 97–142.

Pitzer, K.S. and L. Brewer, 1961, Revised edition of *Thermodynamics*, by G.N. Lewis and M. Randall, 2nd ed. MacGraw-Hill, New York.

Plumlee, G., 1994a, USGS assesses the impact of Summitville. *USGS Office of Mineral Resources Newsletter* **5(2)**, 1–2.

Plumlee, G., 1994b, Environmental geology models of mineral deposits. *Society of Economic Geologists Newsletter* **16**, 5–6.

Plumlee, G.S., K.S. Smith, W.H. Ficklin, and P.H. Briggs, 1992, Geological and geochemical controls on the composition of mine drainages and natural drainages in mineralized areas. In Y.K Kharaka and A.S Maest (eds.), *Water-Rock Interaction*. Balkema, Rotterdam, pp. 419–422.

Plumlee, G.S., M.B. Goldhaber, and E.L. Rowan, 1995, The potential role of magmatic gases in the genesis of Illinois-Kentucky fluorspar deposits, implications from chemical reaction path modeling. *Economic Geology* **90**, 999-1011.

Plummer, L.N., 1992, Geochemical modeling of water-rock interaction: past, present, future. In Y.K Kharaka and A.S Maest (eds.), *Water-Rock Interaction*. Balkema, Rotterdam, pp. 23–33.

Plummer, L.N., D.L. Parkhurst, G.W. Fleming, and S.A. Dunkle, 1988, PHRQPITZ, a computer program incorporating Pitzer's equations for calculation of geochemical reactions in brines. U.S. Geological Survey Water-Resources Investigations Report 88–4153.

Prigogine, I. and R. Defay, 1954, *Chemical Thermodynamics*, D. H. Everett (trans.). Longmans, London.

Reed, M. and N. Spycher, 1984, Calculation of pH and mineral equilibria in hydrothermal waters with application to geothermometry and studies of boiling and dilution. *Geochimica et Cosmochimica Acta* **48**, 1479–1492.

Reed, M.H., 1977, Calculations of hydrothermal metasomatism and ore deposition in submarine volcanic rocks with special reference to the West Shasta district, California. Ph.D. dissertation, University of California, Berkeley.

Reed, M.H., 1982, Calculation of multicomponent chemical equilibria and reaction processes in systems involving minerals, gases and an aqueous phase. *Geochimica et Cosmochimica Acta* **46**, 513–528.

Richtmyer, R.D., 1957, *Difference Methods for Initial-Value Problems*. Wiley-Interscience, New York.

Rimstidt, J.D. and H.L. Barnes, 1980, The kinetics of silica-water reactions. *Geochimica et Cosmochimica Acta* **44**, 1683–1700.

Robie, R.A., B.S. Hemingway, and J.R. Fisher, 1979, Thermodynamic properties of minerals and related substances at 298.15 K and 1 bar (10^5 Pascals) pressure and at higher temperatures. *U.S. Geological Survey Bulletin* **1452** (corrected edition).

Robinson, R.A. and R.H. Stokes, 1968, *Electrolyte Solutions*. Butterworths, London.

Rosing, M.T., 1993, The buffering capacity of open heterogeneous systems. *Geochimica et Cosmochimica Acta* **57**, 2223–2226.

Rowan, E., 1991, Un modèle géochimique, thérmique-hydrogéologique et tectonique pour la genése des gisements filoniens de fluorite de l'Albigeois, sud-ouest du Massif Central, France. Mémoire de DEA, Université Pierre et Marie Curie (Paris IV), Paris.

Rubin, J., 1983, Transport of reacting solutes in porous media, relationship between mathematical nature of problem formulation and chemical nature of reactions. *Water Resources Research* **19**, 1231–1252.

Runnells, D.D., 1969, Diagenesis, chemical sediments, and the mixing of natural waters. *Journal of Sedimentary Petrology* **39**, 1188–1201.

Runnells, D.D. and R.D. Lindberg, 1990, Selenium in aqueous solutions, the impossibility of obtaining a meaningful Eh using a platinum electrode, with implications for modeling of natural waters. *Geology* **18**, 212–215.

Schecher, W.D. and D.C. McAvoy, 1994, MINEQL+, *A Chemical Equilibrium Program for Personal Computers, User's Manual*, version 3.0. Environmental Research Software, Inc., Hallowell, ME.

Shanks, W.C., III, and J.L. Bischoff, 1977, Ore transport and deposition in the Red Sea geothermal system, a geochemical model. *Geochimica et Cosmochimica Acta* **41**, 1507–1519.

Sherlock, E.J., R.W. Lawrence, and R. Poulin, 1995, On the neutralization of acid rock drainage by carbonate and silicate minerals. *Environmental Geology* **25**, 43–54.

Shock, E.L., 1988, Organic acid metastability in sedimentary basins. *Geology* **16**, 886–890 (correction, **17**, 572 573).

Sibley, D.F. and H. Blatt, 1976, Intergranular pressure solution and cementation of the Tuscarora orthoquartzite. *Journal of Sedimentary Petrology* **46**, 881–896.

Sillén, L.G., 1967, The ocean as a chemical system. *Science* **156**, 1189–1197.

Skirrow, G., 1965, The dissolved gases — carbon dioxide. In J.P. Riley and G. Skirrow (eds.), *Chemical Oceanography*, Academic Press, London, pp. 227–322.

Smith, K.S., W.H. Ficklin, G.S. Plumlee, and A.L. Meier, 1992, Metal and arsenic partitioning between water and suspended sediment at mine-drainage sites in diverse geologic settings. In Y.K Kharaka and A.S Maest (eds.), *Water-Rock Interaction*. Balkema, Rotterdam, pp. 443–447.

Smith, W.R. and R.W. Missen, 1982, *Chemical Reaction Equilibrium Analysis: Theory and Algorithms*. Wiley, New York.

Snoeyink, V.L. and D. Jenkins, 1980, *Water Chemistry*. Wiley, New York.

Sorbie, K.S., M. Yuan, and M.M. Jordan, 1994, Application of a scale inhibitor squeeze model to improve field squeeze treatment design (SPE paper 28885). *European Petroleum Conference Proceedings Volume* (vol. 2 of 2), Society of Petroleum Engineers, Richardson, TX, pp. 179–191.

Sørensen, S.P.L., 1909, Enzymstudier, II., Om Maalingen og Betydningen af Brintionkoncentration ved enzymatiske Processer. *Meddelelser fra Carlsberg Laboratoriet* **8**, 1–153.

Sposito, G., 1989, *The Chemistry of Soils*. Oxford University Press, New York.

Sposito, G. and J. Coves, 1995, SOILCHEM on the Macintosh. In R.H. Loeppert, A.P. Schwab, and S. Goldberg (eds.), Chemical Equilibrium and Reaction Models. *Soil Science Society of America Special Publication* **42**, 271–287.

Spycher, N.F. and M.H. Reed, 1988, Fugacity coefficients of H_2, CO_2, CH_4, H_2O and of H_2O-CO_2-CH_4 mixtures, a virial equation treatment for moderate pressures and temperatures applicable to calculations of hydrothermal boiling. *Geochimica et Cosmochimica Acta* **52**, 739–749.

Steefel, C.I. and A.C. Lasaga, 1992, Putting transport into water-rock interaction models. *Geology* **20**, 680–684.

Steefel, C.I. and A.C. Lasaga, 1994, A coupled model for transport of multiple chemical species and kinetic precipitation/dissolution reactions with application to reactive flow in single phase hydrothermal systems. *American Journal of Science* **294**, 529–592.

Steefel, C.I. and P.C. Lichtner, 1994, Diffusion and reaction in rock matrix bordering a hyperalkaline fluid-filled fracture. *Geochimica et Cosmochimica Acta* **58**, 3595–3612.

Steefel, C.I. and P. Van Cappellen, 1990, A new kinetic approach to modeling water-rock interaction, the role of nucleation, precursors, and Ostwald ripening. *Geochimica et Cosmochimica Acta* **54**, 2657–2677.

Steinmann, P., P.C. Lichtner and W. Shotyk, 1994, Reaction path approach to mineral weathering reactions. *Clays and Clay Minerals* **42**, 197–206.

Stoessell, R.K., 1988, 25°C and 1 atm dissolution experiments of sepiolite and kerolite. *Geochimica et Cosmochimica Acta* **52**, 365–374.

Stumm, W., 1992, *Chemistry of the Solid-Water Interface*. Wiley, New York.

Stumm, W. and J.J. Morgan, 1981, *Aquatic Chemistry, An Introduction Emphasizing Chemical Equilibria in Natural Waters*, 2nd ed. Wiley, New York.

Stumm, W. and R. Wollast, 1990, Coordination chemistry of weathering, kinetics of the surface-controlled dissolution of oxide minerals. *Reviews of Geophysics* **28**, 53–69.

Surdam, R.C. and J.R. Boles, 1979, Diagenesis of volcanic sandstones. *Society of Economic Paleontologists and Mineralogists Special Publication* **26**, 227–242.

Sverjensky, D.A., 1984, Oil field brines as ore-forming solutions. *Economic Geology* **79**, 23–37.

Sverjensky, D.A., 1987, The role of migrating oil field brines in the formation of sediment-hosted Cu-rich deposits. *Economic Geology* **82**, 1130–1141.

Sverjensky, D.A., 1993, Physical surface-complexation models for sorption at the mineral-water interface. *Nature* **364**, 776–780.

Sydansk, R.D., 1982, Elevated-temperature caustic/sandstone interaction, implications for improving oil recovery. *Society of Petroleum Engineers Journal* **22**, 453–462.

Taylor, B.E., M.C. Wheeler, and D.K. Nordstrom, 1984, Isotope composition of sulphate in acid mine drainage as measure of bacterial oxidation. *Nature* **308**, 538–541.

Thompson, J.B., Jr., 1959, Local equilibrium in metasomatic processes. In P.H. Abelson (ed.), *Researches in Geochemistry*, Wiley, New York, pp. 427–457.

Thompson, J.B., Jr., 1970, Geochemical reaction and open systems. *Geochimica et Cosmochimica Acta* **34**, 529–551.

Thompson, J.B., Jr., 1982, Composition space: an algebraic and geometric approach. *Reviews in Mineralogy* **10**, 1–31.

Thompson, M.E., 1992, The history of the development of the chemical model for seawater. *Geochimica et Cosmochimica Acta* **56**, 2985–2987.

Thorstenson, D.C., 1984, The concept of electron activity and its relation to redox potentials in aqueous geochemical systems. U.S. Geological Survey Open File Report 84–072, 45 p.

Thorstenson, D.C., D.W. Fisher and M.G. Croft, 1979, The geochemistry of the Fox Hills-Basal Hell Creek aquifer in southwestern North Dakota and northwestern South Dakota. *Water Resources Research* **15**, 1479–1498.

Todd, A.C. and M. Yuan, 1990, Barium and strontium sulphate solid-solution formation in relation to North Sea scaling problems. *SPE Production Engineering* **5**, 279–285.

Todd, A.C. and M. Yuan, 1992, Barium and strontium sulphate solid-solution scale formation at elevated temperatures. *SPE Production Engineering* **7**, 85–92.

Truesdell, A.H. and B.F. Jones, 1974, WATEQ, a computer program for calculating chemical equilibria of natural waters. *U.S. Geological Survey Journal of Research* **2**, 233–248.

Valocchi, A.J., 1985, Validity of the local equilibrium assumption for modeling sorbing solute transport through homogeneous soils. *Water Resources Research* **21**, 808–820.

van der Lee, J., 1993, CHESS, another speciation and complexation computer code. Technical Report no. LHM/RD/93/39, Ecole des Mines de Paris, Fontainebleau.

Van Zeggeren, F. and S.H. Storey, 1970, *The Computation of Chemical Equilibria*. Cambridge University Press, London.

Von Damm, K.L., J.M Edmond, B. Grant, and C.I. Measures, 1985, Chemistry of submarine hydrothermal solutions at 21°N, East Pacific Rise. *Geochimica et Cosmochimica Acta* **49**, 2197–2220.

Warga, J., 1963, A convergent procedure for solving the thermo-chemical equilibrium problem. *Journal Society Industrial and Applied Mathematicians* **11**, 594–606.

Wat, R.M.S, K.S. Sorbie, A.C. Todd, P. Chen, and P. Jiang, 1992, Kinetics of BaSO₄ crystal growth and effect in formation damage (SPE paper 23814). Proceedings Society of Petroleum Engineers International Symposium on Formation Damage Control, Lafayette, Louisiana, February 26–27, 1992, pp. 429–437.

Weare, J.H., 1987, Models of mineral solubility in concentrated brines with application to field observations. In I.S.E. Carmichael and H.P. Eugster (eds.), Thermodynamic Modeling of Geological Materials: Minerals, Fluids and Melts. *Reviews in Mineralogy,* **17**, 143–176.

Wedepohl, K.H., C.W. Correns, D.M. Shaw, K.K. Turekian, and J. Zemann (eds.), 1978, *Handbook of Geochemistry*, volumes II/1 and II/2, Springer-Verlag, Berlin.

Westall, J., 1980, Chemical equilibrium including adsorption on charged surfaces. In M.C. Kavanaugh and J.O. Leckie (eds.), Advances in Chemistry Series **189**, American Chemical Society, Washington, DC, pp. 33–44.

Westall, J.C. and H. Hohl, 1980, A comparison of electrostatic models for the oxide/solution interface. *Advances in Colloid Interface Science* **12**, 265–294.

Westall, J.C., J.L. Zachary, and F.F.M. Morel, 1976, MINEQL, a computer program for the calculation of chemical equilibrium composition of aqueous systems. Technical Note 18, R.M. Parsons Laboratory, Department of Civil and Environmental Engineering, Massachusetts Institute of Technology, Cambridge, MA.

White, W.B., 1967, Numerical determination of chemical equilibrium and the partitioning of free energy. *Journal of Chemical Physics* **46**, 4171–4175.

White, D.E., 1970, Geochemistry applied to the discovery, evaluation, and exploitation of geothermal energy resources. *Geothermics* **2**, 58–80.

White, W.B., S.M. Johnson, and G.B. Dantzig, 1958, Chemical equilibrium in complex mixtures. *Journal of Chemical Physics* **28**, 751–755.

Williamson, M.A. and J.D. Rimstidt, 1994, The kinetics and electrochemical rate-determining step of aqueous pyrite oxidation. *Geochimica et Cosmochimica Acta* **58**, 5443–5454.

Wolery, T.J., 1978, Some chemical aspects of hydrothermal processes at mid-ocean ridges, a theoretical study, I., Basalt-sea water reaction and chemical cycling between the oceanic crust and the oceans, II. Calculation of chemical equilibrium between aqueous solutions and minerals. Ph.D. dissertation, Northwestern University, Evanston, IL.

Wolery, T.J., 1979, Calculation of chemical equilibrium between aqueous solution and minerals: the EQ3/6 software package. Lawrence Livermore National Laboratory Report UCRL-52658.

Wolery, T.J., 1983, EQ3NR, a computer program for geochemical aqueous speciation-solubility calculations: user's guide and documentation. Lawrence Livermore National Laboratory Report UCRL-53414.

Wolery, T.J., 1992a, EQ3/EQ6, a software package for geochemical modeling of aqueous systems, package overview and installation guide (version 7.0). Lawrence Livermore National Laboratory Report UCRL-MA-110662 (1).

Wolery, T.J., 1992b, EQ3NR, a computer program for geochemical aqueous speciation-solubility calculations: theoretical manual, user's guide, and related documentation (version 7.0). Lawrence Livermore National Laboratory Report UCRL-MA-110662 (3).

Wolery, T.J. and L.J. Walters, Jr., 1975, Calculation of equilibrium distributions of chemical species in aqueous solutions by means of monotone sequences. *Mathematical Geology* **7**, 99–115.

Yuan, D. and A.C. Todd, 1991, Prediction of sulphate scaling tendency in oilfield operations. *SPE Petroleum Engineering* **6**, 63–72.

Yuan, M., A.C. Todd, and K.S. Sorbie, 1994, Sulphate scale precipitation arising from seawater injection, a prediction study. *Marine and Petroleum Geology* **11**, 24–30.

Zabala (de), E.F., J.M. Vislocky, E. Rubin, and C.J. Radke, 1982, A chemical theory for linear alkaline flooding. *Society of Petroleum Engineers Journal* **22**, 245–258.

Zeleznik, F.J. and S. Gordon, 1960, An analytical investigation of three general methods of calculating chemical equilibrium compositions. NASA Technical Note D-473, Washington DC.

Zeleznik, F.J. and S. Gordon, 1968, Calculation of complex chemical equilibria. *Industrial and Engineering Chemistry* **60**, 27–57.

Zins-Pawlas, M.-P., 1988, Géochimie de la silice dans les saumares et les milieux évaporatiques. Thèse, Université Louis-Pasteur, Strasbourg, France.

Index